Solitary Wasps

A volume in the
CORNELL SERIES IN ARTHROPOD BIOLOGY

The Tent Caterpillars
by Terrence D. Fitzgerald
Army Ants: The Biology of Social Predation
by William H. Gotwald Jr.
Solitary Wasps: Behavior and Natural History
by Kevin M. O'Neill
The Wild Silk Moths of North America: A Natural History of the Saturniidae of the United States and Canada
by Paul M. Tuskes, James P. Tuttle, and Michael M. Collins

Solitary Wasps

Behavior and Natural History

Kevin M. O'Neill
Montana State University

With illustrations by
Catherine Seibert

Comstock Publishing Associates
a division of
CORNELL UNIVERSITY PRESS
ITHACA AND LONDON

First published 2001 by Cornell University Press

Printed in the United States of America

Library of Congress Cataloging-in-Publication Data
O'Neill, Kevin M.
Solitary wasps : behavior and natural history / Kevin M. O'Neill ; with illustrations by Catherine Seibert.
p. cm.
Includes bibliographical references and index.
ISBN 0-8014-3721-0 (cloth)
1. Solitary wasps—Behavior. 2. Solitary wasps. I. Title.
QL53.O53 2000
595.79'8—dc21 00-010199

Cornell University Press strives to use environmentally responsible suppliers and materials to the fullest extent possible in the publishing of its books. Such materials include vegetable-based, low-VOC inks and acid-free papers that are recycled, totally chlorine-free, or partly composed of nonwood fibers. Books that bear the logo of the FSC (Forest Stewardship Council) use paper taken from forests that have been inspected and certified as meeting the highest standards for environmental and social responsibility. For further information, visit our website at www.cornellpress.cornell.edu.

Cloth printing 10 9 8 7 6 5 4 3 2 1

For Nancy O'Neill,
Howard Evans, and
Mary Alice Evans

Contents

Preface xi

1 **Wasp Diversity and Classification** 1
Order Hymenoptera 2
Suborder Symphyta 3
Suborder Apocrita 4
Conclusion 10

2 **Foraging Behavior of Parasitoids** 11
Characterizing Wasp Foraging Strategies 11
Dryinidae 16
Bethylidae 25
Chrysididae 33
Other Chrysidoidea 40
Mutillidae 40
Tiphiidae and Scoliidae 47
Pompilidae 54
Other Vespoidea 55
Sphecidae 58
Impact of Parasitoid Wasps on Host Populations 60
Biological Control of Pests Using Aculeate Wasp Parasitoids 66
Conclusion 69

3 **Foraging Behavior of Nest-Provisioning Predators** 70
Comparative Survey of Prey 78
Variation in Diets within Species 97
Variation in the Number of Prey per Cell 102
Comparative Survey of Hunting and Stinging Behaviors 102
Comparative Survey of Prey-Carriage Mechanisms 108

Prey Storage 112
Homing and Orientation 113
Comparative Survey of Oviposition Sites 114
Feeding by Adults 115
Impact on Prey Populations 117
Conclusion 118

4 Foraging Behavior of Cleptoparasites 120
Intraspecific Prey Theft 121
Brood Parasitism 122
Conclusion 134

5 Pollen Foraging and Pollination 135
Foraging Behavior of Pollen Wasps 135
The Role of Solitary Wasps in Plant Pollination 143
Conclusion 149

6 Nesting Behavior 152
Ground-Nesting Wasps 153
Pith- and Wood-Burrowing Wasps 168
Cavity-Nesting Wasps 171
Mud-Nesting Wasps 173
Nest Dispersion 177
Nest Usurpation and Supersedure 179
Conclusion 182

7 Natural Enemies and Defensive Strategies 183
A Survey of the Natural Enemies of Solitary Wasps 183
Impact of Natural Enemies on Solitary Wasp Populations 196
Defensive Strategies: Personal Defense 200
Defensive Strategies: Protecting Offspring 206
Conclusion 214

8 Male Behavior and Sexual Interactions 216
Conceptual Background: Sexual Selection Theory 217
Ecological Background 218
Mating at Emergence Sites 221
Male Defense of Individual Nests 237
Mating at Resources 246
Mating at Scent-Marked Territories 252

Mating at Landmarks: Hilltopping 264
Alternative Male Mating Tactics 268
Female Choice and Male Courtship 272
Conclusion 280

9 Thermoregulation, Sleeping, and Overwintering 282
Thermoregulation 282
Sleeping 293
Overwintering 296
Conclusion 296

10 Parental Strategies 298
Reproductive Trade-Offs 299
Parental Investment Decisions 306
The Evolution of Parental Behavior 320
Communal Nesting: A Brief Survey 330
Conclusion 333

Appendix A Superfamilies, Families, and Subfamilies of Solitary Aculeate Hymenoptera 335
Appendix B Solitary Wasp Genera Mentioned in This Book 337
Bibliography 341
Index 397

Preface

Solitary wasps have been a favorite of naturalists and ethologists for well over a century. As pet subjects of Jean Henri Fabre, wasps provided a generous portion of the material for his *Souvenirs Entomologique*, first published in 1879 and later translated into English (Fabre 1915, 1919, 1921). His chapter headings sometimes read like titles of Sherlock Holmes stories: "The Method of the Calicurgi," "The Problem of the Scoliae," and "A Dangerous Diet," for example (Fabre 1921). As was his forensic counterpart, Fabre was a superb detective who used careful observation and deduction in his inquiries. Fabre also asked questions of his subjects that anticipated the interests of later ethologists, addressing topics such as innate behavioral sequences, homing, and sex allocation. His narratives may not be cutting-edge science from our current perspective, but they are certainly informative and entertaining (though often amusingly quaint).

Fabre was just the first of a string of writers fascinated with the lives of solitary wasps—writers such as George and Elizabeth Peckham (1898), Phil and Nellie Rau (1918), Edward Reinhard (1929), John Crompton (1955), Niko Tinbergen (1958), Howard Evans (1963), and John Alcock (1988). Along the way, these and many other observers have contributed to a vast scientific literature on solitary wasps. I like to remind students that, although the Nobel Prize–winning ethologist Niko Tinbergen is most famous for his research on herring gulls and sticklebacks, his doctoral research concerned the behavior of the solitary wasp *Philanthus triangulum*. In doing that research, he provided one of the early successes of ethology. Tinbergen's work is an important link between Fabre's work and modern wasp biology.

Solitary wasps still offer behavioral ecologists and evolutionary biologists model organisms for understanding the evolution of foraging behavior, parental strategies, mating systems, thermoregulation, and eusociality. Fabre, by the way, emphatically rejected the notion of Darwinian evolution, so we can take his inspiration only so far (Pasteur 1994). The future certainly holds much in store as entomologists apply increasingly sophisticated conceptual models and research methods to old and new problems of wasp biology. I hope that this book will encourage others to join in the research, although I must caution that wasp watching can be a nearly addictive habit, not without its minor misadventures and discomforts. Wasp researchers may have to dodge cacti and rattlesnakes while chasing a mating pair or endure heat and wind-blown sand while excavating a nest burrow. Wasps, of course, also share with other animals a frustrating penchant for falsifying pet hypotheses. Nevertheless, one should pay little heed to wasp researchers when they complain. Most wasp biologists of my acquaintance enjoy the physical challenges and surprise observations and may even get satisfaction from doing wasp ethology on a shoestring budget (it's good that they can, because most have no choice in the matter). However, significant contributions to our knowledge of the biology of a wasp species can be made using no more than a pencil, notebook, watch, insect net, pen knife, ruler, and several dollars worth of model airplane paint for giving wasps individual identifying marks. One need only find an unexplored species or population, perhaps dream up a few hypotheses, and settle down to patiently observe and record. Fortunately, there is no shortage of wasps to provide subjects for research and entertainment.

In this book, I have tried to strike a balance between the old and new literature on solitary wasps. Much of the new literature has the obvious merits of being in tune with present-day theory and of taking a quantitative approach. The older literature, though less quantitative, is rich in details on comparative ethology and natural history. Papers published before the rise of modern ethology also have a charm missing from contemporary journals, because our generation has been discouraged from reporting anecdotes and even from placing the research in a personal context. Who today would dare start a journal article with "One fine day in early April I noticed a large, steely blue wasp running awkwardly along the ground on tip toes" (Newcomer 1930). Certainly, such writing can become tedious if overdone, but current papers written in sterile, passive-voiced technical prose can make us forget that the author actually studied real animals in a real world.

Briefly, the plan for the book is as follows. In Chapter 1, I provide a cursory overview of the classification of the Hymenoptera. Chapters 2–5 present a review of the foraging strategies of solitary aculeates, emphasizing the diversity of hosts and prey used and the ways in which prey are located and subdued; I also discuss the role of wasps in biological control. In Chapter 6, I discuss the nests of female solitary wasps. Chapter 7 is a review of information on the natural enemies of solitary wasps and the counterstrategies used to combat them. In Chapter 8, I provide an overview of mating strategies, including courtship. Chapter 9 contains a discussion of thermoregulation and sleeping. Chapter 10 wraps things up with a discussion of the evolution of parental strategies. Throughout the book, I use case studies of individual species to illustrate the details of behavior, while placing much of the comparative information in tables that can be perused to get a sense of the diversity of behaviors of solitary wasps. I have furnished a list of the superfamilies, families, and subfamilies of solitary aculeate Hymenoptera in Appendix I. To avoid having to give a family affiliation each time a genus is mentioned, I have also provided in Appendix II a list with the family affiliations of all genera discussed in the book.

Many thanks go to John Alcock, whose editorial skills and expertise in wasp biology helped polish a rough manuscript. Howard Evans, Bill Wcislo, and Ruth O'Neill reviewed the entire manuscript, and Byron Alexander, Peter Jensen, Rob Longair, Marni Rolston, Catherine Seibert, Bo Turnbow, and several anonymous reviewers commented on various chapters. Special thanks go to Peter Prescott, Nancy Winemiller, and Lynn Coryell of Cornell University Press, to the Interlibrary Loan staff at Montana State University, and to Massimo Olmi.

Originally, this book was planned to be co-authored with Byron Alexander, who would have contributed chapters on classification and nesting. Unfortunately, Byron had not completed his first chapter at the time of his death in 1996. The book is better because of his participation, but not as good as it could have been if he had lived to provide his special touch. I also would have liked to show the book to George Eickwort, who died in 1994. As the original editor of the Cornell Series in Arthropod Biology, it was George (along with Rob Reavill of Cornell University Press) who invited me to write the book and who provided some of the inspiration for its organization and approach.

Solitary Wasps

1/ Wasp Diversity and Classification

For most people, the term *wasp* conjures up images of hornets, yellow jackets, and paper wasps, those common, pesty insects that lurk about our houses and picnics. Among wasps, however, those insects are atypical because they live in social groups that may include thousands of adults (Wilson 1971). Within their societies, there is a division of labor; some females (queens) are the primary egg layers, and the remainder (workers) undertake all of the other tasks in the colony. Males play no role in the colony, other than acting as sperm donors.

While social wasps garner the publicity, the vast majority of wasp species are *solitary*. Adult females of solitary species forage alone and, if they build a nest, it is occupied solely by themselves and their offspring. The offspring are generally abandoned in the egg stage. Usually, the closest solitary species come to a social life is when they nest within dense aggregations, each female going about its activities independently of others (except when they fight or attempt to steal each others' nests and prey). Females within a few wasps, species form closer associations when they share communal nests. Such communally nesting females are still basically solitary because, within the shared nest, each female constructs and provisions her own cells, into which she lays her own eggs (Cowan 1991, Matthews 1991). Thus, *solitary* in the term *solitary wasp* refers to the fact that wasps of these species do not engage in the extended cooperative interactions typical of hornet and paper wasp societies. However, there is also an arbitrary aspect of the term, because not all wasps that live solitary lifestyles have traditionally been called solitary wasps. For example, the ichneumon wasps and chalcid wasps (and their many relatives) are typically called parasitic wasps. To fully understand how the wasps we call solitary wasps are related to other kinds of wasps and to each other, we must survey the classification of the order Hymenoptera.

The standard classification of Hymenoptera partitions it into two suborders, Symphyta and Apocrita, with the Apocrita being further subdivided into the Parasitica and Aculeata (Krombein et al. 1979a,b). The wasps I discuss in this book are members of the latter group and thus are often referred to as the aculeate Hymenoptera. The three groups of aculeates familiar to most people are the bees, the ants, and the social wasps. However, this book will be about the less familiar aculeate Hymenoptera: all aculeates that are *not* bees, ants, or social wasps. No broad reviews of solitary wasps have appeared since *The Wasps* by Howard Evans and Mary Jane West-Eberhard (1970) and *Evolution of Instinct* by Kunio Iwata (1976, originally published in Japanese in 1971). Much work has been done on solitary wasps since Iwata published his book, and the time is ripe for another review. This book cites more than 500 post-1971 papers and books that deal wholly or partially with solitary wasps (and the data in these publications are often interpreted in light of theories not developed or emphasized before the 1970s).

Order Hymenoptera

Wasps, along with ants, bees, and sawflies, are members of the order Hymenoptera, one of the four largest orders of insects, with more than 100,000 described species and an undoubtedly enormous number of undescribed species (Gauld and Bolton 1988, Goulet and Huber 1993). The solitary wasps discussed in this book are part of an evolutionary lineage encompassing 3 of the 14 superfamilies of Hymenoptera: Chrysidoidea, Apoidea, and Vespoidea.

Several aspects of the morphology and physiology of Hymenoptera are critical to an understanding of wasp behavior and life histories: their types of development, mouthparts, and sex determination. Hymenopterans undergo complete metamorphosis during development, passing through egg, larval, and pupal stages before reaching adulthood (and ceasing growth). The larvae of some aculeates, such as wood wasps and sawflies, are similar in appearance and behavior to caterpillars (i.e., the larvae of moths and butterflies). Those of other hymenopterans, including solitary wasps, are usually soft, whitish, legless, grublike creatures. The massive reorganization necessary to transform either type of larva into an adult occurs during the pupal stage; the pupa may or may not develop within a cocoon spun by the larva. Cocoons, however, do not necessarily contain pupae, because many solitary wasps overwinter within their cocoons as diapausing larvae (prepupae) and do not pupate until the following spring (Evans and West-Eberhard 1970). The main function of a hymenopteran larva is to process food. The larva begins this process using a pair of mandibles roughly similar to those of grasshoppers and beetles. During the pupal stage, the mouthparts of hymenopterans undergo a radical change; the adult maxillae and labium become fused into a tube called the proboscis, which is used for imbibing liquids. However, adults still retain their mandibles, which can be used during prey capture and nest building. The combination of a proboscis that sucks liquids and mandibles that bite is unique to the Hymenoptera.

One feature of Hymenoptera important for the topics discussed later is the mode of sex determination, which has had a strong influence on the evolution of parental strategies and social behavior. Male hymenopterans develop from unfertilized eggs, whereas females develop from fertilized eggs. This peculiar mode of sex determination, commonly called *haplodiploidy* because males are haploid and females are diploid, is characteristic of Hymenoptera in general (Whiting 1935, Kerr 1962). A female wasp stores the sperm that she receives during mating in an internal pouch called the spermatheca. Release of sperm from the spermatheca is controlled by a muscle surrounding the duct that leads from the spermatheca to the genital chamber. Thus, a female wasp can readily control the sex of her offspring by releasing or retaining sperm as the egg moves past the spermatheca before oviposition (Flanders 1956). The adaptive consequences of being able to manipulate the sex of offspring are considerable (Charnov 1982) and will be discussed in Chapter 10.

Suborder Symphyta

The fossil record and phylogenetic analyses of extant taxa concur that, of the two hymenopteran suborders, the symphytans (sawflies and horntails) are the oldest (F. Carpenter 1992, Whitfield 1992). Symphytans appeared in the Triassic Period (roughly 200 million years ago) and diversified into 14 families during the Jurassic Period (F. Carpenter 1992). After this heyday of symphytan radiation, however, the fossil record gives no evidence of new families appearing until the Eocene Epoch (55 million years ago), and nine of the families that appeared in the Jurassic are now extinct. Ecologically, symphytans are more similar to most Lepidoptera (butterflies and moths) than to "typical" wasps. Symphytan larvae not only look like lepidopteran caterpillars but also, like caterpillars, feed on plants. The one exception is the family Orussidae, whose larvae feed on the larvae of wood-boring insects (Askew 1971).

Suborder Apocrita

Apocritans are characterized by their so-called wasp waist, a marked constriction between the first and second abdominal segments. The oldest fossils of this lineage are from the middle Jurassic. Although by the late Jurassic nine apocritan families were present, only two of those are known from the Cretaceous Period (about 100 million years ago), (F. Carpenter 1992). However, 26 other apocritan families appeared during the Cretaceous, making the apocritans the major lineage within the Hymenoptera (Gauld and Bolton 1988, F. Carpenter 1992).

The Apocrita are usually partitioned into two groups: Parasitica and Aculeata. Parasitica contains a huge number of species with diverse lifestyles (Askew 1971, Gauld and Bolton 1988, Godfray 1994, Quicke 1997). Most Parasitica, such as the ichneumonid and chalcid wasps, feed upon other insects. Although the Parasitica live solitary lifestyles, they are not referred to as solitary wasps, a common name traditionally reserved for the non-social Aculeata.

ACULEATA

The aculeates take their name from the Latin word *aculeus*, meaning "barb" or "sting." Members of the aculeate branch of the Hymenoptera are characterized by special modifications of the ovipositor, a complex device found only in adult females. In many species of Parasitica, the mechanism serves both as an egg-depositing tube and a sting, allowing a female wasp to temporarily paralyze a host so that an egg can be laid upon or within it (Piek 1985). In aculeates, however, the ovipositor has lost its ancestral function and is used solely for injecting venom. The unusual "ovipositor/sting" of aculeates is the principal evidence that they share a unique common ancestor. Once evolved, however, the sting was then lost independently in several lineages within the Aculeata, such as the stingless bees, two large subfamilies of ants, and most members of the solitary wasp family Chrysididae.

A review and comprehensive analysis of the classification of the aculeates can be found in Brothers and Carpenter (1993), where they recognize three superfamilies within the Aculeata (Fig. 1-1): Chrysidoidea, Apoidea, and Vespoidea. Phylogenetic analyses of extant taxa indicate that Chrysidoidea is the oldest group. In the fossil record, all three aculeate superfamilies date back to the Cretaceous, when the Chrysidoidea was represented by four families (Bethylidae, Chrysididae, Dryinidae, and Scolebythidae); the Vespoidea by at least five families (Formicidae, Pompilidae, Rhopalosomatidae, Tiphiidae, and Vespidae and perhaps the Mutillidae and the extinct Falsiformicidae); and the Apoidea by at least two families (Sphecidae and Apidae) (F. Carpenter 1992).

Superfamily Chrysidoidea

The Chrysidoidea is a relatively small superfamily, with an estimated 4,000 or so species in seven families worldwide (Goulet and Huber 1993) (Fig. 1-1). Although little is known about the biology of the four smallest families, which together have fewer than four dozen known species, the same is not true of the Dryinidae, Bethylidae, and Chrysididae. Dryinids and bethylids are generally minute wasps, whose females are often wingless or short-winged (brachypterous). The forelegs of most female dryinids bear conspicuous pincers (chelae) that they use for subduing hosts. Dryinids are generally black or brown; although bethylids are generally drab in color, some species are blue, red, or green. Most North American chrysidids are uniformly colored metallic green or blue, whereas some Eurasian species have a metallic green head and thorax and a metallic red abdomen. The integument of chrysidids is thick, hard, and often coarsely pitted. The abdomen is often flat or concave ventrally, and, unlike other

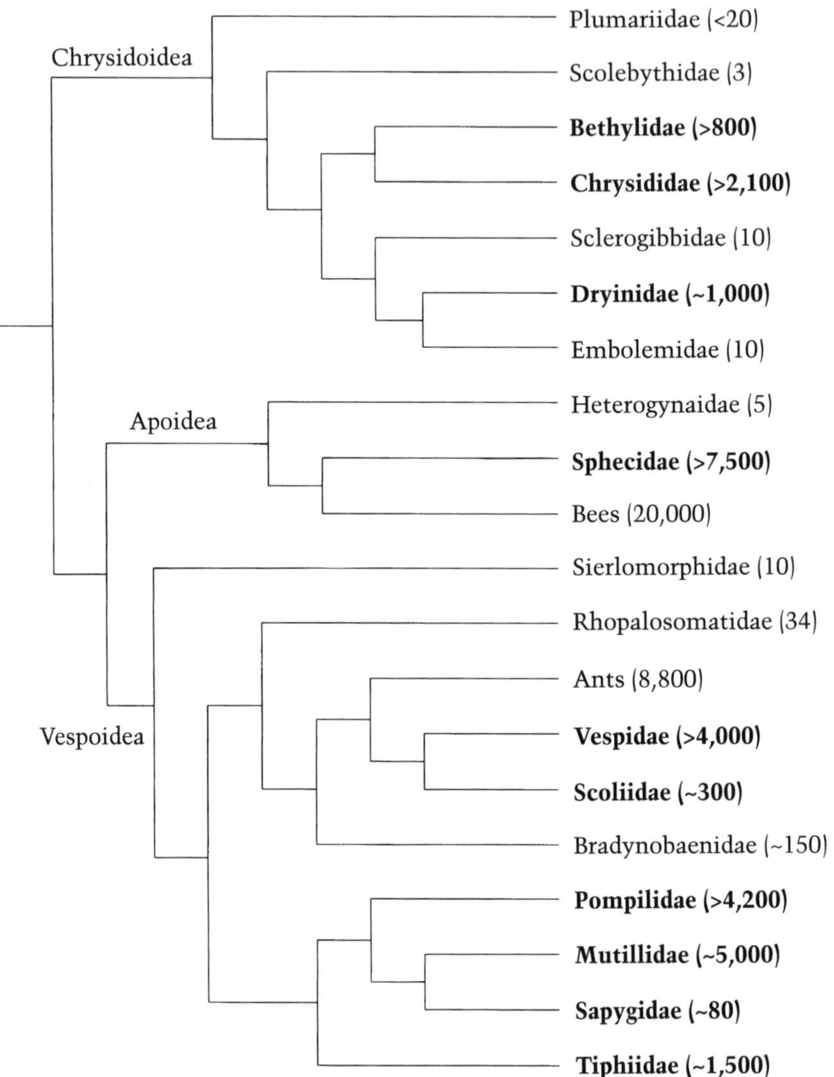

Figure 1-1. Phylogeny of the three superfamilies of Aculeata, based on Brothers and Carpenter 1993. Families listed in bold form the bulk of the material in this book. Numbers following family names are estimates of the number of species in each family (from Bohart and Menke 1976, Kimsey and Bohart 1990, Goulet and Huber 1993).

aculeates, most chrysidids lack a functional sting. Chrysidids do have an ovipositor, though one with a different evolutionary origin than that of the Parasitica: the terminal abdominal segments form an extensible tube that can be withdrawn into the female's abdomen when not in use.

Superfamily Apoidea

The aculeate superfamily Apoidea is currently estimated at about 28,000 species worldwide (Fig. 1-1). Family-level classification within Apoidea has long been a matter of contention, and there is still much debate over whether certain taxa should be assigned family or subfamily status. Over two-thirds of the Apoidea are bees, which belong to a discrete lineage whose females usually provision nests with a mixture of pollen and nectar or floral oils (Michener 1974). Some taxonomists (e.g., Lomholdt 1982, Gauld and Bolton 1988) have proposed that all bees be placed in the family Apidae, whereas others prefer to recognize several families of bees (Michener 1974, Roig-Alsina and Michener 1993, Alexander and Michener 1995). In any case, the bees are universally considered to be a monophyletic group (a group that represents a complete phylogenetic lineage because it contains *all* of the organisms descended from a specified common ancestor). When given a choice, phylogenetic systematists (cladists) prefer a monophyletic group over a paraphyletic group (one that encompasses taxa that share a common ancestor but that excludes some organisms that share that ancestor). To illustrate, the group of *all* of the children of a specific pair of human parents would be analogous to a monophyletic group, whereas a group of *some* of the children of those parents would be comparable to a paraphyletic group. The monophyly of bees has long been accepted because of their distinctive way of life and the large number of unique morphological features found in adults (Alexander and Michener 1995). Thus, the disagreement over the classification of bees is not about monophyletic versus paraphyletic groups.

In contrast, monophyly versus paraphyly *is* an issue in the differences of opinion about how to classify the Apoidea that are not bees. Traditionally, these have been given general common names that reflect the different ranks they have been assigned in the Linnaean system: sphecid wasps (family Sphecidae) or sphecoid wasps (superfamily Sphecoidea). Since the publication of *Sphecid Wasps of the World* (Bohart and Menke 1976), most taxonomists have placed all these wasps in the family Sphecidae. However, preliminary attempts to conduct rigorous phylogenetic analyses of the sphecid wasps suggest that ideas about phylogenetic relationships among sphecids will be changing in the near future (Alexander 1992). Several authors have subdivided the family Sphecidae into multiple families (e.g., Krombein et al. 1979b, Alexander 1992, Goulet and Huber 1993), and this approach is undoubtedly the wave of the future. Nevertheless, because the number of families and the exact subdivisions are not yet finalized, I will adopt the conservative approach of considering a single family (Sphecidae), while acknowledging that this is a temporary convenience unlikely to stand the test of time. My choice of taxonomic schemes has little if any effect on the course of this book. With the exception of a few small subfamilies, the subfamilies of Sphecidae that I take from Bohart and Menke (1976) (see Appendix I) follow basically along the same lines as the families suggested by Alexander (1992).

Thus, as considered here, the Apoidea consists of the bees and two families of solitary wasps: the small family Heterogynaidae and the large paraphyletic family Sphecidae. The Sphecidae contains over 7,500 described species exhibiting great variation in body size and form, from tiny and slender aphid hunters of the subfamily Pemphredoninae to the relatively large and robust cicada killers of the genus *Sphecius*. Sphecids also vary tremendously in color, from a homogeneous black to various combinations of white, yellow, orange, red, and metallic blue or green. Of greater interest here, they also exhibit a great diversity of behaviors that will be featured throughout this book.

Superfamily Vespoidea

The superfamily Vespoidea is a tremendously diverse group that contains about 20,000 described species (Fig. 1-1), although D. Brothers and A. Finnamore (1993) estimate that there are nearly 50,000 in all. The Vespoidea includes two large groups not covered in this book, the ants (Formicidae) and the eusocial wasps (which fall within three of the six subfamilies of the family Vespidae: Stenogastrinae, Polistinae, and Vespinae). However, three other subfamilies of Vespidae—Eumeninae, Masarinae, and Euparagiinae—consist primarily of solitary species.

Roughly 75%, or 3,000, of the species of Vespidae belong to the cosmopolitan subfamily Eumeninae, one of the principal groups of solitary, nest-provisioning wasps. The eumenines are somewhat less morphologically diverse than the sphecid wasps and display a narrower range of body sizes. The approximately 220 species of masarines, or pollen wasps, occur on all continents, but they are most diverse in the Southern Hemisphere, especially southern Africa (S. Gess 1996). Masarines are unique among solitary wasps in that they provision nests with pollen and nectar rather than with paralyzed arthropods (Chapter 5). The final extant vespid subfamily is Euparagiinae, with about 10 carnivorous species that now occur only in arid regions of western North America (Brothers 1992).

Of the remaining eight vespoid families, the Pompilidae, Mutillidae, and Tiphiidae contain the most species. The Pompilidae are commonly known as spider wasps because they feed their young with spiders. Pompilids are generally slender, long-legged, and jet black, though many species have white, yellow, or red markings, and some have translucent orange wings. Members of the large cosmopolitan family Mutillidae are quite striking insects sometimes called velvet ants, because their wingless females resemble large, fuzzy ants. The dense pubescence on their thick integument gives most mutillids black, brown, or orange coloration that may be broken with white, yellow, or red markings (Goulet and Huber 1993).

Wasps of the family Tiphiidae are often black, sometimes with added red or yellow markings (Goulet and Huber 1993). Four of the seven subfamilies are widespread, but the Brachycistidinae occur only in western North America; the Thynninae occur in the neotropics and in the Australian region; and the Diamminae (represented by a single species) occur only in Australia. The Scoliidae is a mostly tropical group of often large, robust, hairy, and colorful wasps, only about two dozen of which occur in North America. The Tiphiidae and Scoliidae were formerly classed together in one family (Scoliidae; Clausen et al. 1932), but the scoliids are now thought to be more closely allied to the Vespidae (Fig. 1-1). Another small family, the Sapygidae, appears to be most closely related to the Mutillidae. Like many mutillids, sapygids attack bees, but unlike their carnivorous relatives, many sapygids live by feeding on the pollen in bee nests. Sapygids are usually predominantly black, though some have white or yellow markings (Goulet and Huber 1993). The three other families of Vespoidea, the Bradynobaenidae, Rhopalosomatidae, Sierlomorphidae, are relatively rare insects whose behavior has been little studied, so they will receive scant attention in this book.

Conclusion

Hymenopteran systematics is in a state of flux, and it seems likely that the venerable taxa such as the Symphyta, Parasitica (Whitfield 1992), and Sphecidae (Alexander 1992) will eventually be broken up and realigned to reflect new knowledge about evolutionary relationships. Such efforts are to be applauded and anticipated with great interest because new systematic analyses will allow better comparative studies of behavior. However, the temporary adoption of paraphyletic groups such as the Sphecidae will not be harmful as long as comparative arguments recognize their existence.

2/ Foraging Behavior of Parasitoids

When a ladybug larva hatches from an egg and starts feeding on aphids, it does so without the assistance of its mother, who at best made the effort to lay eggs in a habitat likely to contain aphids. Thus, to succeed well enough to complete development, the larva must be mobile, efficient at finding prey, and capable of subduing them. Many other insects, for example dragonflies, lacewings, and assassin bugs, also have independent carnivorous larvae. In a sense, these insects have young much like those of precocial birds, such as ducks and shorebirds, that forage for themselves from birth. In contrast, the larvae of aculeate wasps are more like the young of altricial birds, such as hawks and warblers, which cannot forage on their own. Thus, although adult wasps also feed themselves, most of their foraging serves to furnish food for their offspring. All of this parentally motivated foraging is done by the females, which devote most of their activities to locating suitable food and preparing it to support the development of offspring.

Characterizing Wasp Foraging Strategies

Aculeate wasps have evolved a truly astounding variety of foraging adaptations (Askew 1971, Gauld and Bolton 1988, Godfray 1994, Quicke 1997). Before focusing on parasitoids, I will begin by considering 11 ways in which we can characterize aculeate foraging.

1. *Herbivory vs. carnivory.* Although most aculeate wasps are carnivores that feed their young with insects or spiders, all masarine wasps and most sapygid wasps feed their young with pollen and nectar (Goulet and Huber 1993, S. Gess 1996).

2. *Generalists vs. specialists.* Aculeate carnivores vary widely in the taxa and developmental stages of prey they provide offspring. Some wasps specialize on a single species of prey, whereas others have fairly catholic tastes, exploiting prey from many families and, sometimes, from more than one order of insects (Krombein et al. 1979b).

3. *Cleptoparasitism vs. foraging for free-living prey.* Females of most solitary wasps seek out and attack free-living prey (or collect their own pollen). However, females of cleptoparasites provide their young with prey or pollen procured by another solitary wasp species.

4. *Nesting vs. non-nesting species.* The food that female wasps provide their young may be left where it was found or it may be moved to a new location. Many solitary wasps lay their eggs and leave the host in place, often permanently exposed to the elements or to natural enemies. Females of many other wasps, however, deposit the prey in a safe location, often within an elaborate nest whose construction may take many hours or days (Chapter 6).

5. *Number of prey items provided to each wasp larva.* Among the noncleptoparasitic aculeates, most Bethylidae and a few Mutillidae and Dryinidae are gregarious feeders whose females typically lay more than one egg on each host. However, a single prey item is provided for the sole use of each larva of some Sphecidae, most Dryinidae and Mutillidae, and all Pompilidae, Tiphiidae, Scoliidae, and Chrysididae (excluding cleptoparasitic chrysidids). In contrast, most Sphecidae and eumenine Vespidae provide each offspring with more than one prey item.

When multiple eggs are laid on each prey item, the ovipositing female must make decisions concerning the number of eggs and the sexes of eggs to place on a host of a given size (Chapter 10). When one or more prey items are provided for each offspring, females must consider host size in order to be assured that the mass of food provided is sufficient for the rearing of a single larva of a particular sex (female offspring are usually larger than males and so require more food). Females that provide many prey items for each offspring can exploit prey much smaller than that needed if a single prey item is provided, and they can hunt at a relatively great distance from a nest because the prey are small enough to be transported in flight.

At this point, I must digress a bit on the terminology that will be used to distinguish different foraging modes in the Aculeata. All vertebrate carnivores that attack and kill living quarry are referred to as predators; this convention is followed whether we are referring to a lion killing a warthog or a warbler butchering an aphid. However, for insects that feed their young, it is traditional to divide the carnivorous (noncleptoparasitic) species into two categories: parasitoids and predators. The definitions of these terms are often based both on the feeding habits of larvae and on the foraging and parental behavior of adult females (Godfray 1994). I will apply the term *parasitoid* to aculeate wasps on the basis of two criteria. First, an adult female parasitoid provides each larva with a single host that it may have to share with siblings and that dies as a result of the larvae's feeding on it. This criterion distinguishes parasitoids from parasites, which do not necessarily kill their hosts (no solitary aculeate wasps are parasites). Second, parasitoids leave the host where it was captured or, at most, move it only a short distance to deposit it in an existing niche that the females find only after they have captured their prey. The female parasitoid does not modify the niche beyond perhaps just loosely covering the host with soil or debris.

Species of wasps with females that provide many prey items for each young or that construct their own nests or that do both are typically called predators (Godfray 1994). Thus, a species is a predator even if the larva feeds on a single prey item, as long as that prey has been moved from its point of capture to a concealed location (i.e., a nest) prepared, or at least substantially modified, by the mother. The use of a single prey item for each offspring is typical of most Pompilidae and a few sphecid wasps that build nests (Evans and West-Eberhard 1970). Admittedly, applying the nest criterion to the definitions of parasitoid and predator creates a somewhat arbitrary distinction, because there are varying degrees of nest "preparation" among aculeate wasps. In fact, the evolution of nesting behaviors among solitary aculeates has probably been a fairly seamless transition, from species that oviposit on prey left where they were attacked to those that drag the paralyzed prey a short distance to tuck it into a crack in the soil or beneath a leaf (sometimes covering it and sometimes not) to those with more or less elaborate nests made from scratch by the mother before foraging (Chapter 10). Nevertheless, most species of aculeates fit squarely into the categories provided by the definitions, and the use of terms will not affect our descriptions of the foraging adaptations of carnivorous solitary wasps. Parasitoids and predators differ in the way that they treat prey after it is captured, but the foraging behaviors of adult females of parasitoids and predators have much in common.

6. *Means by which food is located by foraging females.* Some wasps specialize on hosts found in the open, whereas others seek hosts hidden in sheltered places. Most aculeate wasps lack specialized ovipositors, but they sometimes attack hidden hosts within plant tissue, in the soil, within nests, or inside cocoons by breaching barriers with their legs or mandibles. Such hosts may be difficult to find, but by laying eggs on cryptic hosts females can provide their young with a refuge from predators and a nursery with a buffered microclimate.

Wherever hosts or prey are located, their efficient use by the foraging female requires strategies that minimize search time. This is not a trivial problem, because prey or hosts may be widely dispersed in the environment (sometimes in cryptic locations) and may have their own tactics for avoiding detection by their enemies. Nevertheless, search patterns need not be random because the prey or hosts may occur in predictable places, such as on host plants or at wasp nesting areas. In addition, potential prey may provide chemical, acoustic, and visual stimuli that betray their presence to foraging wasps.

7. *Means by which hosts and prey are subdued.* Once found, hosts and prey may have to be subdued in order to make them suitable for egg laying and larval feeding. Some aculeate parasitoids that attack resting stages of their hosts and some brood parasites have abandoned stinging altogether; in fact, the sting apparatus itself has become vestigial in most Chrysididae (Kimsey and Bohart 1990). However, most of the solitary aculeate parasitoids and predators paralyze their hosts or prey to some degree. In dedicating the ovipositor to a stinging function, aculeates have evolved sophisticated sting morphologies, stinging behaviors, and venom pharmacologies. In combination, these have permitted wasps to evolve forms of paralysis tailored to their hosts and to the wasps' specific attack and larval feeding strategies. For example, in *Liris nigra*, behavior and morphology coordinate to allow precise stinging in the vicinity of the ganglia of the central nervous system (A. Steiner 1971). In addition, the venoms themselves are often sophisticated enough in their pharmacological effect to rapidly subdue the host or prey, while allowing them to remain in a state of "suspended animation": alive

but immobile so as to avoid movement that could damage the vulnerable egg or early instars of the wasp. The venoms of some wasps may also change their victim's behavior in ways that benefit the wasp (Piek et al. 1984).

8. *Means of transporting prey.* Most parasitoids leave the host where it was attacked, but some parasitoids and all predators transport them elsewhere, either on foot or in flight. Although most wasps use their mandibles or legs to carry prey, a few impale prey on their sting or grip them with specialized claspers on the tips of their abdomen.

9. *Location of the egg and feeding larva on the host or prey.* Aculeate parasitoids and predators usually lay eggs at precise locations within, upon, or near their hosts or prey (Clausen 1940, Evans and West-Eberhard 1970, Iwata 1976). Among the aculeates that lay eggs on the outside of the host or prey, the exact position of the egg varies among species and appears to be an adaptation to several problems. First, the egg of species whose host recovers at least partially from paralysis must be in a position that prevents it from being damaged if the host thrashes about or tries to remove the egg. Second, proper placement of the egg may be critical in aiding the tiny, relatively weak, and immobile first-instar larva to find a point of penetration at vulnerable positions on the often highly sclerotized host. Third, the egg of some brood parasites may be placed where the host adult cannot find it.

The position of the egg is generally correlated with the position of the feeding larva. Parasitoid apocritans may develop within the host (endoparasitoids) or externally on the host (ectoparasitoids). The latter is probably the ancestral form of feeding in most apocritan lineages, with endoparasitism and predation being evolutionarily derived states in the Hymenoptera as a whole (Godfray 1994). The vast majority of aculeate parasitoids feed externally. The major exceptions include the Dryinidae, whose larvae feed internally early in their lives, and the amisegine and loboscelidine Chrysididae, which are endoparasitoids throughout development. Feeding larvae of some aculeate ectoparasitoids and predators may enter the body cavities of their dead hosts during the latter stages of feeding in order to finish off the remaining scraps of food, but they are not considered to be endoparasitoids.

10. *Timing of feeding and killing of host by the wasp larvae.* Most eggs of aculeate parasitoids hatch immediately after they are

deposited so that the larvae can commence feeding on the host. If the host dies quickly (or at least stops eating) because of larval feeding, host resources available at the time of egg laying have to suffice to allow the young wasps to complete development. However, the hosts of some aculeate parasitoids continue to feed and grow long after the wasp's egg has been deposited. Some chrysidid larvae accomplish this by feeding only after the host has completed feeding, grown considerably in size, and spun a cocoon; thus, they are assured an adequate food supply. An alternative solution is exhibited by the Dryinidae, whose larvae hatch out and feed soon after oviposition but initially refrain from feeding on vital organs of the host, which can continue to feed and even fly for a time.

11. *Feeding by adults.* Adult solitary wasps emerge from their cocoons with a store of fat that will fuel their activities and provide materials for developing eggs. However, both sexes also feed on nectar, honeydew (sugary secretions of Homoptera), or sap. Females may also obtain protein by feeding on arthropods. Such host feeding is usually restricted to feeding on hemolymph (blood) exuding from the ovipositor or sting wound or from small holes chewed in the host.

With this beginning, we can now survey the foraging behavior of parasitoid wasps.

Dryinidae

In many ways, dryinids are unique among parasitoid Hymenoptera. First, females of most dryinids have front legs modified in the form of pincers that they use for capturing and holding prey (Fig. 2-1). These "chelae" are constructed of an expansion of the fifth tarsal segment and one enlarged tarsal claw (the other is either missing or vestigial) (Olmi 1984a,b). Second, after prey capture, adult female dryinids often devour much larger portions of tissue than typical host-feeding adults. Third, dryinids are among the few aculeate wasps that lay eggs inside their hosts. Finally, the feeding positions of most dryinid larvae make them endoparasitoids at the beginning of their lives, ectoparasitoids for a brief period in the last larval stage, and somewhat of a combination in the interim.

Facing page
Figure 2-1. Adult female *Gonatopus clavipes* (from Olmi 1984b).

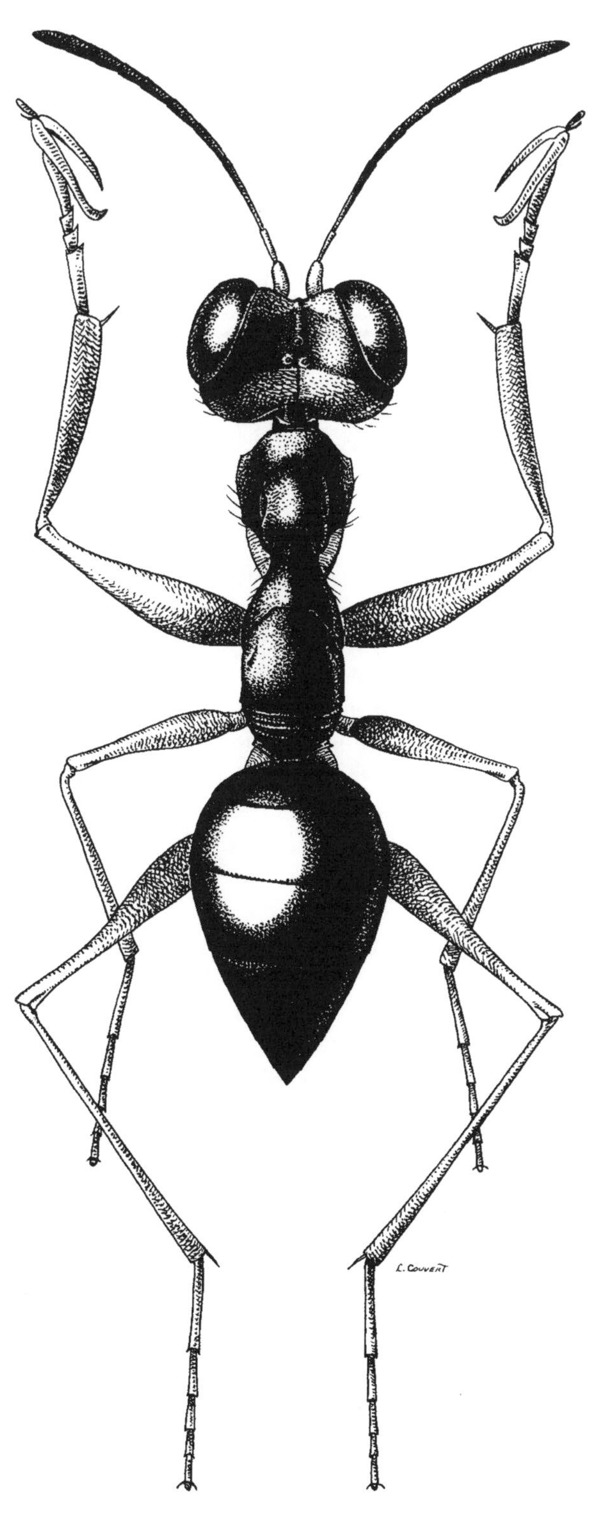
L. COUVERT

Case Study: *Gonatopus clavipes*

Gonatopus clavipes is widely distributed throughout the Palearctic region from Korea to Ireland (Olmi 1984b). Adult female *G. clavipes* are wingless, black with scattered yellow markings, and 2.5–2.75 mm long (Fig. 2-1) (Olmi 1984b). My review of *G. clavipes* biology is based on studies on British grasslands (Waloff 1974, 1975), and on summaries (Waloff 1980, Waloff and Jervis 1987) of studies done elsewhere (Abdul-Nour 1971, Ponomarenko 1971, Currado and Olmi 1972, Olmi 1984b).

A hunting female *G. clavipes* walks about and drums the substrate with her antennae until she finds a leafhopper and lunges forward to grasp it with her chelae (see Fig. 2-3C). When a strike is successful, the wasp holds the leafhopper by the hindlegs and curls her abdomen forward to sting it into transient paralysis. At this point, the wasp has several options. If the prey is the first of the day, the female usually feeds on it, biting out large chunks of tissue and killing the leafhopper (Waloff 1974). Alternatively, the female may forgo feeding and deposit a single egg into the host through an intersegmental membrane. (Curiously, in the lab, the female may also first feed a bit on the leafhopper before ovipositing in it.) *Gonatopus clavipes* females attack leafhoppers only of the subfamily Deltocephalinae, but within that group they have a relatively wide host range, being known to attack 16 species in 11 genera in Britain and another 7 genera and 15 species in other parts of Europe (Waloff and Jervis 1987).

A leafhopper that has not been fed upon, and which harbors a *Gonatopus* egg, recovers from paralysis within several minutes and continues feeding. After 4–5 days, however, a dryinid larva (probably in the second larval stage at this point; Clausen 1940) emerges between two segments on the side of the host abdomen (as in Fig. 2-2A). There, it feeds with its head immersed in the living host's body cavity and with the visible portions of its body covered in a sac composed of cast larval skins. Two to three weeks and several molts later the larva splits the sac, emerges, feeds voraciously on the host for about 1 hour, and moves away from the shriveled host to spin a cocoon at the base of grass stems. If the larva is of the first or early second generation of the year, it pupates and emerges as an adult within about 3 to 6 weeks. If a member of the third generation (or late second generation), it passes the winter in the cocoon as a mature larva before pupating the following spring and emerging as an adult in May.

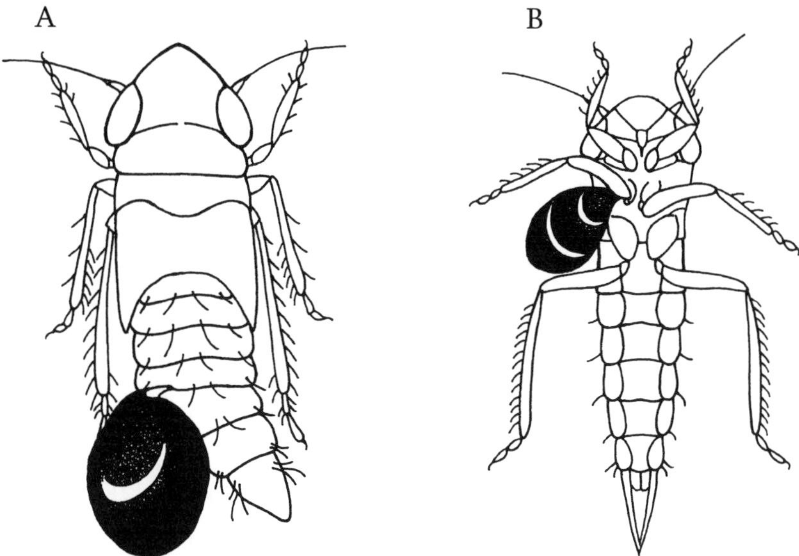

Figure 2-2. Position of larval sac on hosts of Dryinidae. (A) *Gonatopus erythroides* sac on *Deltocephalus inimicus* nymph. (B) *Anteon* sp. sac on *Chlorotettix unicolor* nymph. (A and B redrawn by Catherine Seibert from Fenton 1918.)

COMPARATIVE SURVEY OF DRYINID FORAGING BEHAVIORS

The known hosts of dryinids are restricted to 13 families of the homopteran suborder Auchenorrhyncha referred to by such names as leafhoppers, planthoppers, and treehoppers (Fenton 1918; Baldridge and Blocker 1980; Olmi 1984a,b; Waloff and Jervis 1987; Moya-Raygoza and Trujillo-Arriaga 1993; Guglielmino and Virla 1998). Many species (and even genera) limit themselves to particular families or subfamilies of hosts, but each dryinid may be polyphagous within its host group (Table 2-1). Dryinids also vary in the stages of the host attacked and in the location of the larval sac on the host (Fig. 2-2). Of the species studied, only *Crovettia theliae* has larvae that are completely endoparasitic and do not produce an externally visible larval sac. This species is also notable for being the only aculeate known to be polyembryonic (i.e., a single egg deposited in each host results in multiple young, in this case up to 50) (Kornhauser 1919). In other species, two or more larval sacs sometimes appear on

Table 2-1 Host records, behaviors, and notable features of selected species of Dryinidae

Species	Host family (subfamily)	No. of host genera (no. of species)	Host stage attacked	Location of larval sac on host	Notable observations	References
Aphelopinae						
Aphelopus albopictus	Cicadellidae (Typhlocybinae)	4 (7)	Nymphs	Abdomen or thorax	Mature larvae drop from host on trees to pupate in soil	H. Steiner 1936, 1938; Wilson et al. 1991
A. atratus[1]	Cicadellidae (Typhlocybinae)	8 (22)	Nymphs	Abdomen	Adult females lack chelae	Jervis 1980a,b
A. melaleucus[1]	Cicadellidae (Typhlocybinae)	10 (23)	Nymphs	Abdomen	Adult females lack chelae	Jervis 1980a,b
A. serratus[1]	Cicadellidae (Typhlocybinae)	10 (14)	Nymphs	Abdomen	Adult females lack chelae	Jervis 1980a,b
Crovettia theliae[1]	Membracidae	1 (1)	Nymphs of any stage	Larvae completely endoparasitic	Adult females lack chelae; polyembryonic	Kornhauser 1919
Anteoninae						
Anteon flavicorne[1]	Cicadellidae (Idiocerinae)	4 (9)	—	—	—	Waloff & Jervis 1987

A. pubicorne[1]	Cicadellidae (Macropsinae, Deltocephalinae)	8 (11)	Nymphs, adults	Thorax (lateral)	Adult feeds on host hemolymph without killing the host	Waloff & Jervis 1987
Lonchodryinus ruficornis[1]	Cicadellidae (Deltocephalinae)	10 (15)	Nymphs, adults	Thorax or between head and thorax	Adult feeds on host hemolymph near leg bases without killing host	Waloff 1980, Waloff & Jervis 1987
Bocchinae						
Bocchus europaeus[1]	Issidae	1 (1)	—	—	—	Waloff & Jervis 1987
Dryininae						
Dryinus collaris[1]	Cixiidae, Issidae	3 (4)	—	—	—	Waloff & Jervis 1987
D. tarraconensis[1]	Dictyopharidae	1 (1)	—	—	—	Waloff & Jervis 1987
Megadryinus magnificus[1]	Flattidae	1 (1)	—	—	Chelae >33% of body lengths	Richards 1953
Gonatopodinae						
Gonatopus bicolor[2]	Delphacidae	12 (17)	Nymphs, adults	Abdomen	—	Perkins 1905; Waloff 1974, 1975, 1980

Table 2-1 *Continued*

Species	Host family (subfamily)	No. of host genera (no. of species)	Host stage attacked	Location of larval sac on host	Notable observations	References
G. clavipes[2]	Cicadellidae (Deltocephalinae)	17 (32)	Nymphs, adults	Abdomen	Female may feed and oviposit on same host	Waloff 1974, 1975, 1980
G. distinctus[2]	Delphacidae	8 (9)	Nymphs, adults	Abdomen	Female may feed and oviposit on same host	Waloff 1974, 1980
G. helleni[2]	Cicadellidae	1 (2)	Nymphs, adults	Abdomen	Can disperse as first-instar larvae in migrating hosts	Raatinkainen 1961
G. lunatus	Cicadellidae (Deltocephalinae)	6 (8)	Nymphs, adults	Abdomen	When laid, eggs partially protrude from host abdomen	Guglielmino and Virla 1998
G. unicus[2]	Cicadellidae (Deltocephalinae)	1 (1)	Nymphs	Abdomen	—	Barrett et al. 1965

[1]Female winged. [2]Female wingless.

the same host, probably as a result of two females' coincidently having oviposited on the same host, with the result that both larvae die or emerge undersized (Ainslie 1920, Barrett et al. 1965).

Chelae are present in females of all dryinid subfamilies except the Aphelopinae and Biaphelopinae. The inner margins of the chelae of female dryinids are armed with arrays of lamellae, bristles, and hairs (Fig. 2-3) that presumably help secure the host during capture and stinging. The large and often numerous lamellae, as well as many of the hairs, have blunt, rounded tips that may allow females to grasp their soft-bodied hosts and hold them firmly without puncturing their exoskeleton. However, the larger chelae of dryinids of the subfamilies Dryininae and Gonatopodinae can be used to tear into the host when it is fed upon by the adult (Jervis et al. 1987). The structure of the chelae, including the size, shape, number, and orientation of the lamellae, varies considerably among species (Fig. 2-3). On *G. clavipes*, the chelae are small relative to body size (Fig. 2-1), whereas those of the suitably named *Megadryinus magnificus* are over 40% of the body length of the wasp. On the long, slender, straight chelae of *M. magnificus*, over 150 lamellae in parallel rows face a single row of nearly 80 peglike hairs (Fig. 2-3A). This contrasts with the simpler hooked chelae of *Bocchus europaeus* with its single lamella near the apex (Fig. 2-3B). The chelae of *G. clavipes* and *G. bicolor* have features of both of these species, being both hooked and toothed (Fig. 2-3C,D). Because the homopterans captured by dryinids vary in size, shape, and perhaps strength, some of the interspecific variation in the size and structure of dryinid forelegs may have resulted from selection for efficiency in subduing and manipulating particular types of prey. For example, chela size may be correlated with prey size, and the form of the lamellae may be correlated with the toughness of the host's integument.

After initially subduing hosts with the chelae, dryinids sting them, inflicting complete paralysis that allows them to oviposit on an immobile host. The venom, however, has only a transient effect, so the wasps' larvae can live in hosts that continue feeding (but not molting; Guglielmino and Virla 1998) and providing sustenance for the dryinid. Some dryinid hosts may even migrate with the larval sac attached, thus aiding in the dispersal of the wasps (Waloff 1980).

Feeding on the host, or host feeding, is a common feature of the foraging behavior of adult female dryinids, being absent only in those subfamilies that lack chelae. Host feeding probably provides

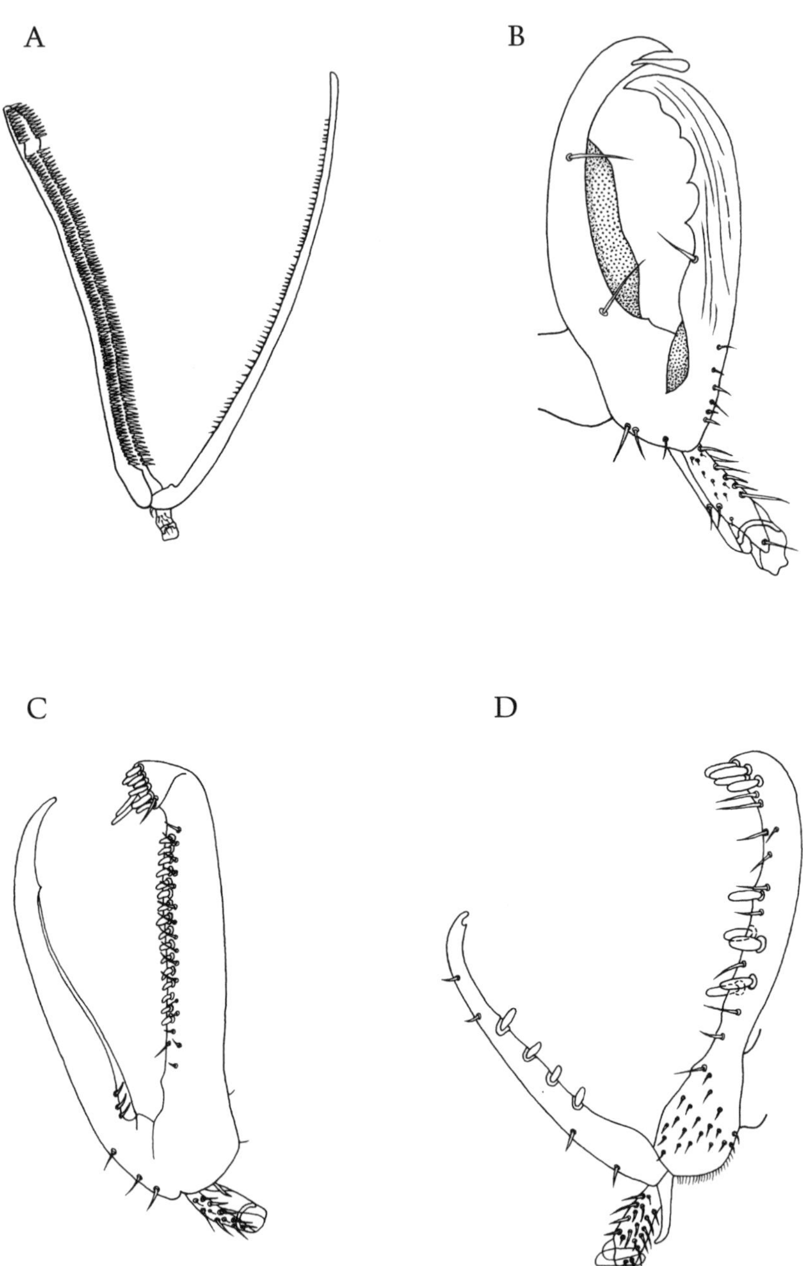

Figure 2-3. Chelae. (A) *Megadryinus magnificus*. (B) *Bocchus europaeus*. (C) *Gonatopus clavípes*. (D) *G. bicolor*. (A–D redrawn by Catherine Seibert from Olmi 1984a,b).

an important source of nutrients for egg-laying females (Jervis et al. 1987). The lifetime egg production of *Gonatopus flavifemur* females increases linearly with the number of hosts fed upon (Sahragard et al. 1991). Females that feed on 100 or more hosts lay over 500 eggs, whereas those that feed on 30 hosts lay only about 10 eggs. Several species, including *Crovettia theliae* (Kornhauser 1919) and *Echthrodelphax fairchildii* (Perkins 1905), are known to feed on honeydew that they take directly from their hosts.

Bethylidae

Most bethylids are gregarious ectoparasitoids whose biology is known primarily from lab studies. Although the front end of feeding larvae may be well submerged in the body cavity of their victim, bethylids (unlike dryinids) do not accumulate cast larval skins.

Case Study: *Laelius pedatus*

Female *Laelius* lay eggs on larvae and, sometimes, on pupae of dermestid beetles of the genera *Anthrenus* and *Trogoderma*, where their own larvae feed as gregarious ectoparasitoids (L. Howard 1901, Evans 1978). One of the best-studied species is *Laelius pedatus*, which attacks pests of stored products and occurs throughout much of North America (Ma et al. 1978; Mertins 1980, 1982; Klein and Beckage 1990; Qi and Burkholder 1990; Qi et al. 1990; Klein et al. 1991; Mayhew 1997, 1998). The winged, shiny black adult females of the wasp average slightly over 3 mm in length.

The basic life-history pattern of *L. pedatus* was first worked out by Mertins (1980), who reared them in the laboratory using varied carpet beetles, *Anthrenus verbasci*, as hosts. Upon contacting an *A. verbasci* larva, a female *L. pedatus* mounts it dorsally and moves to its thorax. The beetle responds by erecting and vibrating long hairs on its abdomen and arching its body to bring the hairs into contact with the wasp. Many of the hairs detach and adhere to the wasp, and the beetle larva gyrates wildly in an attempt to throw off its attacker (Mertins 1982). While taking this wild ride, the wasp manages to work the tip of her abdomen around to the host's venter, where she stings it, usually only once at the base of one of the forelegs.

The beetle is permanently paralyzed within 40 seconds of its discovery, but the wasp delays oviposition. After stinging, she remains with the host for up to a full day, grooming herself to remove hairs shed by the host and repeatedly investigating and biting the host, perhaps to assess its state of paralysis. The female then removes the hairs from the host's venter to create a bare surface to receive the eggs. Because host depilation may take up to 23 hours, two full days may have passed since the host was stung. When the host is finally prepared, the female typically deposits 2–4 eggs (Fig. 2-4), the number depending on host size. Clutches of 1–5 eggs have been reported on *Anthrenus flavipes* (Mayhew 1998). The eggs hatch in 3–4 days, and the larvae feed for 3–7 days, molting twice in the interim and killing the host during the wasps' last larval stage. When the host's body becomes a "hollow chitinous exoskeleton," the larvae spin silken cocoons several millimeters away. In the nondiapausing first generation of each year, adults emerge about 1 month after formation of the cocoons; adults of the second, diapausing generation emerge from cocoons after about 7 months.

Not all *Anthrenus* species are equally suitable hosts for *L. pedatus*. Larvae of *Anthrenus flavipes*, which have longer defensive hairs than *A. verbasci*, have been known to deter attacks (Ma et al. 1978). When a female *L. pedatus* contacts the hairs, which vibrate at ~450 cycles per second, at the tip of an *A. flavipes* abdomen, she retreats and eventually breaks off her attack; however, *L. pedatus* has been reared on *A. flavipes* (Mayhew 1998). Larvae of *Anthrenus fuscus* also resist parasitization by *L. pedatus*, in this case because they recover from paralysis within 1–2 days of being stung and remove the parasite eggs (Mertins 1982). *Anthrenus fuscus* is susceptible, however, to the venom of the *Laelius utilis* (Mertins 1985). Variation in host suitability was also found when *L. pedatus* were given access to larvae and pupae of *Trogoderma glabrum* and *T. variabile*, which can each play host to up to six wasp eggs (Klein and Beckage 1990). *Trogoderma variabile* proved more suitable than *T. glabrum*, because the latter was often rejected as a host by *L. pedatus* females and even if stung was more likely to recover. Even when *T. glabrum* larva remained paralyzed, mortality of the wasp larvae on *T. glabrum* was twice that on *T. variabile*. Thus, the host species more often chosen by the females proved to be more suitable for the wasps' larvae. The differential preference of *Cephalonomia waterstoni* for four species of the cucujid beetle genus *Laemophloeus* is also correlated with host quality (Finlayson 1950b).

Potential host quality for an *L. pedatus* female is also reduced if she comes upon a host previously attacked by another female. The

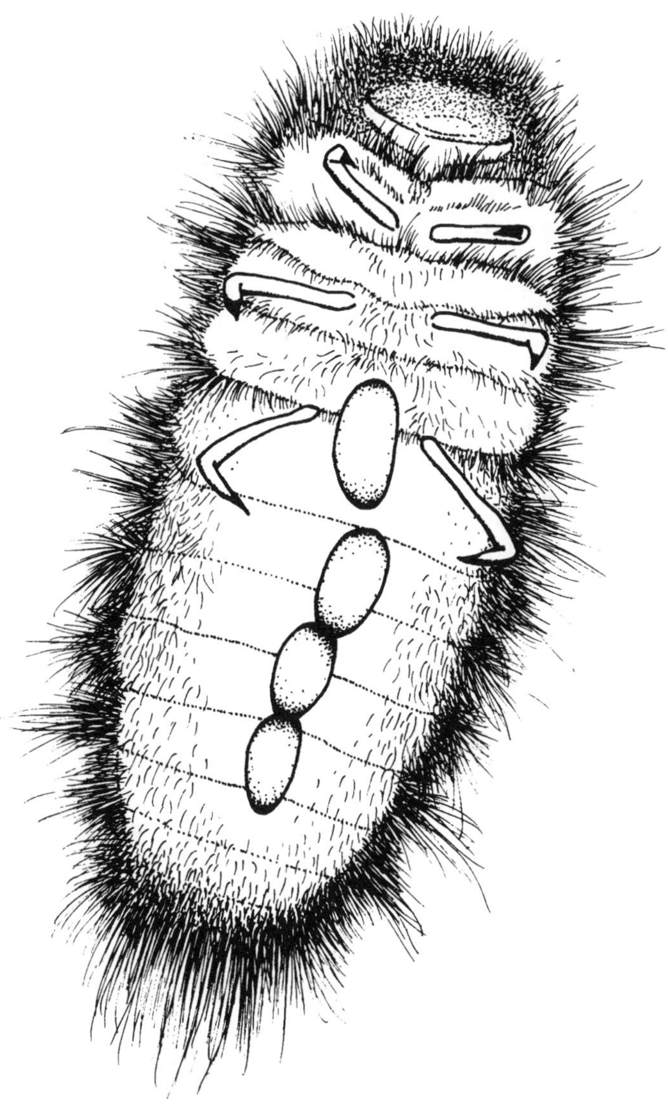

Figure 2-4. Four eggs of *Laelius pedatus* on a host larva (redrawn by Catherine Seibert from Mertins 1980).

second female often eats the eggs of the first, thus reducing competition for her own larvae (Mayhew 1997).

COMPARATIVE SURVEY OF BETHYLID FORAGING BEHAVIORS

Inspection of bethylid host records reveals some broad affinities at the subfamily level (Richards 1939, Evans 1978, Krombein et al. 1979b, Gordh and Móczár 1990). Species of the Bethylinae are parasitoids of moth larvae; Epyrinae attack the larvae and pupae of various beetles; and at least one Old World species of Mesitiinae feed on larvae of chrysomelid leaf beetles (Goulet and Huber 1993) (Table 2-2). Although *Pristocera armifera* has been reared from click beetles (Elateridae), apparently no other host records exist for Pristocerinae. However, bethylids of the genera *Dissomphalus*, *Parascleroderma*, and *Pseudisobrachium* have been collected from ant nests, suggesting that some pristocerines may be parasitoids of ants or their nest associates (Richards 1939, Evans 1978, Gordh and Móczár 1990). In comparisons of closely related species of Bethylidae, the number of eggs laid on hosts tends to increase with host size. However, across the entire family, although larger bethylids tend to attack larger host species, they do not necessarily lay more eggs (Mayhew and Hardy 1998).

Because data on the natural intraspecific host ranges of these small and inconspicuous wasps are not easy to come by, the available records may often underestimate their actual niche breadths (Evans 1978). Indeed, a few field and lab studies reveal that some bethylids have wide host ranges. In Japan, *Goniozus japonicus* exploits leaf-tying and leaf-rolling Lepidoptera from at least 20 species in 5 families that occur on such diverse plants as willow, cherry, oak, sumac, and alder (Iwata 1949, 1961; Kishitani 1961). This flexibility probably allows females to track seasonal changes in host availability (Iwata 1961). Furthermore, bethylids may attack an amazing variety of hosts when given an opportunity in the laboratory. *Sclerodermus immigrans* attacks species in three families of beetles in the field, but in the lab, they also parasitized larvae of three other beetle families (Table 2-2) (Bridwell 1919). Most surprisingly, Bridwell reared them from larvae of bees, wasps, and termites in the lab, suggesting that host range is not just a function of host nutritional suitability.

When near a potential host, bethylid females probably home in using host odors. Females of *L. pedatus* are attracted to airborne

Table 2-2 Host records and behavior of selected species of Bethylidae

Species	Host family	Type of paralysis[1]	Location of egg on host[2]	Disposition of host after attack	No. of eggs per host	References
Bethylinae						
Bethylus fuscicornis	Tineidae, Noctuidae	—	—	Moved	4–6	Richards 1939
Goniozus aethiops	Gelechiidae	P	Lateral (I)	Left in situ	1–8	Gordh & Evans 1976
G. cellularis	Gelechiidae, Olethreutidae, Pyralidae	—	—	Moved	Up to 15	Rude 1937
G. emigratus	Gelechiidae, Pyralidae	—	Dorsal (L)	Left in situ	2–8	Bridwell 1919
G. jacintae	Tortricidae	—	Lateral (I)	Left in situ	1–7	Danthanarayama 1980
G. japonicus	Eucosmidae, Gelechiidae, Gracilariidae, Pyralidae, Tortricidae	P	Dorsal or lateral (I)	Left in situ	1–7	Iwata 1949, 1961; Kishitani 1961
G. legneri	Gelechiidae, Phycitidae	P	Dorsal or lateral	Left in situ	5–20	Gordh et al. 1983

Table 2-2 *Continued*

Species	Host family	Type of paralysis[1]	Location of egg on host[2]	Disposition of host after attack	No. of eggs per host	References
G. longinervus	Gelechiidae	P	Dorsal (I)	Left in situ	1–12	Gordh 1976
G. marasmi	Pyralidae	T	Lateral (I)	Left in situ	1–12	Venkatraman & Chacko 1961
G. natalensis	Pyralidae	T	Dorsolateral (I)	Left in situ	1–40	Conlong & Graham 1988a,b
G. nephantidis	Xylocoridae, Pyralidae, Oecophoridae	—	Lateral	Left in situ	3–18	Antony & Kurian 1960, Kapadia & Mittal 1986, Hardy & Blackburn 1991
G. sensorius	Pyralidae	T	Lateral	—	3–15	Peter & David 1991
G. thailandensis	Pyralidae	T	Dorsal (I)	—	1–17	Witethom & Gordh 1994
Epyrinae						
Allepyris microneurus	Dermestidae	P	Ventral (L)	—	1–4	Yamada 1955
Cephalonomia gallicola	Anobiidae	P?	Dorsal (I)	Left in situ	1–9	Kearns 1934

C. stephanoderis	Scolytidae	P?	Ventral on larval hosts; beneath elytra on pupae	—	1	Abraham et al. 1990
C. utahensis	Scolytidae	P	—	Left in situ	1–6	Schaeffer 1962
C. waterstoni	Cucujidae	P	Ventral (L)	Moved	1–4	Finlayson 1950a,b Rilett 1949
Epyris eriogoni	Tenebrionidae	—	Ventral	Moved	1	Rubink & Evans 1979
E. extraneus	Tenebrionidae	P	Ventral (L)	Moved	1	F. Williams 1919a
Laelius pedatus	Dermestidae	P	Ventral (L)	Left in situ	2–4	Mertins 1980
L. trogodermatis	Dermestidae	P	Ventral	Left in situ	1–6	L. Howard 1901
L. utilis	Dermestidae	P	Ventral	—	2–6	Mertins 1985
Prorops nasuta	Scolytidae	—	Lateral, dorsal, or ventral	—	1	Abraham et al. 1990
Sclerodermus immigrans	6 families of Coleoptera	P	—	Left in situ	10–~40	Bridwell 1920
Pristocerinae						
Pristocera armifera	Elateridae	—	Ventral?	—	1	Hyslop 1916

Source: Modified and expanded from Gordh & Evans 1976.
[1]P, permanent; T, transient. [2]I, intersegmental; L, parallel to longitudinal axis of the host's body and sometimes across several segments.

chemicals produced by *Trogoderma* larvae (Qi and Burkholder 1990, Qi et al. 1990); *Goniozus natalensis* respond to volatile chemicals emanating from the frass produced when the host burrows into sugarcane stalks (G. Smith et al. 1994); and *Cephalonomia waterstoni* orient to chemical trails left on the substrate by *Cryptolestes* larvae (R. Howard and Flinn 1990).

Once a host has been found, the method of attack is fairly standard across the family. The female immediately mounts its victim and quickly stings it in a species-specific location, usually once on or near its neck if it is a larva (e.g., all *Goniozus*), but sometimes farther back on the thorax (L. Howard 1901). Exceptions include *Cephalonomia waterstoni*, which seems to have no particular preferred stinging site (Finlayson 1950a) and *Sclerodermus immigrans*, which stings its host repeatedly in different locations (Bridwell 1920). Stinging is probably imprecise with regard to the location of ganglia in the central nervous system (Piek 1985), but paralysis is usually rapid and permanent (Table 2-2).

Hosts may resist bethylid parasitization by using vibrating body hairs (Ma et al. 1978) or by recovering from paralysis, after which they may dislodge the eggs (Mertins 1982). Some may even molt while partially paralyzed, resulting in the death of the parasitoid larvae (Gordh et al. 1983, Klein and Beckage 1990). Potential hosts also commonly thrash about in an attempt to dislodge the female as she positions herself for stinging (e.g., Rilett 1949), but females usually hang on tightly to the host's integument with their mandibles and legs. Other hosts counterattack more aggressively, attempting to bite the female and occasionally succeeding in severing limbs or the head of their antagonist (Bridwell 1919, 1920; Gifford 1965; Gordh 1976; Gordh and Evans 1976). The ability of hosts to counterattack may explain why female bethylids often mount the victim just behind the head, thereby making it difficult for a host larva to turn around and bite them.

Adult female bethylids often feed on exudations of host hemolymph, sometimes on "hosts" on which they do not oviposit (e.g., Bridwell 1919, Kearns 1934). Females may feed at flowers, and sometimes they eat their own eggs if the host's condition deteriorates before they depart (Bridwell 1920). Females are also known to consume the eggs of conspecific females before laying their own eggs on a host (Mayhew 1997).

Chrysididae

As a group, chrysidids exhibit more diverse foraging habits than either Dryinidae or Bethylidae. The Chrysididae includes brood parasites (Chapter 4), ectoparasitoids, and endoparasitoids, including the only egg parasitoids among the aculeate Hymenoptera. Most species are solitary feeders, but a few feed gregariously as larvae.

Case Study: *Praestochrysis shanghaiensis*

O. Piel (1933) described the female *Praestochrysis shanghaiensis* as a "beautiful insect . . . 11 mm or so in length . . . of a bright green color with metallic tints, adorned in parts with deep blue" (Fig. 2-5). It is widely distributed in the Orient, from India to Japan and Java (Piel 1933, D. Parker 1936, Krombein et al. 1979a). In the early 1900s, entomologists made several unsuccessful attempts to introduce it into the United States to aid in control of the oriental moth (*Monema flavescens*), a shade tree pest also native to the Far East. The biology of *P. shanghaiensis* was studied first by Piel (1933) and Parker (1936) and more recently by Yamada (1987a,b; 1988; 1990) in its native habitat.

Caterpillars of the oriental moth, the only known host of *P. shanghaiensis*, spin 1.5-cm long cocoons on the twigs, branches, or trunks of their host trees. Several days after adult female *P. shanghaiensis* emerge, their eggs mature and they search for moth cocoons. When a female finds a cocoon, she antennates it for several minutes, perhaps to determine if it is alive, unparasitized, and of a suitable age. Cocoons with dead moth larvae or with other *P. shanghaiensis* larvae already present are usually rejected, and only recently formed cocoons are used because those even several hours old are too hard for the wasp to penetrate. At suitable cocoons, the female braces herself against the surface of the tree using five sharp teeth at the tip of her abdomen and chews a round 0.5–0.7 mm diameter hole in the cocoon. She then inserts her flexible 7-mm-long ovipositor and lays a single egg within the cocoon. The chrysidid ovipositor is not homologous with the apocritan ovipositor or the aculeate sting, having independently evolved via modification of the terminal abdominal segments (Kimsey and Bohart 1990). Female *P. shanghaiensis* have fairly low fecundity for a parasitoid. Even in the laboratory with hosts provided ad lib, females rarely laid as many as two or three eggs on

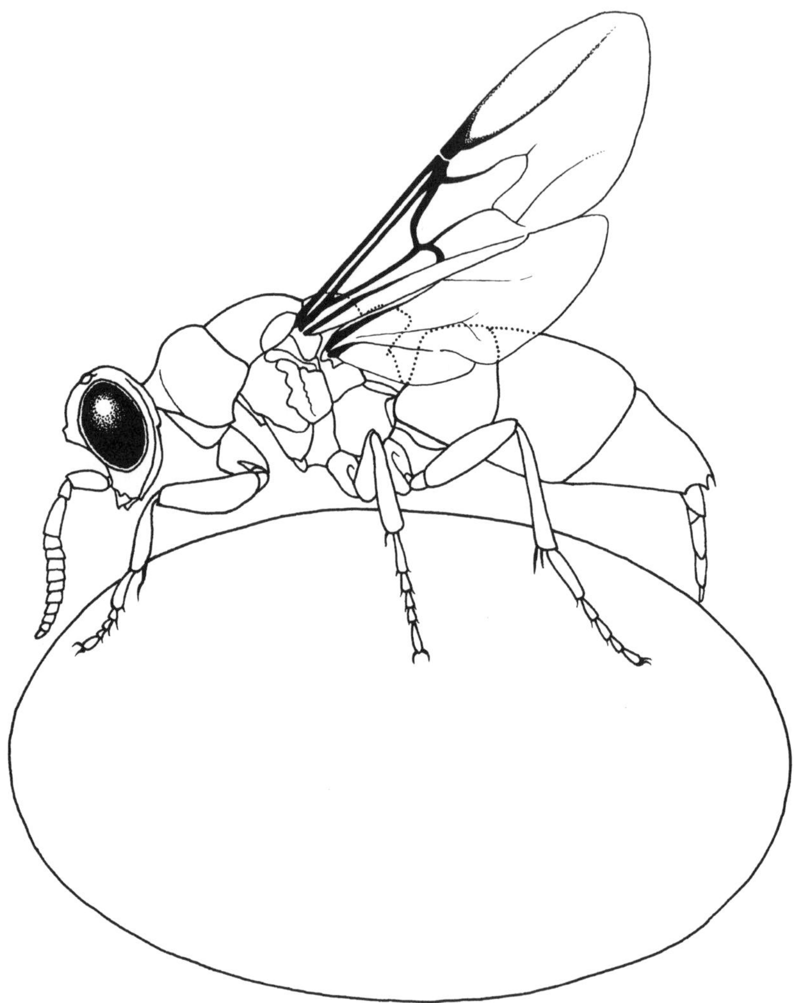

Figure 2-5. Female *Praestochrysis shanghaiensis* showing relative size of host cocoon (Redrawn from Kimsey and Bohart 1990).

a single day and never laid more than 47 during their lives (Yamada 1987b).

Praestochrysis shanghaiensis stings and incompletely paralyzes the host (Piel 1933, Parker 1936, Yamada 1987a), and it has large venom glands that may be over one-third of the wasp's body length

(Buysson 1898, in Piel 1933). The fact that the wasp indeed stings the host is unusual because the "sting is quite reduced and essentially non-functional" in female chrysidids (Kimsey and Bohart 1990:35). After oviposition, the female plugs the hole in the cocoon using saliva mixed with debris generated while chewing the hole. Occasionally, females (probably mistakenly) deposit eggs within cocoons previously attacked by a conspecific female, with the result that the second female's larva is cannibalized by the older larva of the first female (Piel 1933, Yamada 1988). The wasp's egg hatches in 4 days, and the larval development proceeds through five stages (Parker 1936) (or in three, according to Piel [1933]). In China and Japan, *P. shanghaiensis* is bivoltine, with the mature larvae of the second generation overwintering in the moth's cocoon.

COMPARATIVE SURVEY OF CHRYSIDID PARASITOID FORAGING BEHAVIORS

R. Bohart and L. Kimsey (1982) consider the ancestor of the family Chrysididae to be a parasitoid of free-living insects, perhaps beetles, moths, or sawflies. Within the chrysidid lineage, the host preferences and foraging behaviors of parasitoids have diversified, with hosts falling into three orders (Kimsey and Bohart 1990) (Table 2-3): Phasmida, Hymenoptera, and Lepidoptera.

The subfamilies Amiseginae and Loboscelidiinae form a discrete lineage whose larvae feed exclusively on eggs of walkingsticks (order Phasmida) (Heather 1965, Krombein 1983, Kimsey and Bohart 1990). These chrysidids are the only aculeates whose larvae feed strictly on the eggs of their hosts and that are endoparasitoids throughout development. The egg is laid within the host's egg (Fig. 2-6), and the wasp does not emerge until it becomes an adult. Chrysidids of these two subfamilies thus share this habit with wasps of several families of Parasitica, including the Trichogrammatidae and Mymaridae (Askew 1971). Amisegines probably feed on the yolk rather than on the developing embryo (Krombein 1983).

Females of the subfamily Cleptinae search trees or the litter beneath trees for cocoons containing mature larvae of sawflies (Hymenoptera) (Table 2-3). As do *P. shanghaiensis* (Chrysidinae), female cleptines chew holes in the host cocoons, deposit a single egg within, and then seal the opening, sometimes with a secretion from the tip of the abdomen (Iwata 1976). Cleptine larvae complete development and pupate within host cocoons.

Table 2-3 Host records and behavior of selected species of Chrysididae

Species	Host family or subfamily[1]	Oviposition site	Source of food	Notable observations	References
Amiseginae					
Adelphe anisomorphe	Pseudophasmatidae (Phasmida)	Inside egg	Host egg	—	Krombein 1960
Amisega kahlii	Heteronemiidae (Phasmida)	Inside egg	Host egg	—	Milliron 1950
Duckeia cyanea	Pseudophasmatidae (Phasmida)	Inside egg	Host egg	—	Costa Lima 1936 (in Krombein 1983)
Myrmecomimesis rubrifemur	Phasmatidae (Phasmida)	Inside egg	Host egg	—	Hadlington & Hoschke 1959
M. semiglabra	Phasmatidae (Phasmida)	Inside egg	Host egg	—	Riek 1955, Hadlington & Hoschke 1959, Readshaw 1965
Loboscelidiinae					
Loboscelidia spp.	Phasmatidae (Phasmida)	—	Host egg	—	Riek 1970
Cleptinae					
Cleptes purpuratus	Diprionidae	Inside cocoon	Host prepupa	Female chews hole in cocoon; plugs it when done	Dahlsten 1961, 1967

C. semiauratus	Tenthredinidae	Inside cocoon	Host prepupa	Female chews hole in cocoon; plugs it when done	Darling & Smith 1985, Iwata 1976
Chrysidinae					
Allocoelia bidens	Masarinae	—	—	—	F. Gess & Gess 1980
Chrysis angolensis	Sphecinae, Eumeninae	Inside cocoon	Prepupa	Female chews through wall of mud nest and host cocoon; plugs puncture in nest when done	Stage 1960, Bohart & Kimsey 1982
Chrysura kyrae	Megachilidae	Host cell	Prepupa	1–3 eggs per cell, but only one larva survives; some cannibalism; no feeding until host spins cocoons	Krombein 1967
C. pacifica	Megachilidae	Host cell	Prepupa	Larva forgoes feeding until host completes growth	C. Hicks 1933b, Bohart & Kimsey 1982
C. davidi	Megachilidae	On pollen mass in cell	Prepupa	1–4 eggs per host cell, but one larva kills others; larvae forgo feeding until host completes growth	Iwata 1976

Table 2-3 *Continued*

Species	Host family or subfamily[1]	Oviposition site	Source of food	Notable observations	Reference
Chrysurissa densa	Masarinae Megachilidae	Host cell	—	Female chews through hardened cap of host cell	H. Hicks 1927, Hungerford 1937
Parnopes edwardsii	Bembecinae	Host cell	Prepupa	Female digs into host nest	Bohart & MacSwain 1940, Evans 1966c
P. fulvicornis	Bembecinae	Host cell	Prepupa	Female digs into host nest	Bohart & MacSwain 1940
Praestochrysis shanghaiensis	Limacodidae (Lepidoptera)	Inside host cocoon	Prepupa or pupa	Female chews oviposition hole into cocoon and plugs it when finished	Piel 1933, D. Parker 1936, Yamada 1987a,b, 1988, 1990
Stilbum cyanurum	Eumeninae, Sphecinae	On larva inside cell of mud nest	Final-stage larva	Female chews through mud wall of nest to reach host	F. Williams 1919c, Móczár 1961, Kimsey & Bohart 1990, Iwata 1976, West-Eberhard 1987

[1]Subfamily is given for hosts that are solitary aculeates; all hosts in order Hymenoptera, except where indicated.

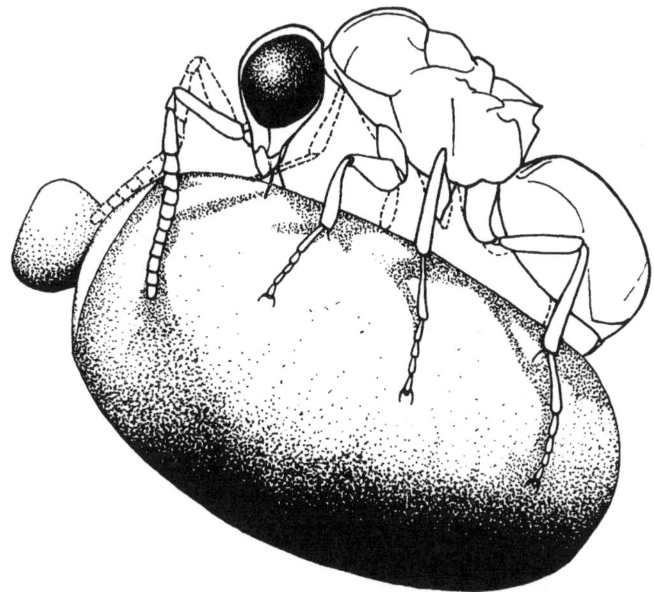

Figure 2-6. Female of *Myrmecomimesis semiglabra* ovipositing in a walkingstick egg (redrawn by Catherine Seibert from Readshaw 1965).

The subfamily Chrysidinae consists of brood parasites (Chapter 4) and ectoparasitoids. With the exception of the few *Praestochrysis* species that are parasitoids of Lepidoptera, the parasitoids of this subfamily attack solitary aculeate bees and wasps (Hymenoptera) within their nests, their larvae feeding on the host's mature larvae or prepupae (Table 2-3). After locating a host nest, a female lays one or more eggs, each in a different nest cell. Gaining access to the host may require laborious penetration of the cell wall of the host's nest, and those chrysidids that attack mud-nesting wasps must breach the nest wall before it hardens.

The timing of oviposition varies among species of parasitoid Chrysidinae, but oviposition generally falls within one of two categories. In one type, represented by *Stilbum cyanurum* and species of *Chrysura*, the females deposit eggs in the host nest over a wide range of times during or after provisioning of the cell by the host female. Although the chrysidid larva hatches immediately, it remains in the first instar, attached to the host but not feeding until the host larva has fully developed. In the second type, of which *Chrysis fuscipennis* is an example, the female oviposits at about the time that the host reaches the prepupal stage, so that the chrysidid can hatch and immediately begin feeding on a large host.

Other Chrysidoidea

The foraging biology of chrysidoid wasps other than the Dryinidae, Bethylidae, and Chrysididae has received little attention because members of the other four families are rare and predominantly tropical in distribution. Almost nothing is known about the natural history of the Plumariidae. Limited information indicates that Scolebythidae are gregarious external parasitoids of the wood-boring larvae of long-horned beetles (Cerambycidae) (Day 1977, Brothers 1981) and that Sclerogibbidae are ectoparasitoids of the nymphs of webspinners (order Embiidina) (Callan 1939). At least one North American Embolemidae, *Ampulicomorpha confusa*, is a parasitoid of the nymphs of leafhoppers, planthoppers, and related insects (Bridwell 1958, Wharton 1989). Brachypterous (short-winged) females of some embolemids have been collected in ant nests (Donisthorpe 1927) and in burrows of small mammals (Heim de Balsac 1935), but their hosts are unknown.

Mutillidae

For naturalists interested in the spectacular, it is perhaps disappointing that the common name sometimes applied to mutillids, "cow-killers" (Mickel 1928), actually has nothing to do with their foraging habits. Apparently, it was not until the mid-1800s that entomologists realized that mutillids (velvet ants) are parasitoids; previous writers had concluded that they were social insects or commensals within nests of social insects (Mickel 1928). Mutillids do live within the nests of solitary bees and wasps, but as parasitoids.

Case Study: *Pseudomethoca frigida*

Females of *Pseudomethoca frigida* are brown, wingless wasps, just 4–6 mm long (Fig. 2-7). The bee *Lasioglossum zephyrum* is one of its major host species (Brothers 1972). A typical nest of *L. zephyrum*, which is often inhabited by several female bees, consists of a

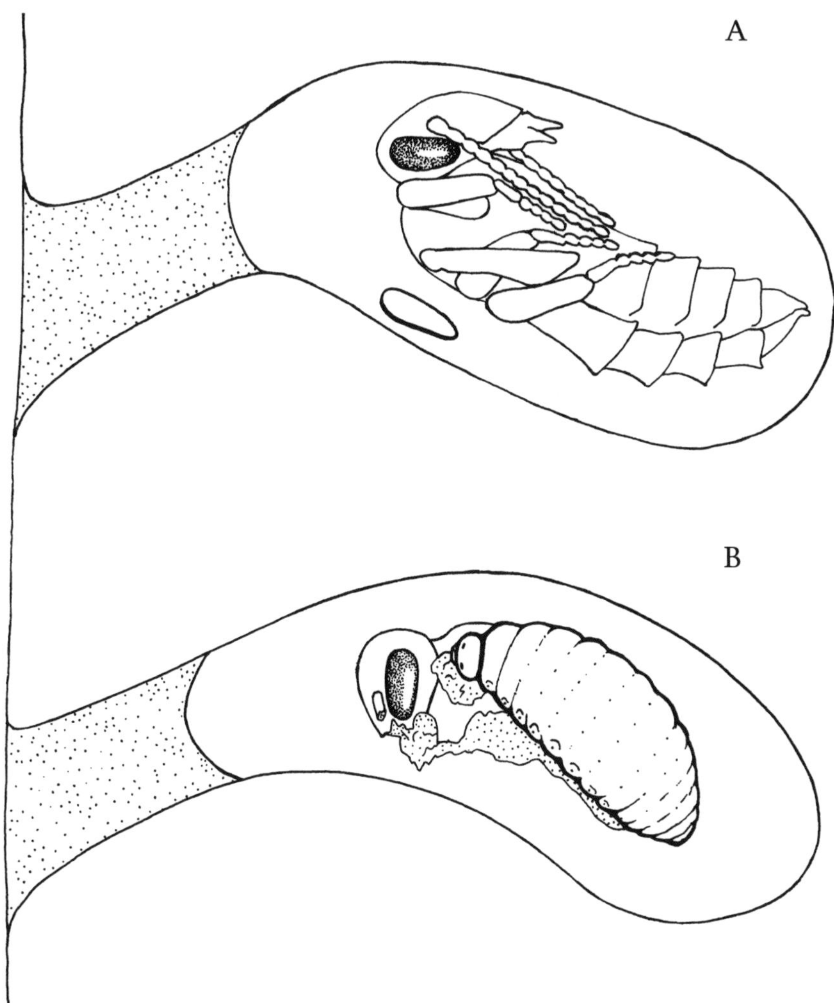

Figure 2-7. (A) Position of *Pseudomethoca frigida* egg in nest cell of *Lasioglossum zephryum* containing a pupa. (B) Larva of *P. frigida* feeding on pupa of *L. zephryum* (stippled area on bee pupa indicates tissue exposed by feeding). (A and B redrawn from Brothers 1972.)

branching set of tunnels descending from a single nest entrance into a nearly vertical soil bank. Adult female *P. frigida* emerge later than their hosts so that they can attack *L. zephyrum* when host prepupae and pupa are present. A *P. frigida* attempting to gain entrance to a nest must often deal with the guard bee, which uses its abdomen to block the entrance. A female *P. frigida* may try to bypass the guard, sometimes by digging around it, but most such attempts are unsuccessful and the female moves on to another nest. Some guards are more aggressive, emerging from their posts to wrestle, bite, and sting the intruder. Such attacks by *L. zephyrum* guards may be successful but sometimes result in the death of the guard when the mutillid counterattacks (Melander and Brues 1903, Krombein 1938, Batra 1965, Brothers 1972).

It is critical to the bees that the female velvet ant be prevented from entering the nest, because bees deep within the burrow usually do not hinder the wasp's movements. When a female *P. frigida* does enter a nest, she moves to the bottom of the burrow system to search for cells containing immature bees at an advanced stage of development. Upon locating a cell, the female first removes the soil barrier that walls off the cell. After determining whether the occupant is a mature larva, prepupa, or pupa, the female wasp inserts her abdomen into the cell, probes for a few minutes with her sting, and dumps a single egg somewhere onto the host or the wall of the cell. The mutillids sting only hosts that have reached the pupal stage, delaying host development for the 4 days it takes for the mutillid larva to hatch and begin feeding. After laying an egg, the wasp loosens soil particles with her front legs and mandibles, passes them beneath her body, and uses the tip of her abdomen to pack the soil into place and reseal the cell.

Once inside a nest, some females attack more than one cell, sometimes up to six within 36 hours. Because females apparently do not recognize or remember cells they have already attacked until they reopen the cell, they occasionally lay a second egg in a cell, probably by mistake.

After taking about 4 days to hatch, *P. frigida* larvae pass through five larval stages in about 8–9 days, while consuming the entire host. During the final stage, the larva spins a silken cocoon. In all but the final generation of the season, the prepupa molts to a pupa and remains in that stage for about 2 weeks. Development in the overwintering generation is halted at the prepupal stage until the following summer.

COMPARATIVE SURVEY OF MUTILLID FORAGING BEHAVIORS

Although the Mutillidae are most closely related to the Tiphiidae, their foraging habits are more similar to the ectoparasitoid Chrysidinae (Chrysididae). Most Mutillidae are solitary ectoparasitoids of the mature larvae, prepupae, or pupae of their hosts (Table 2-4). Species of at least 5 families of bees and 10 families of wasps play host to mutillids (see also Mickel 1928, Krombein et al. 1979b, Krombein 1992), but a few mutillids are ectoparasitoids of the resting stages of Lepidoptera, Coleoptera, and Diptera (Mickel 1928, Clausen 1940, Brothers 1972). Unlike those of the Bethylidae and Chrysididae, however, host preferences do not follow presently recognized subfamily divisions of the Mutillidae. Thus, *Mutilla clythrae,* which attacks chrysomelid beetle larvae inside ant nests, and *Chrestomutilla glossiniae,* which attacks pupae of tsetse flies (*Glossinia morsitans*), are both within subfamilies whose members usually attack aculeates. Host-parasitoid relationships are unknown for several subfamilies of Mutillidae (Brothers 1972).

Not enough information is available to determine the specificity of mutillid host-parasitoid relationships. *Chrestomutilla glossiniae* attacks pupae of *G. morsitans* but apparently not of the related and larger *G. brevipalpus* (Lamborn 1916). However, species such as *Dasymutilla vesta* and *Sphaeropthalma orestes* attack many different aculeate hosts. Although other mutillids have been found on a narrow range of hosts, many studies report very small sample sizes from a single site, so it is possible that many species have wider host ranges than we are aware of. Females of mutillids that attack Hymenoptera enter host nests and oviposit either within the cocoon or within the host cell if the host larva does not spin a cocoon. *Chrestomutilla glossiniae,* however, oviposits within tsetse fly puparia found on the soil surface. Female mutillids commonly reseal the oviposition puncture in host cocoons either with a secretion or with extrinsic materials such as soil.

Most Mutillidae probably lay just a single egg on each host (Brothers 1984). Multiple oviposition by *Pseudomethoca frigida* apparently occurs only when the female fails to detect an egg already laid by another female (Brothers 1972). W. Ferguson (1962) found up to seven *Sphaeropthalma orestes* eggs or larvae within single nest cells of its bee host. The *S. orestes* larvae from eggs laid after the first egg may develop as hyperparasitoids of conspecifics, or the oldest

Table 2-4 Host records and behavior of selected species of Mutillidae

Species	Larval host family or subfamily[1]	Oviposition site	Stages attacked	Notable observations	Reference
Mutillinae					
Ephuta slossonae	Pompilidae	In nest cell	On prey (araneid spider)	Thought to be an obligate brood parasite of pompilids	Krombein & Norden 1996
Mutilla attenuata	Halictidae	In host nest cell	Pupa	—	Janvier 1933 (in Clausen 1940)
M. clythrae	Chrysomelidae	—	Larva	Hosts live inside ant nests	Iwata 1976
M. europaea	Apidae	In closed nest cell	Prepupa	Adult females also feed on adult honey bees	Mickel 1928
Mutilla sp.	Tiphiidae	In host cocoon	Prepupa	—	F. Williams 1919c
Myrmosinae					
Ephutomorpha ignita[2]	Sphecinae	In host cocoon	Prepupa or pupa	Four larvae per host cocoon	Brothers 1984
Myrmosa bradleyi	Crabroninae	—	—	—	Court 1961, Bohart & Menke 1976
M. unicolor	Tiphiidae Halictidae	—	—	—	Melander & Brues 1903

Myrmosula parvula	Halictidae	In closed nest cell	Pupa	Female closes cell after ovipositing	Brothers 1978
Sphaeropthalminae					
Chrestomutilla glossinae	Muscidae	In host puparium on soil surface	Pupa	Female chews puncture in puparium, seals the hole with a secretion, and covers puparium with soil when finished; multiple parasitism common	Lamborn 1915, 1916; Heaversedge 1968, 1969
Dasymutilla bioculata	Bembecinae	In host cocoon	Prepupa	After oviposition, female seals hole with mixture of sand grains and saliva	Cottrell 1936 (in Brothers 1972)
D. nigripes	Philanthinae	—	—	—	Krombein et al. 1979b
D. occidentalis	Apidae	—	—	—	Fattig 1943
D. scaevola	Philanthinae	In host cocoon	Pupa or prepupa	—	Hook & Evans 1991
D. vesta	Bembecinae, Halictidae, Sphecinae	—	—	—	Fattig 1943, Krombein et al. 1979b
Pseudomethoca frigida	Halictidae	In closed cell within host nest	Mature larva, prepupa, pupa	Female plugs hole after oviposition	Brothers 1972

Table 2-4 *Continued*

Species	Larval host family or subfamily[1]	Oviposition site	Stages attacked	Notable observations	Reference
Sphaeropthalma orestes	Crabroninae, Eumeninae, Megachilidae, Mutillidae	In host cell within host nest	"Mature larva"	One cell attacked up to 3 times; later larvae feed on pupae of earlier females	W. Ferguson 1962
S. pensylvanica	Crabroninae	In host cocoon in host nest	Prepupa	Female chews through mud cap, partition, and cocoon to lay egg, then seals it with mud	Krombein 1967, Matthews 1997
S. unicolor	Anthophoridae, Megachilidae, Sphecinae	In host nest cell	"Mature larva"	—	Ferguson 1962

[1]For hosts that are solitary aculeates. [2]May actually be a member of some other undescribed genus (Brothers 1984).

larva may kill the younger larvae. Similarly, newly parasitized puparia of tsetse flies usually contain two or three eggs of *Chrestomutilla glossiniae*, but the first larva to hatch invariably kills its potential competitors (Heaversedge 1969). However, several species of *Ephutomorpha* are apparently true gregarious parasitoids, rearing up to four young on a single host (Brothers 1984).

Adult females of some mutillids prey upon adult insects. Female *Pseudophotopsis continua* enter burrows of the sand wasp *Bembix olivacea* and prey on adult females (Mellor 1927). Similarly, *Mutilla europaea* females, whose larvae are parasitoids of bumble bees, also invade colonies of honey bees and kill huge numbers of adults, leaving the immature stages untouched (Mickel 1928). Like tiny vampires, females of both species bite their victims in the neck and draw out body fluids. C. Clausen (1940) reported that one observer had seen as many as 200 worker bees being killed in a single day by mutillids.

Tiphiidae and Scoliidae

In the past, the families Tiphiidae and Scoliidae were classed together either in the superfamily Scolioidea (Krombein et al. 1979b) or in the family Scoliidae (Clausen et al. 1932). Recent taxonomic treatments, which recognize a more distant relationship, place both families in the superfamily Vespoidea but within the two major lineages of that group (see Fig. 1-1). Here, however, I will consider them together because larvae of most Tiphiidae and Scoliidae are solitary ectoparasitoids of scarab beetles.

Case Study: *Tiphia popilliavora*

Tiphia popilliavora was one of a dozen or so *Tiphia* species introduced into the United States from Japan, China, and Korea during the 1920s and 1930s with the goal of controlling scarab beetle pests (Clausen et al. 1932). By the late 1940s, several had become well established in the eastern United States, with mixed results as to their effectiveness (Krombein 1948, Clausen 1956). *Tiphia popilliavora*, introduced to control the Japanese beetle (*Popillia japonica*), has been a poor control agent but has become one of the best-studied tiphiids (McColloch et al. 1928; King and Holloway 1930; Holloway 1931; Clausen et al. 1932; Brunson 1934, 1938; Clausen 1940, 1956;

White 1943). Besides *P. japonica*, it also attacks *Phyllopertha* species, as well as three other *Popillia* species in China and Korea.

The shiny black, 1-cm-long females of this tiphiid (Fig. 2-8) hunt by tunneling their way into soil and searching for scarab larvae residing in feeding cells. When attacking *P. japonica*, *T. popilliavora* will accept both second- and third-instar larvae, but they prefer the larger third instars (Brunson 1938). Upon locating a grub, a female mounts it from behind and crawls forward to curl her abdomen beneath the host and sting it in the thorax. Stinging may be repeated until the larva is completely paralyzed, although it will recover within 20–40 minutes. After kneading the ventral surface of the quiescent larva with her mandibles for several minutes, the female applies the tip of her abdomen to the groove between the fifth and sixth abdominal

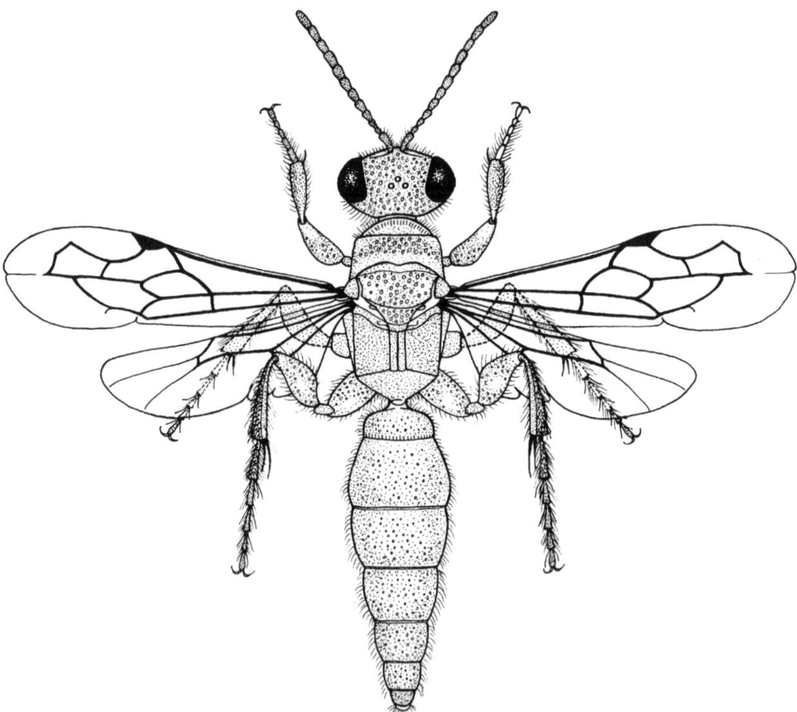

Figure 2-8. Adult female *Tiphia popilliavora* (modified and redrawn by Catherine Seibert from Clausen et al. 1932).

segments. Her doing so may broaden the groove and "rasp" the integument to prepare it for attachment of the egg and penetration by her own larva (Clausen 1940). A single egg is then deposited within the groove and attached by a "mucilaginous material." After oviposition, the female may chew one of the legs of the larva until hemolymph exudes and is fed upon. Females also feed on nectar at flowers and extrafloral nectaries as well as on honeydew.

Within 5–7 days, the anterior end of the wasp's egg shell splits, and the larva extends its head and feeds, but it does not completely split out of the egg until it grows larger. After each molt, the larva makes a few feeding punctures on the living host until the last stage, when it completely devours the host (exclusive of the legs and head capsule). After 18–30 days, the wasp larva spins a cocoon in the host's feeding cell and passes the winter underground as a mature larva (McColloch et al. 1928, Clausen 1940).

COMPARATIVE SURVEY OF TIPHIID AND SCOLIID FORAGING BEHAVIORS

Most Tiphiidae attack the larvae of either scarab beetles (Scarabaeidae) or tiger beetles (Cicindelidae) (Table 2-5). Exceptions include *Myzinum andrei*, *Hylomesa* spp., and *Diamma bicolor*, but all tiphiid hosts live cryptically in soil or in plant tissue. Scoliid wasps also attack scarab beetle larvae, sometimes those of the same genera used by tiphiids (Table 2-6). However, A. Molumby (1995) reared three *Scolia* species from cocoons of the sphecid *Trypoxylon politum*, concluding that they were parasitoids of the wasp. Females of each tiphiid and scoliid species lay a single egg on each host at a species-specific location (Fig. 2-9).

Females of *Methoca* and *Pterombrus* are parasitoids of tiger beetle larvae (Cicindelidae) that live in the soil in vertical burrows from which they strike at passing prey with their sharp, curved mandibles. Surprisingly, females of both genera conduct a frontal attack directly into the formidable jaws of the much larger beetle larvae, using their small size, body shape, and armor to avoid injury. Female *Methoca californica* enter the burrow of a *Cicindela* larva and quickly mount its head as the larva lunges forward and snaps it mandibles closed (Burdick and Wasbauer 1959). The slender, wingless, heavily sclerotized wasp thus gets inside the defenses of the larva, in the space between the beetle's mandibles and its head. From this position, the female quickly crawls beneath the beetle and stings it between the

Table 2-5 Host records and behavior of selected species of Tiphiidae

Species	Host family	Stinging and paralysis[1]	Location of egg on host	Disposition of host after attack	References
Anthoboscinae					
Anthobosca chilensis	Scarabaeidae	—	—	—	Janvier 1933 (in Clausen 1940)
Tiphiinae					
Tiphia lucida	Scarabaeidae	T	Venter between thorax and abdomen	—	F. Williams 1919c
T. popilliavora	Scarabaeidae	M-T	Venter of abdomen	Left in host feeding cell	King & Hollway 1930, Clausen et al. 1932, Clausen 1940
T. segregata	Scarabaeidae	—	Near ventral tip of abdomen	—	Williams 1919c
T. vernalis	Scarabaeidae	M-T	Venter of thorax or abdomen	Left in host feeding cell	Clausen et al. 1932
Methocinae					
Methoca californica	Cicindelidae	M-T	Near a hind coxa	Left in burrow, buried by wasp	Burdick & Wasbauer 1959
M. ichneumonoides	Cicindelidae	M-T	Near a hind coxa	Left in burrow, buried by wasp	F. Williams 1916

M. striatella	Cicindelidae	S-P	Venter of abdomen	Left in burrow, buried by wasp	F. Williams 1919c, Iwata 1976
M. stygia	Cicindelidae	M-T	Near hind coxae	Left in burrow, buried by wasp	F. Williams 1916, E. Wilson & Farish 1973
M. yasumatsui	Cicindelidae	S-P	Venter of abdomen	Left in burrow, buried by wasp	Iwata 1936
Myzininae					
Hylomesa sp.	Cerambycidae				Krombein 1968
Myzinum andrei	Tenebrionidae	—	—	—	Iwata 1976
M. quinquecinctum	Scarabaeidae	S-P	Venter of abdomen	Left in situ	J. Davis 1919
Pterombrus cicindelus	Cicindelidae	P	Venter of abdomen	Left in burrow, buried by wasp	F. Williams 1928
P. iheringi	Cicindelidae	P	Venter of abdomen	Left in burrow, buried by wasp	F. Williams 1928
P. piceus	Cicindelidae	M-P	Venter of abdomen	Left in burrow, buried by wasp	Palmer 1976
P. rufiventris	Cicindelidae	M-P	Venter of abdomen	Left in burrow, buried by wasp	Knisley et al. 1989
Thynninae					
Epactothynnus opaciventris	Scarabaeidae	P	Venter of abdomen	—	F. Williams 1919b
Diamminae					
Diamma bicolor	Gryllotalpidae	M-T	—	—	Turner 1907 (in Iwata 1976)

[1]S, single sting per host; M, multiple stings per host; T, paralysis transient; P, paralysis permanent.

Table 2-6 Host records and behavior of selected species of Scoliidae

Species	Host family	Stinging and paralysis[1]	Location of egg on host	Disposition of host after attack	References
Campsomerinae					
Campsomeris annulata	Scarabaeidae	P	Venter of abdomen	Stung, then moved to deeper cell	Iwata 1976, Clausen et al. 1932
C. dorsata	Scarabaeidae	—	—	—	Iwata 1976, Clausen et al. 1932
C. marginella	Scarabaeidae	—	—	—	Lai 1988
C. plumipes	Scarabaeidae	—	Venter of abdomen	—	Kurczewski 1963a, 1966b, 1967a
Scoliinae					
Megascolia flaviphrons	Scarabaeidae	M-P	—	—	Piek 1985
Scolia japonica	Scarabaeidae	P	Venter of abdomen	Stung, then moved to deeper cell	Clausen et al. 1932
S. manilae	Scarabaeidae	M-P	Venter of abdomen	Left in host feeding cell	F. Williams 1919c
S. ruficornis	Scarabaeidae	—	—	—	Iwata 1976

[1]S, single sting per host; M, multiple stings per host; T, paralysis transient; P, paralysis permanent.

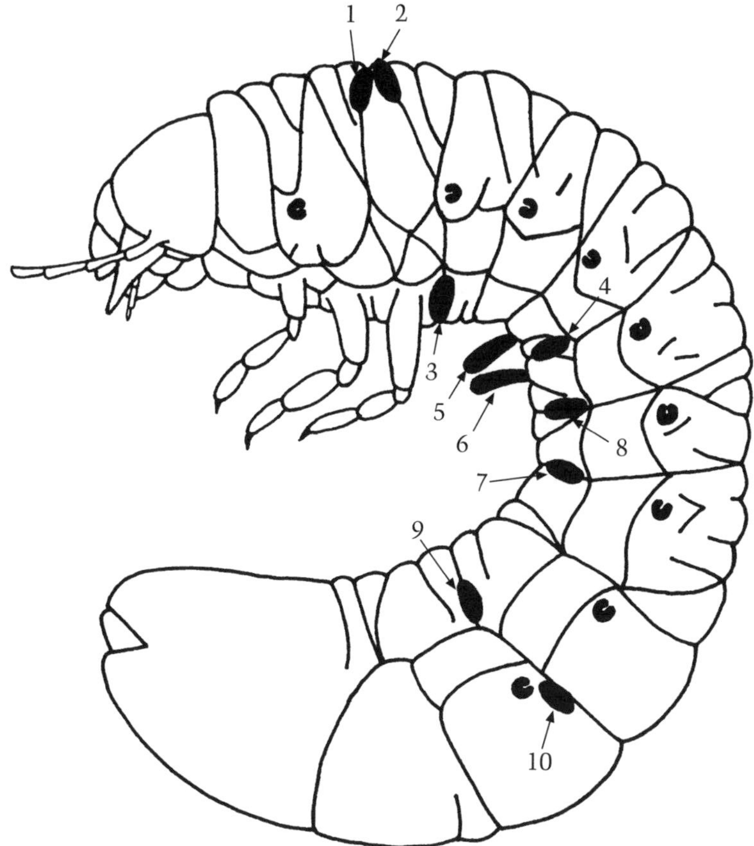

Figure 2-9. Species-specific egg positions of Tiphiidae and Scoliidae. (1) *Tiphia tegitiphaga.* (2) *T. koreana.* (3) *T. vernalis.* (4) *T. agilis.* (5) *Campsomeris annulata.* (6) *Scolia japonica.* (7) *T. popilliavora.* (8) *T. ovinigris.* (9) *T. notopolita.* (10) *T. asericae.* (1–10 redrawn by Catherine Seibert from Clausen 1940.)

neck and thorax. After numerous stings, the number apparently related to the beetle's initial degree of paralysis, the wasp lays a single egg near the hind coxa. The host of *M. californica* and several other species of *Methoca* eventually recovers from the multiple stings, but the host of other species is permanently paralyzed after a single sting (Tables 2-5, 2-6).

Host feeding by adult females is common among the Tiphiidae (Clausen et al. 1932, Burdick and Wasbauer 1959, E. Wilson and Farish 1973, Clausen 1940). Adult scoliids and tiphiids also feed on carbohydrates from flowers, extrafloral nectaries, and aphids and scale insects. In Australian Thynninae, feeding is a component of courtship; the wingless females mate only with males that provide them with nectar or honeydew (see Chapter 8).

Pompilidae

Most pompilid wasps deposit prey in nests that the female wasps build from scratch or create by substantially modifying an existing cavity. These predators will be considered in Chapter 3, and the cleptoparasitic Pompilidae will be covered in Chapter 4. Here, I will discuss the ectoparasitoid pompilids, many of which simply deposit an egg on a temporarily paralyzed host before quickly departing. Others, however, stash the paralyzed host in the host's own burrow or in the nearest available crevice.

Case Study: *Homonotus iwatai*

Several species of *Homonotus* are parasitoids of clubionid spiders of the genus *Chiracanthium* and have lifestyles clearly analogous to those of the Bethylidae, Chrysididae, Mutillidae, and Tiphiidae that attack hosts in the hosts' own refuges or nests. In this case, the refuges consist of grass leaves bound together by the spider's silk. In Japan, *Homonotus iwatai* females search for these refuges and enter them, either by chewing a hole through the leaves or by pulling apart the leaves at the silk bindings (Iwata 1932, 1976). The wasp then stings the spider between the legs, injecting a venom that paralyzes the host for just 5–10 minutes but that permanently prevents the spider from laying eggs. The spider's future is very brief in any case, because the pompilid egg hatches in 2–3 days, producing a larva that devours the spider in another 5 days. The wasp larva then spins a cocoon within the spider's refuge, which has not been modified by its mother.

COMPARATIVE SURVEY OF POMPILID PARASITOIDS

Several other parasitoid Pompilidae are also known to attack spiders within host refuges either made of plant parts or dug in the soil and covered with a trap door (Table 2-7). Usually, the spider is either stung within the refuge and left there, or it may be flushed from the refuge, stung, and then carried back in. The spider's lair is not modified by the wasp, and the spider is not even repositioned in any particular way (as occurs in nest-provisioning Pompilidae). Complete paralysis is usually only transient, but the venom often has lingering effects that serve to keep the spider from wandering and damaging the eggs or young larvae.

Although details are scant, some parasitoid pompilids, such as *Minagenia osoria*, apparently attack free-living spiders (those that do not build refuges), laying an egg on the spider after inflicting a temporary paralysis that wears off and allows the spider to continue its activities (Kaston 1959). In other cases, such as *Agenioideus cinctellus*, the wasp may take the precaution of depositing the paralyzed spider in a safe location. The exact placement of the paralyzed host and egg can influence host survival. Female *Pseudopompilus humboltdi* sting their spider host within its silken tube-shaped retreat and then drag the paralyzed host to the entrance of the tube. At this location, it is more likely to be found by foraging birds. However, the trade-off appears worth it, because wind blowing across the entrance of the tube keeps the spider up to 15°C cooler than in the inner end of the tube, where temperatures may reach 50°C, a temperature that would kill both the wasp larva and its paralyzed host (Ward and Henschel 1992).

Other Vespoidea

To this point I have examined the most common families of parasitoid Vespoidea: the Mutillidae, Tiphiidae, Scoliidae, and Pompilidae. The solitary Vespidae include nest provisioners Eumeninae and Masarinae (Chapters 3 and 5) and the brood parasitic Sapygidae (Chapter 4). Little is known about the remaining families, though all are probably parasitoids. One species of Bradynobaenidae is an ectoparasitoid of wind scorpions (Arachnida: Solifugae) (Brothers and Finnamore 1993). Larvae of Rhopalosomatidae are ectoparasitoids of immature crickets and have larvae that form saclike structures resembling those of dryinids and embolemids in the Chrysidoidea (Gurney 1953). Nothing is known about the biology of the 10 species of Sierlomorphidae (Goulet and Huber 1993).

Table 2-7 Host records and behavior of selected species of parasitoid Pompilidae

Species	Host family	Attack method	Form of paralysis[1]	Oviposition site	Disposition of spider after attack	References
Notocyphinae						
Notocyphus tyrannicus	Theraphosidae	Enters spider's refuge; stings spider	T	Anterior dorsum of abdomen	—	F. Williams 1928
Pompilinae						
Agenioideus cinctellus	—	—	P	—	Deposited in old bee tunnels, snail shells, or crevice and covered with debris	Richards & Hamm 1939
Allochares azureus	Filistatidae	Flushes spider from its silken retreat; stings it	T	Venter of abdomen	Returned to web and entangled in silk	Deyrup et al. 1988
Anospilus orbitalis	Ctenizidae	Bites apart lid of spider's trap-door tunnel; flushes spider out; stings it between legs	T	Posterior dorsum of abdomen	Paralyzed and dragged back into tunnel before oviposition	Iwata 1976

Aporus hirsutus	Ctenizidae	Drives spider from its trap-door nest; stings it between legs	T	Dorsolaterally on abdomen	Paralyzed and dragged back into tunnel before oviposition	F. Williams 1928
Homonotus iwatai	Clubionidae	Enters spider's refuge; stings it between legs	T	Abdomen	Left in place in its refuge	Iwata 1932 (in Clausen 1940), Iwata 1976
H. sanguinolentus	Clubionidae	Enters spider's refuge; stings it between legs	P	Abdomen	Left in place in its refuge	Nielsen 1936 (in Kaston 1959, Iwata 1976)
Pseudopompilus humboltdi		Enters spider's silken tube to sting it	—	Abdomen	—	Ward & Henschel 1992
Psorthaspis planata	Ctenizidae	Wasp chews trap-door lid; enters tunnel of spider; stings spider	P	Dorsum of abdomen	Left in place in its tunnel	Davidson 1905, Jenks 1938
Pepsinae						
Minagenia osoria	Lycosidae	—	T	Right side of abdomen	Left to resume its activities after it recovers from paralysis	Kaston 1959

[1]T, transient; P, permanent.

Sphecidae

Like the family Pompilidae, the family Sphecidae includes a large number of nest-provisioning predators (Chapter 3), a few cleptoparasitic genera (Chapter 4), and small number of ectoparasitoids. Some of the parasitoids belong to subfamilies that include only ectoparasitoids (e.g., the Ampulicinae), whereas others belong to groups comprising primarily predators.

Case Study: *Ampulex bantuae*

Information on foraging biology is available for only a handful of the over 150 known species of the Ampulicinae. One of the best studies is that of *Ampulex bantuae* in southern Africa by F. Gess (1984). The approximately 1.5-cm-long, metallic blue females of this species are very specialized in their host and habitat preferences, taking only wingless adult females and larger nymphs of the derocalymmid cockroach *Bantua dispar.* The female wasps generally reject smaller nymphs and never take the small winged males. *Bantua dispar* itself is restricted to the bark and old beetle galleries of *Acacia karoo* trees and shrubs.

When a female wasp encounters a roach, she grabs it by its pronotal shield and brings the tip of her abdomen forward to sting it between the legs or in the neck or in both places (Fig. 2-10). The resulting paralysis, which is neither permanent nor complete, allows the wasp to solve the problem of having to transport a much larger host on a vertical surface. It does so by walking backward after grasping the dazed, but still ambulatory, roach by the antennae. By this means, it leads the roach over the surface of the bark into an unoccupied gallery of a wood-boring beetle, where the wasp lays a single egg near the base of a hind leg. Finally, the wasp plugs the entrance to the gallery with whatever appropriately size objects are available, including materials as diverse as bark fragments, leaf debris, rodent droppings, a dead tick, and a moth pupa. Although no other modifications of the cavity are made, the whole process from host contact to final closure takes about 2.5 hours. During the sequence, the female may sever the tip of an antenna and feed on oozing roach blood, a behavior common in *Ampulex* (F. Williams 1929, Piek et al. 1984).

The offspring, now left to its own devices, hatches from the egg in 3–9 days and feeds for another 4–11 days near the oviposition site on the roach. The larva molts but does not move from its position until

A

B

Figure 2-10. *Ampulex bantuae* female (A) grabbing and (B) stinging a cockroach (from Gess 1984).

it reaches about 4.5 mm in length, after which it enters the still-living roach's body cavity through the feeding wound. After the larva has fed (as an endoparasitoid) for another 2–6 days, the host dies and the wasp spins a cocoon and pupates within the roach's hollow and desiccated exoskeleton. Adults of the spring-summer generation emerge 9–10 weeks after oviposition, whereas those of the overwintering generation have an extended prepupal diapause.

COMPARATIVE SURVEY OF SPHECID PARASITOIDS

Ampulicines are probably the most primitive sphecids, as indicated by their morphology (R. Bohart and Menke 1976, Alexander 1992) and parasitoid lifestyle. All ampulicines feed on cockroaches, and all inflict temporary paralysis on victims that are placed singly in existing cavities and quickly covered with debris (Table 2-8); some species make simple nests, with multiple cells separated by partitions made of plant debris (Krombein 1967). As does that of *Ampulex bantuae*, the venoms of *A. assimilis*, *A. compressa*, and *A. canaliculata* exhibit a very sophisticated pharmacology that enables the wasps to lead away their docile victims. Although stung cockroaches remain submissive "zombies," they are capable of running (often at normal speed) if stimulated. Thus, the venoms appear to affect the centers in the central nervous system that are important in "regulating and activating behavior" (Piek et al. 1984), without having an effect on the peripheral motor and sensory neurons and muscles that allow reflex responses and locomotion.

True parasitoids are also present in the subfamilies Sphecinae and Crabroninae (Table 2-8). However, these parasitoids reside in groups that include species with more complex nesting strategies. The genus *Chlorion* includes not only two true parasitoids (Table 2-8) but also *C. aerarium*, whose females deposit multiple prey within each cell in a multicellular nest of their own construction. Similarly, although *Larropsis chilopsidis* and all *Larra* spp. are true parasitoids, wasps of other closely related genera are primarily nest builders.

Impact of Parasitoid Wasps on Host Populations

Some studies of parasitoid aculeate wasps have gone beyond behavioral observations to assess the impact of the wasps on host population sizes. Information on this subject is uneven among groups for several reasons. First, the motivation to obtain the data is higher when the host has an economic impact on humans, as in the case of tsetse flies and various agricultural pests. Second, reliable estimates of rates of parasitism are easier to obtain for some groups than for others. Here, I review information on rates of parasitism by parasitoids and brood parasites in the Dryinidae, Chrysididae, and Mutillidae (see also Chapter 7).

DRYINIDAE

The levels of impact of dryinids on their host populations are relatively well known compared with those of other aculeates because rates of parasitism can be estimated by checking leafhoppers for the

Table 2-8 Host records and behavior of selected species of parasitoid Sphecidae

Species	Host family	Attack method	Form of paralysis	Oviposition site	Disposition of host after attack	References
Ampulicinae						
Ampulex amoena	Blattidae	Stings host laterally between legs or in neck; bites off host's antennae; feeds on hemolymph	Transient	—	Carried to concealed location and covered with plant debris and soil	Iwata 1976
A. bantuae	Derocalymmidae	Stings host multiply between legs and/or in neck	Transient	Hind coxa	Is "led" to existing nest cavity (a beetle burrow)	F. Gess 1984
A. canaliculata	Blattidae	Stings host between legs, in the neck, and venter of thorax; bites off antennae; feeds on hemolymph	Transient	Mid-coxa	Is "led" to existing nest cavity (a beetle burrow) and covered with plant debris and soil	F. Williams 1929

Table 2-8 *Continued*

Species	Host family	Attack method	Form of paralysis	Oviposition site	Disposition of host after attack	References
A. compressa	Blattidae	Stings host in thorax and neck	Transient	—	Is "led" to existing nest cavity (a beetle burrow) and covered with plant debris and soil	F. Williams 1942
Dolichurus stantoni	Blattidae	Stings host in venter of thorax	Transient	Mid-coxa	Is carried to hole or crack in soil or twig and is subsequently buried	F. Williams 1919c
Sphecinae						
Chlorion lobatum	Gryllidae	Drives host from burrow and stings it	Transient	Behind a forecoxa	Host burrows into soil, or wasp drags host back into its burrow and seals entrance	Iwata 1976, Bohart & Menke 1976

C. maxillosum	Gryllidae	Attacks within or near host burrow	Transient	Near first abdominal spiracle	Remains in burrow or digs another (if stung outside of burrow)	Valdeyron-Fabre 1955 (in Bohart & Menke 1976)
Larrinae						
Larra amplipennis	Gryllotalpidae	Enters host burrow; usually drives cricket out before stinging it between fore- and mid-coxae	Transient	Membrane behind a forecoxa	Left in its burrow, where it continues to feed	Iwata & Tanihata 1963
L. analis	Gryllotalpidae	Enters host burrow; drives cricket out before stinging it between the thorax and abdomen, and in the neck; adult feeds on host hemolymph; removes eggs of previous females	Transient	On side of body just behind a hindleg	Left in place, but after recovering from paralysis, cricket burrows into soil	C. Smith 1935

Table 2-8 *Continued*

Species	Host family	Attack method	Form of paralysis	Oviposition site	Disposition of host after attack	References
L. bicolor	Gryllotalpidae	Enters host burrow and drives cricket out before stinging it at bases of legs and then near mandibles	Transient	Between pro- and mesothoracic legs	Left in place, but after recovering from paralysis, cricket burrows into soil	Castner & Fowler 1987, Hudson et al. 1988
Larropsis chilopsidis	Gryllacrididae	Apparently stings host within the cricket's burrow	Transient	Venter of thorax near prothoracic legs	Left in place in its burrow	Gwynne & Evans 1975

presence of the dryinid larval sac. Mean rates of parasitism (i.e., the percentage of a sample of potential hosts that are parasitized) ranged as high as 32% on *Dicranotropis hamata* (Waloff and Jervis 1987), 37% on *Dichoptera hyalinata* (Swaminathan and Ananthakrishan 1984), and 59% on *Eupteryx aurata* (Stiling 1994). However, low levels of parasitism were the norm, with 80% of the 70 reported values being ≤10%. The rates sometimes varied dramatically both among sites and between seasons for a single host species (Peña and Shepard 1986, Stiling 1994).

At this time, it is difficult to say whether dryinids are generally important as regulating influences on their host populations, because the relationship between host density and dryinid parasitism appears to be density-independent in most cases (Cronin and Strong 1994). For several reasons, however, many assessments of the impact of dryinids based on the percentage of hosts with larval sacs are probably underestimates. First, the presence of first instars, which feed internally, cannot be detected without dissection (Waloff and Jervis 1987). Perhaps more sensitive techniques of detecting the presence of dryinids in leafhoppers, such as the use of electrophoresis, will help provide more accurate estimates (DeMichelis and Manino 1998). Second, dryinid females also kill large numbers of leafhoppers by host feeding, thus reducing the host population in another, less easily quantifiable way (Kidd and Jervis 1989). For example, in the laboratory, *Gonatopus clavipes* females consumed one in four leafhoppers without ovipositing on them (Waloff 1974). However, one should not assume that a host-feeding parasitoid is a better biological control agent than are species that do not host feed (Jervis et al. 1996).

OTHER PARASITOIDS AND BROOD PARASITES

Hosts of Bethylidae, Tiphiidae, Scoliidae, and some Chrysididae are often more difficult to sample than are dryinid hosts because they are often permanently paralyzed and quickly eaten within cryptic locations. Thus, information on parasitism rates for these groups is much less extensive than for dryinids; two exceptions, however, are the chrysidids and mutillids that can be found in the cells of bee and wasp nests.

Although different methods and sample sizes preclude detailed comparisons among species, several points are clear from examination of estimates of parasitism rates for aculeate parasitoids. First, rates of parasitism are often low (e.g., Linsley and MacSwain 1942,

W. Ferguson 1962, Krombein 1967). For example, while monitoring populations of host and parasitoid for 16 generations, W. Danthanarayana (1980) found parasitism rates of 0% to 8% on the light brown apple moth and concluded that *Goniozus jacintae* was an "insignificant mortality factor." Second, rates of parasitism by a single species can be highly variable in time and space, revealing that single-season studies at one site may be misleading (Knisley 1987, Yamada 1987a, Conlong and Graham 1988b). Rates of parasitism may also vary within a season. In a study of *G. natalensis* at one site over a 2-year period, rates varied from 0% to 29% in approximately weekly surveys (Conlong 1990). Similarly, in an eight-generation study of *Praestochrysis shanghaiensis*, 17% to 70% of the oriental moth cocoons were parasitized; rates of parasitism were even higher, often over 80%, among overwintering cocoons of the moth that managed to escape parasitism earlier in the fall (Yamada 1987b).

Biological Control of Pests Using Aculeate Wasp Parasitoids

Two forms of biological control based on release of parasitoid and predatory insects have been used to control pest populations (Pedigo 1996). In classical biological control, natural enemies are imported from the native habitat of an exotic pest species and released in relatively small numbers. Populations of the natural enemy are then allowed to build up on their own, in the hope that they will control the pest without the need for repeated local releases. In augmentative biological control, native or exotic natural enemies are reared and released in huge numbers in an attempt to overwhelm the pest. Augmentative control may require repeated releases in order to maintain a high population of a natural enemy, which may normally have a low population due to climatic factors (DeBach and Rosen 1991). Most examples of successful classical or augmentative biological control using insects have involved parasitoids, but only a few of the notable successes have used aculeate parasitoids.

CLASSICAL BIOLOGICAL CONTROL USING ACULEATE PARASITOIDS

There have been some successful biological control programs using aculeate wasps. The three examples I give here occurred in Hawaii and California.

Control of the oriental beetle (*Anomala orientalis*), a sugar cane pest that entered Hawaii from the Orient, was achieved by two aculeate wasps imported from the Philippines. *Campsomeris marginella* was released in 1916, and *Tiphia segregata* was released in 1917 (Krombein 1948, Funasaki et al. 1988, Lai 1988). Releases of *C. marginella* to control sugar cane pests in Taiwan, however, proved unsuccessful (Cheng 1991).

Larra luzonensis, which was imported into Hawaii from 1921 to 1925, brought the mole cricket *Gryllotalpa africana*, a pasture and turf pest, under complete control (Funasaki et al. 1988, Lai 1988). *Larra bicolor* also became established in Puerto Rico (in the 1930s) and in Florida (in the 1980s) after its importation from Brazil, but little data exist on its effectiveness against *Scapteriscus* mole crickets in these locations (Hudson et al. 1988, J. Frank et al. 1995, Parkman et al. 1996).

The navel orangeworm, *Amyelois transitella*, is a native of the neotropics that entered California in the early 1940s and became a major pest of almonds within a decade (Legner and Gordh 1992). Dissatisfaction with various control techniques led to the importation of three species of parasitoid wasp. One, an encyrtid wasp (Chalcidoidea) introduced in the 1960s, attacks mature larvae. The other two, *Goniozus emigratus* (from Texas) and *G. legneri* (from Argentina and Uruguay), both bethylid species, attack and permanently paralyze all larval instars of the moth. Both *Goniozus* species were liberated into unsprayed orchards in 1979 and 1980, but *G. legneri* proved more effective (Legner 1983, Legner and Silveira-Guido 1983, Legner and Warkentin 1988, Legner and Gordh 1992). In combination with the encyrtid, *G. legneri* reduced the proportion of attacked almonds in late summer to <0.5% (Legner and Gordh 1992). Several other species of the genus *Goniozus* have been used in classical biological control attempts, although their effectiveness has not been fully explored (Gordh and Evans 1976, Gordh and Medved 1986, Funasaki et al. 1988, Berry 1998).

Unfortunately, the notable successes in the use of aculeate parasitoids in classical biological control are few, while the list of disappointments is long. The reasons for the failure of imported biological control agents are sometimes known or suspected. Numerous attempts have been made to introduce species of *Campsomeris* and *Tiphia*, but, with the possible exception of *Tiphia vernalis*, most have proved ineffective, often because the wasps do not establish self-sustaining populations (Krombein 1948, Clausen 1956). The failure of *T. popilliavora* imported to control the Japanese beetle (*Popillia japonica*) in the United States has been attributed to an insufficient supply of food for adults, which need honeydew and nectar sources close to their hunting sites (Clausen 1940). *Tiphia popilliavora* may also have been unable to withstand the cold northeastern U.S. winters (D. Parker 1936). Similarly, when *Praestochrysis shanghaiensis* was imported into the eastern United States to control the oriental moth, *Monema flavescens*, the rate of parasitism was just 6% in 1919, after which the wasp disappeared.

Apparently, several factors were responsible for the failure of *Haplogonatopus vitiensis* to control the sugarcane planthopper (*Perkinsiella saccharida*) after the species was introduced into Hawaii from Fiji in 1906. At first, the wasps reduced leafhopper densities, but they eventually became scarce; perhaps the species became a victim of its own initial success and that of other imported enemies of the leafhopper. After leafhopper populations dropped, the wasp did not recover from low host densities. In addition, populations of *H. vitiensis* and another dryinid, *Pseudogonatopus hospes*, introduced from China in 1906 and 1907, were suppressed by their own parasitoids (Clausen 1956).

AUGMENTATIVE BIOLOGICAL CONTROL USING ACULEATE PARASITOIDS

In programs proposed to control pests by releasing huge numbers of parasitoid aculeates into the habitat, several technical difficulties must be overcome. First, biological studies must identify candidate control agents that appear to be ineffective only because of their inability to build up populations on their own (DeBach and Rosen 1991). Second, efficient and low-cost techniques for mass rearing and release of an agent must be developed. Most of the attempts to develop augmentative release programs using aculeate parasitoids have involved dryinids and bethylids (Paul et al. 1979, Chua et al. 1986, Graham and Conlong 1988, Betbeder-Matibet 1990, Cheng 1991, Hill 1994). In the United States, one of the major attempts to assess the effectiveness of a mass-release program employing a parasitoid aculeate has been the use of *Cephalonomia waterstoni* against the rusty grain beetle (*Crytolestes ferrugineous*), a stored-grain pest. Although lab studies and simulations suggest that the program could work (Flinn 1991, Flinn and Hagstrum 1995), results showing real control within grain bins have been scarce (Flinn et al. 1996). One problem has been that the wasp is hard to keep out of grain bins that are used as controls in experiments testing for the impact of the wasp. This difficulty may actually bode well for the future if it reflects the ability of the wasp to invade pest habitat.

Conclusion

The parasitoid way of life of most aculeates is undoubtedly a holdover from the parasitoid lifestyle of their nonaculeate ancestors. However, aculeate parasitoids have evolved tremendous behavioral diversity within the bounds of the definition of parasitoid. There are specialists and generalists, solitary and gregarious species, and at least one polyembryonic species. There are also species that exhibit simple forms of parental behavior in which the female moves the host to a nearby, quickly found, and somewhat safer location before ovipositing. Those species set the stage for two evolutionary developments typical of the so-called predatory aculeates: complex nests and nesting behaviors (Chapter 6) and an ability to exploit smaller prey by providing multiple prey items for each offspring, using a nest as a home base (Chapter 3).

3/ Foraging Behavior of Nest-Provisioning Predators

The parasitoid lifestyle common among wasps has no counterpart among vertebrates. In contrast, the foraging strategies of nest-provisioning solitary wasps detailed in this chapter have much in common with those of breeding songbirds. Many species of both wasps and songbirds are central-place foragers that feed their sedentary offspring by making repeated trips between hunting sites and nests. Each wasp nest may contain more than one nest cell, each of which provides a home for a single offspring (Chapter 6).

The building of nests adds several dimensions to foraging that are absent among parasitoids: the carriage of prey from the hunting grounds to the nest, the provision of multiple prey items for each offspring (in most species), and the ability to learn and home to nest sites. This chapter begins with three detailed case studies of foraging behavior, one each for the three families considered: Sphecidae, Pompilidae, and Vespidae.

Case Study: *Philanthus triangulum* (Sphecidae)

The European beewolf, *Philanthus triangulum*, a yellow-and-black wasp approximately 1 cm long, nests on partially vegetated sand dunes or in sandy woodlands throughout much of Europe and North Africa (Simonthomas and Simonthomas 1980). During their adult lives, which probably last no longer than about 6 weeks, female *P. triangulum* dig one to three multicellular nests. The account of the foraging by *P. triangulum* provided here is derived primarily from Hamm and Richards (1930); Tinbergen (1932, 1935, 1972); Hirschfelder (1952); Rathmayer (1962); Simonthomas and Simonthomas (1972); and Piek (1985).

Philanthus triangulum is a specialist, preferring to prey on worker honey bees (*Apis mellifera*). Prey also include a few bees of other families, perhaps taken either by mistake or when honey bee numbers are low. When foraging, beewolves first choose a general hunting locale, guided visually and perhaps by experience to flowers or bee hives. Upon arriving at flowers, a wasp flies slowly from plant to plant, not reacting to bees sitting on flowers (or to bees tied fast to twigs by Tinbergen), but responding rapidly to flying insects within about 30 cm. After the wasp notices a bee, it hovers 5–20 cm downwind, facing its potential prey. According to Tinbergen, however, it is "unlikely that *Philanthus* is able to distinguish honeybees visually from other insects" (Tinbergen 1972:123). Rather, the wasp apparently makes its final decision on whether to pounce by analyzing the odor wafting downwind from the potential victim. In experiments with various moving, wasp-sized "dummies," hunting females responded readily to freshly killed honey bees or to small objects given the scent of a bee, but they responded less readily to bees descented with a solvent.

Upon contacting its intended prey and further analyzing its odor, the beewolf grasps the bee's head with its legs and mandibles, curls its abdomen forward, and inserts its sting into the membrane at the base of a foreleg (Fig. 3-1). The sting is probably guided by tactile hairs at the tip of the wasp's abdomen. The sting itself does not reach the bee's central nervous system to cause physical damage. However, the venom contains a cocktail of pharmacologically active components that disrupt normal function in the central nervous system and at neuromuscular junctions in skeletal muscle, traveling to those sites via the blood (Piek 1985). The physiological effect of the venom spreads outward from the sting site; the front and middle legs become paralyzed at once, the hindlegs after about 60 seconds, and the tips of the legs after 30–40 minutes. The rapid muscle paralysis caused by the initial sting protects the wasp from counterattack by the bees, which are capable of killing beewolves. The first sting is often followed by a second delivered near the subesophageal ganglion in the neck. Although the legs and wings of the bee are paralyzed, the gut and heart continue to function, maintaining the prey in fresh condition for many days.

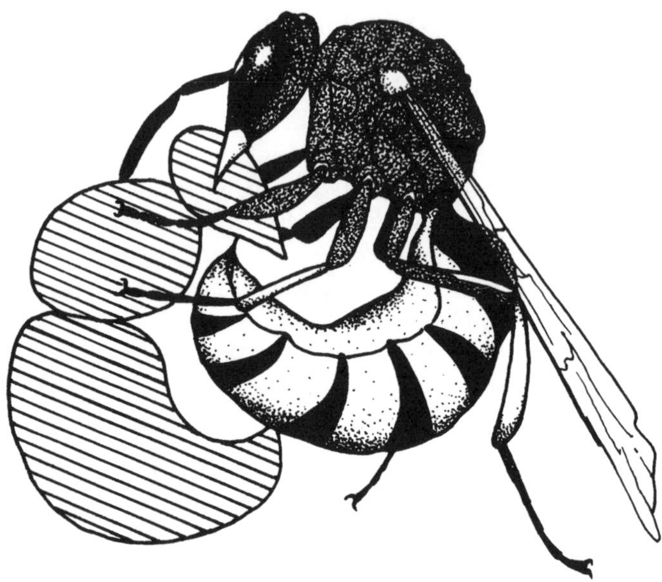

Figure 3-1. Position taken by female *Philanthus triangulum* while stinging prey (redrawn by Catherine Seibert from Rathmeyer 1962).

When subdued, the bee is carried in flight back to the nest, held upside down and gripped by the wasp's middle legs. Upon landing at the nest, the beewolf opens the entrance with her forelegs, shifts her burden to her hindlegs, and enters. The female places from one to six prey items in each nest cell and then loosely attaches an egg by its anterior end near a forecoxa of the topmost prey. R. Simonthomas and A. Simonthomas (1972) found that females brought home an average of four prey items each day, sufficient to fully stock one or more cells.

When a beewolf returns to the nesting area with prey, it must relocate its own nest among the many within the aggregation. The means by which beewolves solve this problem are worth considering in detail. In the 1930s, Tinbergen studied homing and orientation by *P. triangulum* in Holland; the results of his doctoral research and later studies with W. Kruyt and R. J. van der Linde are summarized in Tinbergen (1951, 1958, 1972). Tinbergen (1958:23) describes how female *P. triangulum*, upon leaving their nests to hunt, "circled a little while over the burrow, at first low above the ground, soon higher, describing ever widening loops; then flew away, but returned to cruise once more low over the nest." Such complex departure flights are typical of all *Philanthus*, though there is some variation in form and height (Evans and O'Neill 1988). The Peckhams (1898) and the Raus (1918) suggested that the information required to learn the location of nests is acquired when a wasp makes a "locality study" during these flights, now commonly called orientation flights or learning flights.

Prey-laden females reentering the nesting area use the pattern of objects on the ground to recognize the position of their nests. In his earliest experiments, Tinbergen surrounded a nest entrance with a ring of pine cones and allowed the resident to make two or three foraging trips. After this brief period of "training" (Fig. 3-2A), he then shifted the ring 30 cm away to surround a sham nest that included a fake mound of soil (Fig. 3-2B). Returning females invariably flew directly to the sham nest, supporting the hypothesis that females use local landmarks (rather than just the nest) for learning the position of their nests. Another set of experiments suggested that odor was not an important cue in homing.

Later, Tinbergen and Kruyt developed a technique to determine which of a pair of similar cues females were most likely to use in homing. They again started by training females, this time to a ring consisting of alternating objects of two types, for example flat black discs and black hemispheres placed flat side down (Fig. 3-2C). They then separated the intermeshed rings, moving them to sham nests on either side of the true nest (Fig. 3-2D). In this and similar experiments, female responses to the two rings gave a clear indication that they are most likely to learn the location of objects projecting above the soil surface. In the end, Tinbergen and Kruyt concluded that configurations consisting of relatively large, three-dimensional, patterned objects contrasting with background color and close to the nest are more important to females than are arrays of smaller, uniformly colored, flat objects that match the substrate and sit at a greater distance away. G. van Beusekom (1948) later demonstrated that *P. triangulum* are capable of such feats as distinguishing a circular array of pine cones from square or triangular arrays (their skills as geometers break down when they are asked to distinguish circles from ellipses). While provisioning, females may revise their cognitive maps during further orientation flights so that they can respond to changes in landmark patterns (Tinbergen 1972).

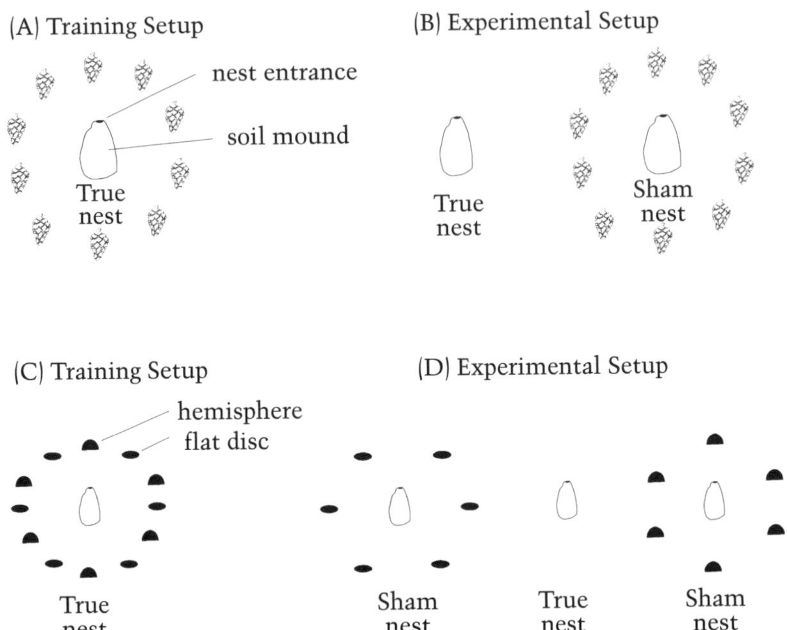

Figure 3-2. Protocol for Tinbergen's experiments on stimuli important in orientation by the European beewolf, *Philanthus triangulum*. (A) Training setup for the first experiment: a ring of pine cones was placed around the true nest. (B) Experimental setup: the ring of cones was shifted from the real nest to surround the sham nest. (C) Training setup in the experiment to determine whether a wasp responds better to two- or three-dimensional stimuli: intermeshed rings of flat black discs (ellipses) and black hemispheres (half-circles), both with 2-cm radii, were placed around the nest. (D) Experimental setup: rings of the discs and hemispheres were separated and shifted to two sham nests near the true nest. (A–D redrawn from Tinbergen 1972.)

Philanthus triangulum females obtain nectar either at flowers or by compressing a paralyzed bee's abdomen telescopically, causing the bee to regurgitate nectar that can be lapped up. Some of these prey may be discarded. Although most prey are used only for feeding larvae, a female may also bite some prey just behind the first pair of legs, where she then takes a blood meal.

Case Study: *Pepsis thisbe* (Pompilidae)

Female tarantula hawks of the species *Pepsis thisbe* are striking insects, up to 4 cm in long, with reddish brown wings and a black body covered with fine hairs that give the wasps a bluish green or purplish hue. *Pepsis thisbe* occurs in deserts of the U.S. Southwest. In California (F. Williams 1956) and in Texas (Punzo 1994a,b), the wasps provision shallow, single-celled nests with a single prey item, always a tarantula of the genus *Aphonopelma* (Theraphosidae). *Aphonopelma* of both sexes are longer and more massive than the wasp and live in burrows beneath entrances covered with silken sheets.

Female *P. thisbe* hunt by running over the surface of the ground, tapping the soil in search of an occupied tarantula burrow as they go; alternatively, the wasp may encounter a male spider outside of a burrow. Upon locating a burrow, the wasp cuts through the silk covering with its mandibles and enters. The complete sequence of attack occasionally occurs within the burrow, but the wasp usually forces the tarantula into the open before attacking. Surprisingly, although the tarantula may exhibit a threat display at this time, it makes no attempt to counterattack or to escape. After antennating the tarantula, the wasp moves off a short distance to groom, perhaps as a way of ensuring that its antennal sense receptors are in peak condition during the upcoming attack. The attack occurs rapidly (Fig. 3-3): the wasp rushes at the spider, grabs a leg, flips the spider onto its back, and stings it, usually through the membrane at the base of a foreleg. With this single sting, the prey remains paralyzed until it is killed by the fourth-stage wasp larva. During the attack, the tarantula may finally counterattack, though it rarely succeeds in killing the wasp. Experience gained in successive encounters apparently increases the foraging efficiency of the wasp. In series of staged encounters, the average duration of the initial approach and antennation of prey declined from about 26 to 6 minutes after 10 encounters, and the duration of the actual attack dropped from 17 to 6 minutes (Punzo 1994b).

If the tarantula is found in a burrow, the wasp deposits it back in its burrow, which is modified to serve as a nest. If the prey is a wandering male, the wasp digs a nest from scratch. In either case, the female moves the prey to the nearby burrow by dragging it over the ground while walking backward. Before oviposition, the female removes some of the dense hairs from the venter of the spider's abdomen, where the egg will be glued to a position just behind the book lungs.

Immediately after paralyzing the tarantula, the wasp often feeds on blood oozing from the sting wound or on liquid within the mouth

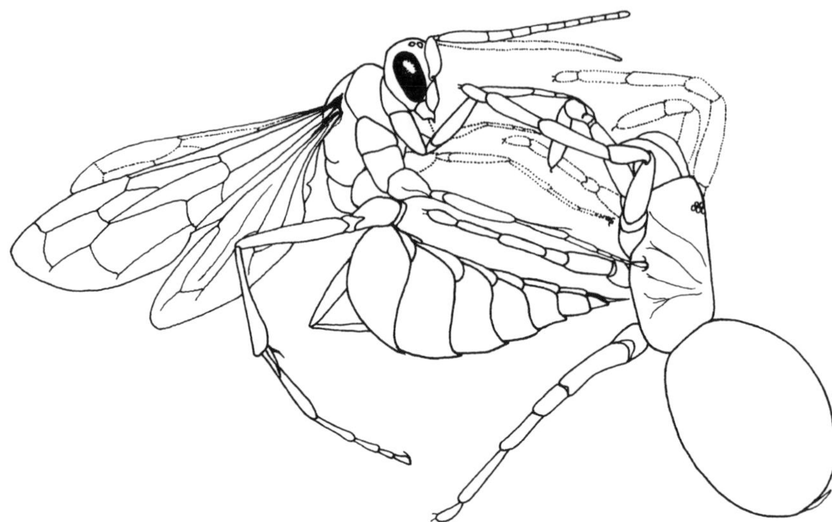

Figure 3-3. *Pepsis* female stinging a tarantula (redrawn by Catherine Seibert from Williams 1956).

of the spider. The probability that host feeding will occur apparently increases with increased time searching for prey, so such feeding may function to replenish recently used energy stores. Adults of both sexes also feed at flowers.

Case Study: *Euodynerus foraminatus* (Vespidae: Eumeninae)

Euodynerus foraminatus occurs throughout much of the United States and southern Canada, where it nests in hollow twigs, in cavities in wood, and in abandoned mud nests of other species. A. Steiner (1983b, 1984) provided detailed information on prey searching and stinging behaviors. Although *E. foraminatus* sometimes provision with sawfly larvae (A. Steiner 1984), they most often take lepidopteran caterpillars of the families Gelechiidae, Olethreutidae, Pyralidae, Pyraustidae, Thyridae, and Tortricidae (Rau and Rau 1918, Medler 1964, Krombein 1967).

The range of prey taken and the fact that individual nests often contain multiple prey families suggest that females forage in a variety of microhabitats (Medler 1964). Like *Philanthus triangulum*, *E. foraminatus* use different sensory modalities in sequence to locate and subdue prey. Females begin their search by investigating indi-

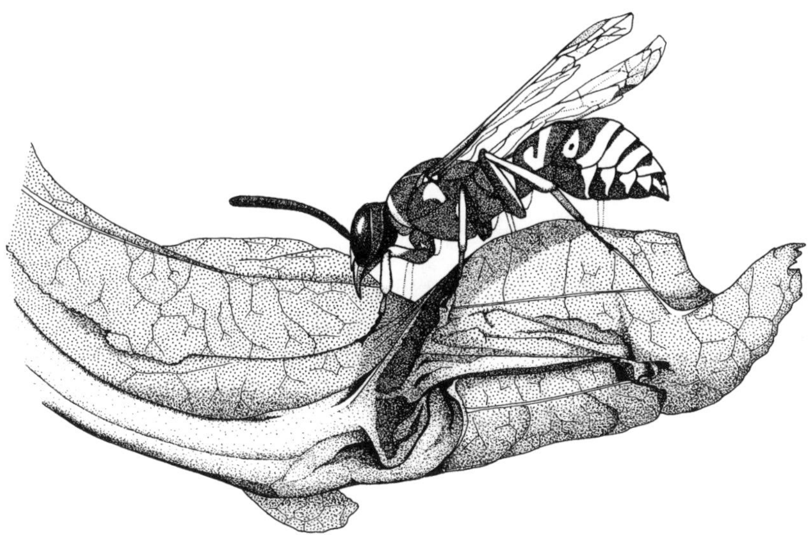

Figure 3-4. *Euodynerus foraminatus* female searching for prey in leaf tied together with silk of a caterpillar (redrawn by Catherine Seibert from Steiner 1984).

vidual leaves on trees to which they are probably attracted visually (Steiner 1984). They move quickly from leaf to leaf until they detect odors associated with caterpillars located within refuges, which consist of one or more leaves bound together with silk (Fig. 3-4). Odors emanating from the caterpillar, frass within the refuge, and silk cause an abrupt shift in the behavior of the wasp, which then chews its way into the refuge and extracts the caterpillar.

The number and location of stings are variable, but *E. foraminatus* always delivers more stings than do beewolves and tarantula hawks. Typically, the wasp approaches its prey from above and behind, grasps it with its mandibles, curls its own abdomen forward, and stings near the caterpillar's neck one or more times. Moving backward, she usually delivers the next sting on or near the metathorax. She may then repeat the process, stinging the prey up to eight times. Paralysis is usually incomplete because it takes effect only in the vicinity of the wound, but the wasp packs the caterpillar so tightly into the nest cell that it has little room to move anyway. The wasp holds the paralyzed caterpillar in her mandibles and carries it in flight to the nest, where 2–24 prey items are placed in each cell.

The egg is not laid on the prey; before provisioning, the wasp had suspended the egg from the ceiling of the cell by a slender thread (Hartman 1944a).

Euodynerus foraminatus adults feed on nectar of such flowers as *Melilotus* (sweetclover) and *Penstemon*, robbing nectar from *Penstemon* by chewing a hole in the side of the corolla. Female *E. foraminatus* also feed on caterpillars, but unlike *Philanthus triangulum*, they do not always sting before feeding and they devour large portions of the prey rather than just blood. Such high-protein–high-fat meals appear to be vital for nesting females as a source of nutrients for producing eggs. The egg volumes of females that had fed on prey were about 2.5 times those of females that had been allowed access only to sugar water (Chilcutt and Cowan 1992).

Comparative Survey of Prey

In surveying the extensive literature on prey of nest provisioners, I have selected primarily species for which prey have been identified to genus and for which large numbers of prey have been collected at a single site or set of sites that were near one another. (The exceptions are mainly the specialist predators.) Of the data that were available for different sites, I selected data from the site with the largest sample. Data in Tables 3-1 through 3-8 undoubtedly reflect a bias for recent studies (which tend to be more quantitative than older studies) and those published in English. Information on wasp diets is also reviewed in Iwata (1976), Bohart and Menke (1976), and Krombein et al. (1979b).

The data on wasp prey should be interpreted with the following provisos. First, with regard to the behavioral and physiological processes that affect prey choice, a wasp that takes 10 genera of aphids may not actually have a broader diet than one that preys on only 5 genera of flies. If the fly prey were more ecologically, behaviorally, and morphologically diverse than the aphids, they would require greater plasticity in foraging behavior. Second, sample size may affect apparent niche breadth. For example, an analysis of more than 3,000 prey items of *Philanthus sanbornii* uncovered 108 species of bees and wasps (Stubblefield et al. 1993). However, even after 5 years of study, new prey species were still being discovered. In addition, the small prey samples reported in many studies were probably not sampled in a statistically random fashion (either deliberately or accidentally). Even a large prey sample may represent the activities of a small number of females; for instance, the prey may have been collected from a few multicelled nests with many prey items in each cell. Third, prey records are notoriously site- and time-specific; even a large sample may not accurately characterize the diet of a species if it is taken at a single place and time.

PREY OF NEST-PROVISIONING SPHECIDAE

Prey of Sphecinae

All wasps of the tribe Sphecini (subfamily Sphecinae) provision nests with Orthoptera. Most species stock their nests with prey from a single family of grasshoppers or crickets (Table 3-1). The Ammophilini are likewise restricted, taking either lepidopteran caterpillars or the morphologically and ecologically similar sawfly larvae (Hymenoptera). The only exception is *Eremochares dives*, a predator of Orthoptera. The Sceliphrini display the greatest diversity of diets among the Sphecinae. For example, *Chalybion* and *Sceliphron* take spiders; *Chlorion* takes crickets; *Podium*, *Penepodium*, and some *Trigonopsis* take cockroaches; and *Stangeella* takes nymphal and adult mantids (reviewed by Iwata 1976, Bohart and Menke 1976).

Among the most specialized Sphecinae are *Palmodes laeviventris*, which preys only on Mormon crickets (*Anabrus simplex*), and several species of *Podalonia* known to take the larvae of a single species or genus of moths. Intermediate degrees of preference are exhibited by species such as *Prionyx atratus*, which restricts itself to late-instar nymphs and adults of various acridid grasshoppers. Females in some populations of *Ammophila*, *Chalybion*, and *Sceliphron*, in contrast, may take prey from a large number of families. Sphecines that use large caterpillars or orthopterans may provide only one prey item for each cell; few species provide more than 12 (Iwata 1976).

Prey of Pemphredoninae

Members of the subfamily Pemphredoninae are small wasps that take small prey, most often Homoptera (Table 3-2). Species in the tribe Psenini provision primarily with leafhoppers and planthoppers of the homopteran suborder Auchennorryncha (but also Psyllidae, suborder Sternorrhyncha), whereas Pemphredonini usually take either aphids or nonhomopteran prey. The major exception to the preference for Homoptera is the use of Thysanoptera (thrips) or

Table 3-1 Prey diversity of selected species of nest-building Sphecidae, subfamily Sphecinae

Species	No. of prey items	No. of prey genera	No. of prey families	No. of prey items per cell	Most common prey[1] (%)	References
Sphecini						
Isodontia auripes	55	2	1	9–18	Orth: Gryllidae (100)	Krombein 1967
I. mexicana	371	6	2	up to 21	Orth: Tettigoniidae (51)	Medler 1965
I. pelopoeiformis	556	≥3	1	1–10	Orth: Tettigoniidae (100)	F. Gess & Gess 1982
Palmodes laeviventris	211	1	1	1–4	Orth: Tettigoniidae (100)[2]	LaRivers 1945
Prionyx atratus	18	6	1	1	Orth: Acrididae (100)	Evans 1958c
Sphex argentatus	217	4	1	2–7	Orth: Tettigoniidae (100)	Tsuneki 1963, Iwata, 1976
S. cognatus	>100	1	1	4–5	Orth: Tettigoniidae (100)	Ribi & Ribi 1979
Ammophilini						
Ammophila azteca	44	≥5	3	7	Hym: Tenthredinidae (64)	Evans 1965
A. dysmica	128	≥3	2	1–2	Lep: Geometridae (>99)	Rosenheim 1987b
A. harti	196	≥12	7	5–12	Lep: Geometridae (44)	Hager & Kurczewski 1986

Podalonia luctuosa	32	≥8	1	1	Lep: Noctuidae (100)	O'Brien & Kurczewski 1982a
P. valida	64	1	1	1	Lep: Arctiidae (100)[2]	O'Neill & Evans 1999
Sceliphrini						
Chalybion caerulum	927	—	7	—	Ara: Theridiidae (68)	Muma & Jeffers 1945
C. spinolae	28	1	1	3–11	Ara: Theridiidae (100)	F. Gess et al. 1982
Chlorion aerarium	~60–70	1	1	2–9	Orth: Gryllidae (100)[2]	Peckham & Kurczewski 1978
Penepodium goryanum	52	1	1	1–4	Blat: Epilampridae (100)[2]	Garcia & Adis 1993
Podium rufipes	41	3	1	3–14	Blat: Blattidae (100)	Krombein 1967
Sceliphron caementarium	2,021	—	9	6–15	Ara: Argiopidae (56)	Muma & Jeffers 1945

[1]Ara, Araneae; Blat, Blattodea; Clm, Collembola; Col, Coleoptera; Dip, Diptera; Hem, Hemiptera; Hom, Homoptera; Hym, Hymenoptera; Lep, Lepidoptera; Orth, Orthoptera; Thy, Thysanoptera. [2]All prey were of a single species.

Table 3-2 Prey diversity of selected species of nest-building Sphecidae, subfamilies Pemphredoninae and Astatinae

Species	No. of prey items	No. of prey genera	No. of prey families	No. of prey items per cell	Most common prey[1] (%)	References
Pemphredoninae: Pemphredonini						
Diodontus occidentalis	160	4	1	23–30	Hom: Aphididae (100)	J. Powell 1963
Microstigmus comes	1,197	4	2	31–58	Clm: Entomobryidae (86)	Matthews 1968
M. thripoctenus	<300	2	1	up to 171	Thy: Thripidae (100)	Matthews 1970
Passaloecus cuspidatus	9,618	6	1	14–74	Hom: Aphididae (100)	Fricke 1993
P. insignis	181	4	1	up to 50	Hom: Aphididae (100)	Danks 1970
Pemphredon lethifer	2,016	5	1	up to 89	Hom: Aphididae (100)	Danks 1970
Spilomena subterranea	46	5	2	up to 14	Hom: Psyllidae (65)	McCorquodale & Naumann 1988
Pemphredoninae: Psenini						
Mimesa basirufa	85	3	1	2–6	Hom: Cicadellidae (100)	Kurczewski & Lane 1974
M. cressoni	294	11	3	9–17	Hom: Cicadellidae (99)	Kurczewski & Lane 1974

Mimumesa littoralis	111	2	2	up to 10	Hom: Cicadellidae (59)	Tsuneki 1959
M. mixta	100	8	2	up to 19	Hom: Cicadellidae (99)	Rosenheim & Grace 1987
Pluto littoralis	72	2	2	7–17	Hom: Delphacidae (92)	Grissell 1979
Psen barthi	46	2	1	up to 5	Hom: Cicadellidae (100)	Barth 1907
Psenulus schencki	307	1	1	up to 23	Hom: Psyllidae (100)	Danks 1970
Astatinae: Astatini						
Astata occidentalis	122	5	1	3–6	Hem: Pentatomidae (100)	Evans 1957a
A. unicolor	—	2	1	2–4	Hem: Pentatomidae (100)	Evans 1957a
Diploplectron peglowi	34	2	1	4–6	Hem: Lygaeidae (100)	Kurczewski 1972

[1]See Table 3-1 for prey order abbreviations.

Collembola by *Microstigmus*, *Xysma*, and some *Spilomena* (Bohart and Menke 1976). Most often, prey records are restricted to one prey family, but some species may use two or three families. *Spilomena subterranea*, however, preys on an odd mixture of Psyllidae (Homoptera) and Eulophidae (Hymenoptera) (McCorquodale and Naumann 1988). Although small themselves, pemphredonines often stock each nest cell with huge numbers of prey items. Up to 89 aphids have been found in *Pemphredon lethifer* cells, and 171 thrips were found in a *Microstigmus thripoctenus* cell (Iwata 1976).

Prey of Astatinae

Species of the small subfamily Astatinae prey solely upon Hemiptera (Table 3-2). Females of the best-studied genus, *Astata*, take only Pentatomidae (stink bugs), whereas *Diploplectron* provision with Lygaeidae (seed bugs). *Dryudella*, in contrast, take at least seven families of Hemiptera, and *Dinetus* prey on either Nabidae or Lygaeidae. Both nymphal and adult prey are used, depending on the wasp species, and relatively few prey items are provided in each cell.

Prey of Larrinae

In the subfamily Larrinae there are several notable predator-prey affinities at the tribe level (Table 3-3). Most Larrini are predators of grasshoppers and crickets, although some *Tachytes* take geometrid caterpillars, some *Tachysphex* take cockroaches and mantids, and *Prosopigastra* take small hemipterans (Bohart and Menke 1976). All Trypoxylonini, as well as *Miscophus* of the Miscophini, prey on spiders. Other Miscophini take insects: *Sericophorus* take flies; *Lyroda* take crickets and pygmy grasshoppers; *Solierella* take hemipterans, psocopterans, and grasshoppers; *Nitela* take psocopterans and aphids; and *Plenoculus* take caterpillars and hemipterans (Bohart and Menke 1976). Females of the genus *Palarus* (the sole genus of the tribe Palarini) have relatively broad prey preferences, though they restrict themselves to either Hymenoptera or Diptera (Gayubo et al. 1992). The number of prey items in each cell varies among species. Some Larrini provide only one item, but Trypoxylonini that prey on small spiders may provide several dozen or more.

Prey of Crabroninae

Members of the subfamily Crabroninae prey on at least 12 orders of insects (Bohart and Menke 1976, Iwata 1976). The most common

Table 3-3 Prey diversity of selected species of nest-building Sphecidae, subfamily Larrinae

Species	No. of prey items	No. of prey genera	No. of prey families	No. of prey items per cell	Most common prey[1] (%)	References
Larrini						
Ancistromma distincta	~45	1	1	1–3	Orth: Gryllidae (100)[2]	Evans 1958b
Tachysphex pechumani	25	2	1	1	Orth: Acrididae (100)	Kurczewski & Elliott 1978
T. terminatus	984	6	2	3–7	Orth: Acrididae (>99)	Kurczewski 1966c
Tachytes intermedius	112	1	1	2–6	Orth: Tridactylidae (100)	Kurczewski & Kurczewski 1984
T. validus	202	1	1	1–5	Orth: Tettigoniidae (100)	Kurczewski & Ginsberg 1971
Trypoxylonini						
Pison koreense	101	1	1	20–31	Ara: Dictynidae (100)	Sheldon 1968
P. rufipes	47	3	1	4–9	Ara: Salticidae (100)	Evans et al. 1980c
Pisonopsis clypeata	41	3	2	9–17	Ara: Theridiidae (98)	Evans 1969a
Trypoxylon antenuatum	231	≥16	7	6–31	Ara: Thomisidae (27)	Asis et al. 1994
T. figulus	198	≥4	3	up to 32	Ara: Argiopidae (97)	Danks 1970
T. politum	792	3	1	2–17	Ara: Araneidae (100)	Rehnberg 1987
T. superbum	265	≥33	4	6–22	Ara: Salticidae (97)	Coville & Griswold 1984

Table 3-3 *Continued*

Species	No. of prey items	No. of prey genera	No. of prey families	No. of prey items per cell	Most common prey[1] (%)	References
T. tenoctitlan	836	11	6	11–44	Ara: Araneidae (47)	Coville & Coville 1980
T. tridentatum	183	1	1	7–12	Ara: Araneidae (100)[2]	O'Brien 1982
Bothynostethini						
Bothynostethus distinctus	38	1	1	4–9	Col: Chrysomelidae (100)	Kurczewski & Evans 1972
Miscophini						
Lyroda subita	67	2	1	1–7	Orth: Gryllidae (100)	Kurczewski & Peckham 1982
Miscophus kansensis	138	7	6	9–29	Ara: Theridiidae (93)	Kurczewski 1969
Nitelopterus californicus	33	4	1	4	Ara: Salticidae (100)	J. Powell 1967
Plenoculus cockerelli	28	5	4	up to 6	Lep: Gelechiidae (64)	Rubink & O'Neill 1980
Sericophorus viridis	97	1	1	3–9	Dip: Calliphoridae (100)	Matthews & Evans 1970
Palarini						
Palarus almariensis	39	7	4	8–13	Dip: Stratiomyiidae (56)	Gayubo et al. 1992
P. variegatus	41	11	7	5–12	Hym: Halictidae (41)	Gayubo et al. 1992, Iwata 1976

[1]See Table 3-1 for prey order abbreviations. [2]All prey were of a single species.

prey orders are listed in Table 3-4. In addition, some *Crossocerus*, *Lindenius*, *Rhopalum*, and *Tracheliodes* take adult Hymenoptera; the *Tracheliodes* are specialists on worker ants (Bohart and Menke 1976). Adult Diptera seem to be particularly favored by both the Oxybelini (especially *Oxybelus*) and the Crabronini (especially *Ectemnius* and *Crabro*); 49 families of Diptera are known to be taken by Crabronini (Iwata 1976). Each tribe has a spectrum of species ranging from relative specialists to generalists that take over 15 families of prey. Most crabronines provide many prey items for each cell, nearly 100 in the case of *Crabro argusinus*.

Prey of Bembicinae

The subfamily Bembicinae is a morphologically diverse group with a corresponding diversity of diets (Table 3-5). All Alyssonini and Gorytini, as well as *Bembecinus* of the Stizini, take nymphal and adult Homoptera ranging in size from tiny leafhoppers to massive cicadas. *Mellinus arvensis*, sometimes placed in the small family Mellinidae and sometimes in the bembicine subfamily Mellininae, is a generalist predator, taking adult flies of more than 10 families (Hamm and Richards 1930).

The Bembicini include the only true detritivore among the Hymenoptera (exclusive of the ants). *Microbembex* provision nests with dead or disabled arthropods, as well as with arthropod parts; nearly all common terrestrial insects are included in prey records (Evans 1966c). A *Microbembex monodonta* nest examined by H. Evans contained dead insects of seven orders and a dead spider. However, most Bembicini are predators and, as a group, take insects of at least seven orders, the five mentioned in Table 3-5 as well as Odonata (damselflies), and Neuroptera (ant lions and lacewings) (Evans and Matthews 1973c). With the exception of *Bicyrtes*, which take both nymphal and adult Hemiptera, Bembicini provision nests only with adult insects. Many *Bembix* have quite broad prey records, and *Glenostictia scitula* has probably the most catholic diet among predatory aculeate wasps, taking prey of at least 27 families of Hemiptera, Diptera, and Hymenoptera; all three orders may be represented in the prey of a single female (Evans 1966c).

The number of prey items in each cell varies. Some cicada-killers, *Sphecius speciosus* and *Exeirus lateritus* (both in the tribe Gorytini) provide only one item (Evans 1966c), whereas some Bembicini provide at least three dozen. However, because many Stizini and

Table 3-4 Prey diversity of selected species of nest-building Sphecidae, subfamily Crabroninae

Species	No. of prey items	No. of prey genera	No. of prey families	No. of prey items per cell	Most common prey[1] (%)	References
Oxybelini						
Belomicrus forbesii	—	1	1	5–9	Hem: Miridae (100)	Evans 1969a
Enchemicrum australe	91	6	3	6–21	Dip: Ephydridae (95)	Peckham & Hook 1994
Oxybelus bipunctatus	787	34	14	2–12	Dip: Stratiomyiidae (28)	Peckham et al. 1973
O. emarginatus	672	44	21	4–38	Dip: Milichiidae (30)	Peckham et al. 1973
O. exclamans	238	5	1	1–9	Dip: Sarcophagidae (100)	Peckham 1985
O. fossor	221	4	1	7–20	Dip: Scenopinidae (100)	Peckham 1985
O. sericeus	903	10	7	3–14	Dip: Otitidae (77)	Hook & Matthews 1980
O. subulatus	492	2	1	3–11	Dip: Therevidae (100)	Peckham et al. 1973
O. uniglumis	367	33	11	2–13	Dip: Anthomyiidae (38)	Peckham et al. 1973
Crabronini						
Anacrabro ocellatus	174	2	1	4–9	Hem: Miridae (100)	Kurczewski & Peckham 1970
Crabro arcadiensis	206	41	18	8–33	Dip: Dolichopodidae (66)	Miller & Kurczewski 1976
C. argusinus	470	38	15	5–92	Dip: Ephydridae (44)	Matthews et al. 1979
C. maculiclypeus	203	12	6	9–20	Dip: Empididae (62)	Kurczewski et al. 1969

C. rufibasis	443	53	15	3–29	Dip: Muscidae (26)	Miller & Kurczewski 1976
Crossocerus annulipes	648	7	3	14–30	Hom: Cicadellidae (>99)	Kurczewski & Miller 1986
Dasyproctus westermanni	53	9	7	14–26	Dip: Simuliidae (32)	F. Gess 1980b
Ectemnius paucimaculatus	308	16	9	7–31	Dip: Ephydridae (55)	Krombein 1964
E. sexcinctus	127	16	≥3	—	Dip: Muscidae (83)	Hamm & Richards 1926
E. spiniferus	53	3	3	up to 10	Dip: Acroceridae (77)	Bechtel & Schlinger 1957
Entomognathus memorialis	220	1	1	3–9	Col: Chrysomelidae (100)	Miller & Kurczewski 1972
Lindenius armaticeps	445	4	1	3–15	Dip: Chloropidae (100)	Miller & Kurczewski 1975
Moniaecera asperata	42	6	4	~20	Hom: Psyllidae (62)	Evans 1964
Podagritus parrotti	—	1	1	5–15	Col: Chrysomelidae (100)[2]	Harris 1998
Rhopalum atlanticum	30	5	4	31–41	Hom: Psocidae (60)	Kislow & Matthews 1977

[1]See Table 3-1 for prey order abbreviations. [2]All prey were of a single species.

Table 3-5 Prey diversity of selected species of nest-building Sphecidae, subfamily Bembicinae

Species	No. of prey items	No. of prey genera	No. of prey families	No. of prey items per cell[1]	Most common prey[2] (%)	References
Alyssonini						
Alysson melleus	31	9	1	3–23	Hom: Cicadellidae (100)	Evans 1966c
Gorytini						
Austrogorytes bellicosus	218	4	1	5–27	Hom: Cicadellidae (100)	Evans & Matthews 1971
Gorytes canaliculatus	135	1	1	6–19	Hom: Cicadellidae (100)	Evans 1966c
Hoplisoides nebulosus	<150	9	1	up to 20	Hom: Membracidae (100)	Evans 1966c
Ochleroptera bipunctata	13	12	4	6–18	Hom: Cicadellidae (69)	Evans 1966c
Sphecius speciosus	703	1	1	1–4	Hom: Cicadidae (100)	Dambach & Good 1943
Stizini						
Bembecinus hirtulus	183	11	3	P	Hom: Cicadellidae (97)	Evans & Matthews 1971
B. nanus	71	11	3	P	Hom: Dictyopharidae (61)	Evans & O'Neill 1986
B. quinquespinosus	757	2	1	P	Hom: Cicadellidae (100)	Evans & O'Neill 1986
Stizus pulcherrimus	117	4	2	P	Orth: Acrididae (98)	Tsuneki 1965

Bembicini						
Bembix americana	114	29	12	P	Dip: Muscidae (30)	Lane et al. 1986
B. amoena	209	45	10	>30	Dip: Tabanidae (35)	Evans 1966c
B. bubalus	78	14	7	P	Dip: Bombyliidae (62)	S. Gess & Gess 1989b
B. moma	301	27	12	P	Hym: Halictidae (43)	Evans & Matthews 1973c
B. sayi	193	15	7	P	Dip: Tabanidae (65)	Evans 1966c
B. texana	125	9	6	P	Dip: Tabanidae (94)	Evans 1966c
B. tuberculiventris	85	4	2	P	Hym: Halictidae (59)	Evans & Matthews 1973c
B. u-scripta	99	26	9	up to 40	Dip: Otitidae (26)	Evans 1960
Bicyrtes quadrifasciata	96	8	2	4–14	Hem: Coreidae (71)	Krombein 1955, 1958
Glenostictia scitula	113	22	19	≥40	Hym: Andrenidae (57)	Gillaspy et al. 1962
Steniolia obliqua	60	3	2	P	Dip: Bombyliidae (97)	Evans & Gillaspy 1964
Stictia carolina	123	6	3	25–35	Dip: Tabanidae (97)	Evans 1966c
Stictiella formosa	33	5	3	up to 11	Lep: Libytheidae (82)	Gillaspy et al. 1962

[1]P refers to progressive provisioners where total number of prey items per cell is unknown. [2]See Table 3-1 for prey order abbreviations.

Bembicini are progressive provisioners (i.e., females continue to bring in prey after their offspring have begun feeding), it is often difficult to determine how many prey items have been provided unless females are observed continually for prolonged periods.

Prey of Philanthinae

Females of two of the three tribes of the subfamily Philanthinae take only adult Hymenoptera as prey, but the range of prey of the Cercerini includes adults of both Hymenoptera and Coleoptera (Table 3-6). The small tribe Aphilanthopini includes some the most extreme specialists among the aculeate wasps. Female *Aphilanthops* prey only on alate queens of the ant genus *Formica*, whereas *Clypeadon* take only worker ants of the genus *Pogonomyrmex*. Species of Cercerini and Philanthini are usually much less specialized, although *Cerceris halone* apparently takes only acorn weevils, *Curculio nasicus*. Most Cercerini, including all North American species studied, take beetles as prey, but several species in Europe, South Africa, and Japan prey on bees and wasps. All Philanthini are predators of bees and wasps. Some species of the well-studied genus *Philanthus* include in their diets more than a dozen families and many dozens of species of bees and wasps (Evans and O'Neill 1988). *Aphilanthops*, as well as the species of *Philanthus* and *Cerceris* that take relatively large prey, commonly provide fewer than a half dozen prey items in each cell. However, *Philanthus* species that use small sweat bees, *Cerceris* species that use small leaf beetles, and *Clypeadon* species may provide dozens of prey items to each offspring.

PREY OF NEST-PROVISIONING POMPILIDAE

All members of the family Pompilidae are called spider wasps because all species forage for spiders (Table 3-7). Four of the six pompilid subfamilies contain only parasitoids or cleptoparasites (Chapter 2), but the Pepsinae and Pompilinae also include many species that provision nests of their own construction. There are no major differences in prey of Pepsinae and Pompilinae: 71% of the 28 spider families used by Pepsinae are also used by Pompilinae, which take at least 23 families of prey (Shimizu 1994). In fact, this broad variation in prey use may occur within a single genus. Ten families of prey are taken by various *Agenioideus*, 13 by *Auplopus*, 15 by *Priocnemis*, and 15 by *Anoplius*. Furthermore, a single species may have a diet breadth nearly as broad (Field 1992a). According to Evans and Yoshimoto

Table 3-6 Prey diversity of selected species of nest-building Sphecidae, subfamily Philanthinae

Species	No. of prey items	No. of prey genera	No. of prey families	No. of prey items per cell	Most common prey[1] (%)	References
Aphilanthopini						
Aphilanthops frigidus	~75	1	1	2–3	Hym: Formicidae (100)[2]	Evans 1970a
A. subfrigidus	23	1	1	3–4	Hym: Formicidae (100)[2]	O'Neill 1990
Clypeadon laticinctus	175	1	1	15–26	Hym: Formicidae (100)[2]	Evans 1962
Cercerini						
Cerceris antipodes	100	10	2	10–53	Col: Chrysomelidae (91)	Evans & Hook 1986b
C. arenaria	269	6	1	5–12	Col: Curculionidae (100)	Hamm & Richards 1930
C. australis	51	4	1	2–11	Col: Scarabaeidae (100)	Evans & Hook 1986b
C. californica	1,127	4	1	5–18	Col: Buprestidae (100)	Linsley & MacSwain 1956
C. flavofasciata	263	7	1	12–30	Col: Chrysomelidae (100)	Kurczewski & Miller 1984
C. fumipennis	310	7	1	2–32	Col: Buprestidae (100)	Evans 1971
C. holconota	13	8	6	—	Hym: Tiphiidae (54)	F. Gess 1980a

Table 3-6 *Continued*

Species	No. of prey items	No. of prey genera	No. of prey families	No. of prey items per cell	Most common prey[1] (%)	References
C. rybiensis	40	2	2	4–8	Hym: Halictidae (60)	Hamm & Richards 1930
Eucerceris ruficeps	188	2	1	6–22	Col: Curculionidae (100)	Linsley & McSwain 1954
Philanthini						
Philanthus crabroniformis	406	15	5	7–24	Hym: Halictidae (94)	Evans & O'Neill 1988
P. inversus	168	3	2	5–13	Hym: Halictidae (99)	O'Neill & Evans 1982
P. multimaculatus	107	4	2	5–23	Hym: Halictidae (97)	Evans & O'Neill 1988
P. pulchellus	407	32	14	4–9	Hym: Halictidae (43)	Asis et al. 1996
P. sanbornii	3,138	43	15	3–15	Hym: Halictidae (61)	Stubblefield et al. 1993
Trachypus petiolatus	14	7	4	—	Hym: Halictidae (36), Andrenidae (36)	Evans & Matthews 1973b

[1]See Table 3-1 for prey order abbreviations. [2]All prey were of a single species.

Table 3-7 Prey diversity of selected species of nest-building Pompilidae

Species	No. of prey items	No. of prey genera	No. of prey families	Most common prey family[1] (%)	References
Pepsinae					
Dichragenia pulchricoma	39	5	4	Lycosidae (54)	F. Gess & Gess 1974
Dipogon sayi	114	4	2	Thomisidae (99)	Fye 1965
Hemipepsis ustulata	—	2	1	Theraphosidae (100)	F. Williams 1956
Pepsis thisbe	—	2	1	Theraphosidae (100)	F. Williams 1956, Punzo 1994a,b
Priocnemis cornica	73	7	5	Lycosidae (64)	Kurczewski & Kurczewski 1968a,b, 1973
P. germana	30	6	4	Amaurobiidae (67)	Kurczewski & Kurczewski 1968a, 1973
P. minorata	86	9	6	Amaurobiidae (49)	Kurczewski & Kurczewski 1973
Pompilinae					
Anoplius apiculatus	14	1	1	Lycosidae (100)	Evans et al. 1953
A. cleora	21	3	1	Lycosidae (100)	Kurczewski & Kurczewski 1968a,b, 1972
A. eous	11	1	1	Lycosidae (100)	Shimizu 1992

Table 3-7 *Continued*

Species	No. of prey items	No. of prey genera	No. of prey families	Most common prey family[1] (%)	References
A. marginatus	15	8	7	Salticidae (40)	Kurczewski & Kurczewski 1968a,b, 1972
A. tenebrosus	38	11	5	Lycosidae (63)	Alm & Kurczewski 1984
A. semirufus	63	6	5	Lycosidae (75)	Kurczewski & Kurczewski 1968a,b, 1972
A. splendens	32	12	5	Lycosidae (38)	Kurczewski & Kurczewski 1968a,b, 1972
Aporinellus wheeleri	194	5	2	Salticidae (>99)	Kurczewski et al. 1988
Episyron quinquenotatus	210	5	1	Araneidae (100)	Kurczewski & Kurczewski 1968a,b, 1972
Poecilopompilus algidus	45	4	1	Argiopidae (100)	R. Martins 1991
Pompilus scelestis	8	1	1	Lycosidae (100)	Gwynne 1979
Sericopompilus apicalis	14	6	4	Salticidae (64)	Evans & Yoshimoto 1955
Tachypompilus xanthopterus	35	1	1	Heteropodidae (100)	R. Martins 1991

[1]All prey are in the order Araneae.

(1962), many pompilids are poor "spider taxonomists" in that they hunt spiders in a particular size range without restricting themselves to a narrow set of prey taxa. On the other hand, tarantula hawks (both *Pepsis* and *Hemipepsis*) and some of the *Anoplius* that hunt wolf spiders have a relatively narrow diet, perhaps because they need specialized hunting techniques to subdue their large and dangerous prey. Although both immature and adult spiders are hunted (depending on the species of wasp), all nest-provisioning pompilids provide each nest cell with a single prey item.

PREY OF NEST-PROVISIONING VESPIDAE

Despite the diversity of their nesting behaviors (Chapters 6 and 10), all members of the subfamily Eumeninae provision with insect larvae, primarily larval Lepidoptera (Table 3-8). The prey records of the 110 eumenine species reviewed by Iwata (1976) include 24 families of Lepidoptera. Many of the eumenines have just one or two known families of prey listed, and only three species took as many as six families of prey. This niche breadth is comparable to that of sphecids (Ammophilini) that take caterpillars, but it is a much more restricted diet than that of many Sphecidae. Nonlepidopteran prey are used by a few Eumeninae: *Paragymnomerus spiricornis* take sawfly larvae (Móczár et al. 1973); *Symmorphus* prey primarily on the larvae of leaf beetles; and *Raphiglossa zethoides* take weevil larvae (Iwata 1976). Species of the related subfamily Euparagiinae, which includes just 10 species, are also predators of larval beetles (Krombein et al. 1979b).

Variation in Diets within Species

Variation in prey use within species of wasps can be partitioned among various sources: spatial variation within and between nests and between populations and temporal variation across seasons and years. Such variation must often result from spatial and temporal variation in prey availability, but it could also be caused by variation in genetically based prey preferences or from body size constraints that restrict smaller females to smaller prey (O'Neill 1985).

At the largest spatial scale, prey records vary among populations. In samples of *Euodynerus leucomelas* prey from two sites in Maine in 1984, 100% of the prey at Penobscot were tortricid caterpillars,

Table 3-8 Prey diversity of selected species of nest-building Vespidae, subfamily Eumeninae

Species	No. of prey items	No. of prey genera	No. of prey families	No. of prey items per cell[1]	Most common prey[2] (%)	References
Ancistrocerus antilope	112	1	1	up to 12	Lep: Pyralidae (100)	Jennings & Houseweart 1984
Antepipona scutellaris	41	—	1	up to 13	Lep: Pyralidae (100)	F. Gess & Gess 1991
Anterhychium flavomarginatus	1,135	1	3	3–21	Lep: Pyralidae (95)	Itino 1992
Discoelius japonicus	46	—	2	up to 10	Lep: Pyralidae (61)	Itino 1992
Euodynerus dantica	478	1	1	19–50	Lep: Pyralidae (100)	Itino 1992
Monobia quadridens	90	1	1	4–19	Lep: Pyralidae (100)	Frost 1944
Odynerus dilectus	711	1	1	8–31	Col: Curculionidae (100)[3]	G. Bohart et al. 1982
Orancistrocerus drewseni	1,615	—	5	P	Lep: Pyralidae (87)	Itino 1992
Parachilus major	54	1	1	5–11	Lep: Psychidae (100)	F. Gess & Gess 1988a

Paraleptomenes mephitus	104	≥8	≥7	up to 15	Lep: Gelichiidae (53)	Krombein 1978
Parancistrocerus fulvipes	24	—	2	8–13	Lep: Pyraustidae (88)	Krombein 1967
Pterocheilus texanus	61	1	1	3–9	Lep: Noctuidae (100)	Grissell 1975
Rhynchium marginellum	>400	—	1	4–31	Lep: Pyralidae (100)	F. Gess & Gess 1991
Symmorphus cristatus	~75	1	1	2–7	Col: Chrysomelidae (100)[3]	Fye 1965
Zethus otomitus	98	—	1	≥40	Lep: Olethreutidae (100)	Calmbacher 1977

[1]Prefers to progressive provisioners where maximum number of prey items is unknown. [2]See Table 3-1 for prey order abbreviations. [3]All prey were of a single species.

whereas 93% of prey at Piscataquis were pyralids. The diet of *Ancistrocerus adiabatus* also varied from 100% tortricids at Penobscot to 79% gelechiids at Piscataquis (Collins and Jennings 1987a,b). Similarly, in 1971, the prey of *Lindenius columbianus* consisted of 61% chironomid midges and fewer than 0.1% scaptopsid flies at Auburn, New York, but 51% scatopsids and fewer than 0.1% chironomids at nearby Sennett (Miller and Kurczewski 1975).

Because the abundance of potential prey is likely to change within and between years, it should not be surprising that diets may also vary temporally. In Michigan, H. Brockmann (1985a) observed both yearly and seasonal changes in the subfamilies of prey taken by *Sphex ichneumoneus*, which she attributed to seasonal changes in the availability and vulnerability of different prey (Fig. 3-5). Seasonal variation in provisions has also been reported for *Chlorion californicum* (Landes et al. 1987), *Episyron arrogans* (Endo 1976), *Pachodynerus nasidens* (Jayasingh and Taffe 1982), *Poecilopompilus* algidus (R. Martins 1991), and *Trypoxylon politum* (Rehnberg 1987). Yearly variation was found in the diet of *Ancistrocerus catskill*, which took 56% Gelechiidae (Lepidoptera) and 8% Pyralidae (Lepidoptera) in 1977, but 23% Gelechiidae and 40% Pyralidae in 1978 at one site in Maine (Jennings and Houseweart 1984). D. Gwynne (1981) found that *Philanthus bicinctus*, normally a bumble bee predator, responded to a local dearth of its preferred prey in 1978 by including various other large Hymenoptera in its diet. Thus, although the data presented in Tables 3-1 through 3-8 provide a starting point for considering variation in prey use among wasp taxa, diets are often more variable than such data indicate.

Facing page

Figure 3-5. Temporal variation in the diet of *Sphex ichneumoneus*. (Top) Yearly variation at a site in Michigan. (Bottom) Seasonal variation (for 5-day intervals beginning on 5 July) in diet (different years and sites combined). Symbols for yearly variation correspond to those for seasonal variation. The four major groups of prey included the subfamilies Copiphorinae (cone-headed katydids), Conocephalinae (meadow katydids), and Phanopterinae (false katydids) of the family Tettigoniidae and Oecanthinae (tree crickets) of the family Gryllidae. (All data from Brockmann 1985a.)

100
80
60
40
20
0
1972
1973
1974
1975

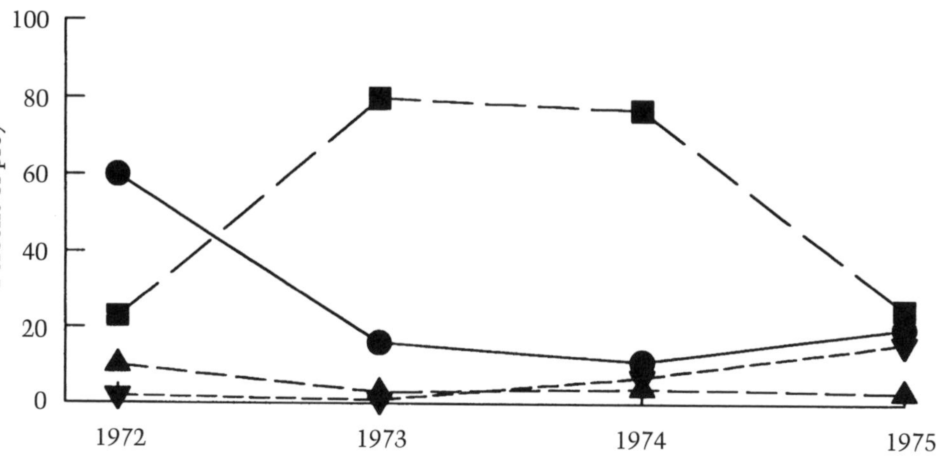

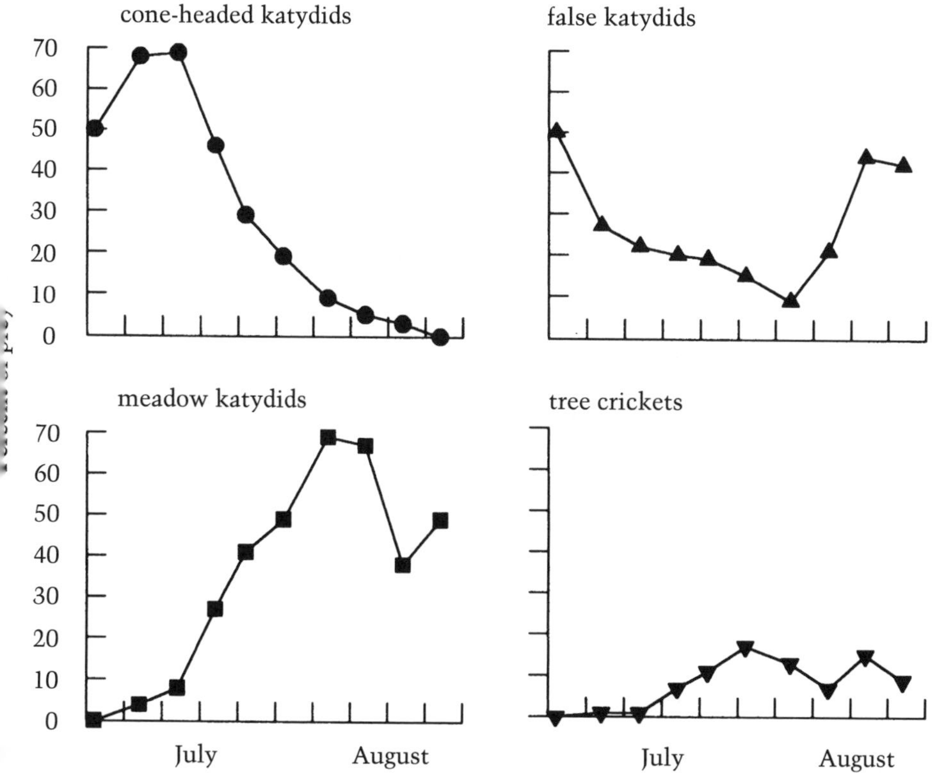
cone-headed katydids
false katydids
meadow katydids
tree crickets
70
60
50
40
30
20
10
0
70
60
50
40
30
20
10
0
July
August
July
August

Variation in the Number of Prey per Cell

Interspecific variation in the number prey items in each cell (Tables 3-1 to 3-8) can be attributed primarily to the relative sizes of predator and prey. Species taking relatively large prey, such as all pompilids and sphecines of the genera *Prionyx* and *Podalonia*, need never stock a nest cell with more than a single prey item. At the other extreme are the pemphredonines that provide each offspring with many dozens of small aphids or thrips. Variation in the number of prey items per cell within a species can result from differences in allocation of resources to male and female offspring (Chapter 10) and to variation in the availability of prey of different sizes. For example, as the summer progressed and the relative abundance of larger adult spider prey increased, the mean number of prey items that *Trypoxylon politum* placed in each cell dropped from about 10 to 6 (Rehnberg 1987).

Comparative Survey of Hunting and Stinging Behaviors

HUNTING SITES AND SEARCHING BEHAVIOR

Observations on *Philanthus triangulum, Pepsis thisbe,* and *Euodynerus foraminatus* suggest that searches for prey move from general to specific sites as the predator narrows down the location of prey and that cues used in locating prey may require the use of multiple sensory modalities. Most studies do not address hunting in such detail, but a few observations on other wasps suggest a similar sequence of foraging. Some species may search first for conspicuous landmarks associated with prey, such as ant mounds, livestock, spider webs, dung, or host plants (Table 3-9). *Oxybelus emarginatum,* for example, first search for cattle, then for dark spots on the surface of the animal, and then they investigate objects closely with their antennae (Snoddy 1968).

To find their prey, many wasps make restricted searches in particular habitats where prey may occur in cryptic niches that are difficult to reach (Table 3-9). Some wasps seek prey in locations that change over time, such as moving livestock, fresh dung that remains only ephemerally attractive to dipteran prey, and ant mating swarms that appear only on certain days. Once wasps reach specific hunting

Table 3-9 Common hunting locations of selected species of solitary predatory wasps

Predator	Prey	Hunting sites	References
Sphecidae			
Aphilanthops subfrigidus	Alate queen ants	Ant mating swarms	O'Neill 1990, 1994, 1997
Clypeadon laticinctus	Worker ants	Ant nests	Alexander 1985, 1986
Bembix spp., *Oxybelus emarginatus*, *Stictia carolina*, *S. signata*	Flies	On and around livestock	Evans 1966c, Snoddy 1968, Hine 1906 (in Bohart & Menke 1976), Sheehan 1984, Philippi & Eberhard 1986
Bembix spp., *Mellinus* spp., *Oxybelus similis*	Flies	Animal droppings	Hamm & Richards 1930, Evans 1966c, Evans 1989
Chalybion californicum, *Trypoxylon* spp.	Spiders	Spider webs	Coville 1976, Rayor 1996
Chlorion spp., *Liris* spp.	Crickets	Refuges in burrows, bark crevices, under rocks, grass clumps	Peckham & Kurczewski 1978 (and references therein), Steiner 1976
Palarus latifrons	Honey bees	Apiaries	Clauss 1985
Palmodes laeviventris	Mormon crickets	Migrating bands	LaRivers 1945
Pemphredon lethifer	Aphids	Aphids colonies on plants	Danks 1970
Philanthus spp.	Bees	Flowers, natural bee nests, apiaries	Tinbergen 1972, Lin 1978a, Simonthomas & Simonthomas 1980, Evans & O'Neill 1988

Table 3-9 *Continued*

Predator	Prey	Hunting sites	References
Prionyx atratus	Grasshoppers	Bunch grasses	Evans 1958c
P. crudelis	Migratory locusts	Locust swarms	Haskell 1955
Sphecius speciosus	Cicadas	Trees	Dambach & Good 1943
Tachytes mergus, T. minutus, Palmodes carbo	Crickets	Subterranean tunnels	Kurczewski 1966a, Dodson & Gwynne 1984
Pompilidae			
Anoplius depressipes, A. eous	Aquatic spiders	Beneath or on water surface	Roble 1985, Shimizu 1992
Anoplius spp.	Wolf spiders	Spider burrows	Gwynne 1979, McQueen 1979
Episyron arrogans	Spiders	Orb webs	Endo 1976
Pepsis thisbe	Tarantulas	Tarantula burrows	Punzo 1994b
Vespidae: Eumeninae			
Ancistrocerus antilope, Euodynerus foraminatus, Pachodynerus nasidens	Caterpillars	Inside leaves rolled or tied with silk by caterpillar; inside leaf mines	Cooper 1953, A. Steiner 1984, Jayasingh & Taffe 1982
Odynerus dilectus	Alfalfa weevils	Surface of alfalfa leaves	G. Bohart et al. 1982
Pterocheilus texana	Cutworms	Inside flowers	Habeck et al. 1974
Zethus spinipes	Pine tip moths	Inside tips of branches infested by larvae	Lashomb & Steinhauer 1975

Note: Some of these species may also hunt elsewhere for prey.

grounds, they may have to flush prey hidden within burrows (A. Steiner 1976, Peckham and Kurczewski 1978, Gwynne 1979), grass clumps (Evans 1958c), or spider webs (Coville 1976). Alternatively, a female may have to enter the tight confines of plant tissue (Lashomb and Steinhauer 1975), leaf rolls (Cooper 1953, A. Steiner 1984), flower blossoms (Habeck et al. 1974), or subterranean tunnels (Kurczewski 1966a) to extract hidden prey. The oddest wasp is, perhaps, *Anoplius depressipes*, which hunts under water for its aquatic spider prey (Evans 1949, Roble 1985).

STINGING AND PARALYSIS

Once a potential prey has been located, the next step is to subdue it. Predatory wasps usually sting prey through the venter, most often through membranous areas near the bases of legs, in the thorax, or in the neck. A. Steiner (1971, 1976, 1978, 1979, 1981, 1983a,b, 1986), who has done more than anyone to improve our understanding of stinging behavior, provides strong evidence that stings are delivered in the vicinity of ventral ganglia of the central nervous system that control movement of legs and mouthparts. Although lesions caused by stings are commonly found on the body surface (providing a lasting record of the attack), stings rarely penetrate the ganglia. Rather, venom travels to these nerve centers in the blood.

Because the effect of the venom is only on the nearest ganglion, the number and location of stings inflicted appear to be correlated primarily with the number of ganglia that need to be inactivated and secondarily to the location of vulnerable soft spots (membranes) on well-sclerotized prey (A. Steiner 1981, 1986). *Podalonia luctuosa* typically stings its caterpillar prey once in the head (corresponding to the subesophageal ganglion that controls the mandibles, maxillae, and labium), once in each thoracic segment near ganglia that control the true legs, and once each in abdominal segments 1–6 near ganglia that control those segments. Eumenines often sting caterpillars only twice, perhaps because they prey on smaller, weaker species (A. Steiner 1986). Species of *Isodontia*, *Liris*, *Palmodes*, *Prionyx*, and *Tachysphex* that attack Orthoptera sting in the head and thorax only (Fig. 3-6), and *Oxybelus uniglumis* sting flies only once in the thorax, near the single large ganglion that controls all three pairs of legs (A. Steiner 1978, 1979). Among wasp species, the number of stings needed apparently decreases with the increased consolidation of ganglia in the central nervous system of prey that has occurred during

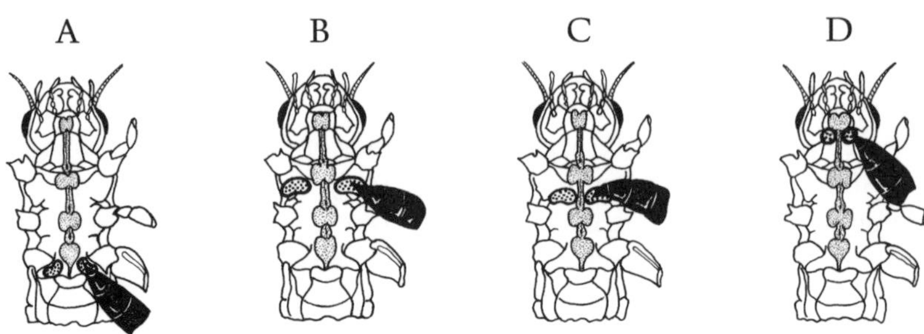

Figure 3-6. Sting sequence of *Liris nigra* attacking a cricket; ventral view of cricket, showing ganglia and ventral nerve cord of the central nervous system (fine stippling) and regions where sting is usually inserted (coarse stippling). Stinging near the (A) metathoracic ganglion, (B) prothoracic ganglion, (C) mesothoracic ganglion, and (D) subesophageal ganglion in the neck. (A–D redrawn by Catherine Seibert from A. Steiner 1971.)

insect evolution (A. Steiner 1981). However, the number of stings delivered may be variable within a species, increasing with decreased responsiveness of prey to early stings (Steiner 1983a, Truc and Gervet 1983) and with increased prey size (Steiner 1986, Kurczewski 1987a).

The order in which some species sting may be related to which appendages need to be subdued first to effect an efficient and safe attack. *Podalonia* stings first in the third thoracic segment, inhibiting the defensive curling response of the cutworm, which makes it difficult for the wasp to reach the venter for further stings and impossible for it to store in the narrow confines of the nest (A. Steiner 1983a, 1986). Other wasps target potential defenses on the head or thorax of prey. To inhibit counterattacks or escape responses of prey, pompilids often sting first near the ganglia that control the spiders' fangs (A. Steiner 1986). Similarly, *Tachysphex costai* and *Stizus ruficornis* first sting their mantid prey near the base of the front (raptorial) legs, whereas *Liris* species make their first penetration at the base of the hind (jumping) legs of crickets (A. Steiner 1976, 1986). *Prionyx parkeri*, which attacks grasshoppers, strikes first in the neck (rather than near the hindlegs), but this wasp has the advantage of physically overpowering its prey with its long mandibles and long, spiny legs (A. Steiner 1981). Delivering the sting in the neck may also inhibit oral secretion of defensive compounds by the grasshoppers (A. Steiner 1982).

The degree and duration of paralysis vary considerably among species (Piek 1985). At one extreme are the wasps that kill their prey, either with their sting or with their mandibles. The stings of certain *Alysson*, *Bembix*, *Glenostictia*, *Rubrica*, and *Steniolia* apparently kill their prey, though not always instantly (Evans 1966c, Piek and Spanjer 1986); however, many reports of mortal stings given by other wasps are probably erroneous (Piek and Spanjer 1986). *Passaloecus* kill their aphid prey by crushing them with their mandibles (J. Powell 1963), and *Calligaster cyanoptera* subdue caterpillars by biting off the head (F. Williams 1919c).

More commonly, stinging paralyzes the prey without killing them. Most pompilid and eumenine venoms tend to have incomplete and temporary effects on prey. The effects of the venoms of nest-provisioning sphecids, in contrast, range from incomplete and temporary to complete and permanent paralysis (Piek 1985, Piek and Spanjer 1986). When the venom paralyzes prey incompletely, the movements of legs, wings, and mandibles are inhibited, while the heart and sometimes the gut continue to function; the prey may continue to defecate and some species even transform from a larva to a pupa (Rau and Rau 1918, Cooper 1953, Piek 1985). Paralysis preserves prey so that they remain fresh and immobilizes prey so that transport is easy and damage to sensitive eggs and larvae is minimized. Many authors have noted that paralyzed prey remain alive in the lab for several weeks and may recover much muscular function (Brockmann 1985a and references therein, Piek 1985).

The sting apparatus itself is a complex organ adapted to deliver venom rapidly and precisely, often during a vigorous struggle between predator and prey. This task is facilitated by the presence of venom reservoirs, muscles that power the sting thrust, structures that allow extrusion and rotation of the sting, numerous minute sensory hairs that guide the sting (Rathmayer 1962), and the sting itself, a sharp and relatively stiff hypodermic needle. A functional morphologist could probably make much of the variation in sting morphology among species attacking different prey. The curvature of the sting is greatest in sphecids that hunt highly mobile insects such as flies, bees, and wasps, whereas the stings are straight or slightly curved in species with less mobile (perhaps weaker) prey such as slow-moving spiders and caterpillars (Radović 1985). The curved sting of some sphecids probably aids in inserting the sting into the venter of prey (Hermann and Gonzalez 1986).

Comparative Survey of Prey-Carriage Mechanisms

Parasitoid wasps usually do not move hosts (Chapter 2), although some Bethylidae, Tiphiidae, and parasitoid Pompilidae and Sphecidae drag them several centimeters and deposit them in an unmodified crevice or inside the prey's own burrow. However, the predators discussed in this chapter often attack prey far from nests and transport them considerable distances while flying or walking. During hunting, the added burden of carrying prey may add a significant energy cost and limit prey size.

One of the major sets of constraints on foraging efficiency in optimal foraging models for central-place foragers consists of the time and energy costs of transporting prey, both of which increase with prey size and distance to hunting grounds (Stephens and Krebs 1986). Wasps that transport arthropods approaching or exceeding their own body weight must be severely constrained in their ability to travel great distances quickly and efficiently. Most prey of wasps that stock just a single item for each cell have to be larger than the adult female. The spider prey of pompilids and the caterpillar and orthopteran prey of many sphecines commonly range in size from two to eight times the body mass of the predator (Iwata 1942, Dodson and Gwynne 1984). Even when several prey items are given to each offspring, large prey, such as the cicada prey of *Sphecius speciosus* (Lin 1979b, Hastings 1986) and the larger bees transported by some *Philanthus* (Strohm and Linsenmair 1997a), obviously impose a considerable burden. The difficulty of transporting prey is reflected in positive correlations between the size of females of many wasps and the size of their prey, correlations that undoubtedly reflect the difficulty small females have in carrying (and perhaps subduing) large prey (O'Neill 1985). For example, larger *Philanthus triangulum* females can carry larger honey bee prey than can smaller females (and for longer periods in continuous flight) (Strohm and Linsenmeir 1997a), and *Ammophila sabulosa* transport small caterpillar prey faster than large prey (Field 1992d).

For species that move prey great distances or that stock nest cells with many small items, the distance between the nest and hunting grounds is probably a major factor in foraging success, even when the prey itself may add relatively little to travel time and expense. Unfortunately, the distance between hunting sites and nests is usually unknown for wasps (because most observations are limited to nesting areas). Many wasps probably hunt close to their nest. For example, *Tachytes* may hunt for nymphal grasshoppers within 1–5 m of their nests (e.g., Kurczewski and Spofford 1986); *Aphilanthops* take many ants within 5–10 m of their nests (O'Neill 1994); and *Passaloecus* take aphids within 15 m of their nests (Corbet and Backhouse 1975). Nevertheless, the stocking of a single cell in a nest may require many dozens of provisioning flights, totaling hundreds of meters in length. Furthermore, there are several potential problems associated with hunting near nests. First, prey could become rapidly depleted near large nesting aggregations. Second, suitable nest sites and prime hunting grounds may not always be adjacent. Perhaps for these reasons, some wasps transport prey considerable distances. Even species that drag large prey along the ground may travel 10 or more meters to reach the nest site. *Palmodes laeviventris*, for example, may carry their Mormon cricket prey more than 25 m (La Rivers 1945), and *Podalonia luctuosa* haul caterpillars up to 20 m (O'Brien and Kurczewski 1982a). For those species that carry prey along the ground (particularly in dense vegetation), the actual distance traveled while weaving around and through barriers exceeds the "beeline" distances traveled by species that carry prey in flight. Terrestrial prey carriage, which occurs in species that use relatively large prey, is supplemented in some species by short flights made after the prey is carried up onto short vegetation.

Mechanisms of prey carriage have probably evolved under constraints of load-carrying ability and balance, the effect of prey on the wasp's ability to see, and the need to keep the front legs free for other activities, such as opening the nest to bring prey in. H. Evans (1962) categorized the general mechanisms of prey carriage on the basis of the way prey are held and the mode of transport (Table 3-10, Fig. 3-7). Each category also includes several variations on the theme, with most of the variation being exhibited by the Sphecidae. All Eumeninae carry their prey in their mandibles while flying forward, whereas all Pompilidae carry prey in their mandibles while dragging them across the ground or (rarely) over the surface film of water (Evans 1949, Shimizu 1992). (After stinging, some pompilids amputate the legs of their prey to more fully immobilize them and to make them easier to transport and store in the nest.) Mandibular forms of carriage are also practiced by many sphecids (as well as some Bethylidae and Tiphiidae). Many Sphecidae, however, use more evolutionarily derived forms of prey carriage that involve the use of their legs and various structures on their abdomen.

Species that carry prey with their mandibles (Fig. 3-7A) or legs (Fig. 3-7B) have not evolved special structures for prey transport. However,

Table 3-10 Prey-carriage mechanisms of selected species of solitary nest-provisioning wasps

Mechanism	Description	Sphecidae	Pompilidae	Vespidae: Eumeninae
Mandibular 1	Prey held in mandibles only and usually dragged backward over ground.	All Ampulicinae (parasitoids)	Most species, including *Pepsis*, *Episyron*	—
Mandibular 2	Prey held in mandibles and carried forward over ground; wasp may make short flights after climbing short vegetation. Prey must be lifted and so may partially block wasp's view.	Many Sphecinae; some Larrinae *(Tachysphex)*	Some species of *Pompilus*, *Priocnessus*, and *Auplopus*	—
Mandibular 3	Prey held in mandibles (sometimes with aid of legs) and carried forward in flight.	Many Pemphredoninae; many Sphecinae; some Larrinae *(Palarus, Tachysphex, Tachytes, Trypoxylon)*; some Philanthinae (most *Cerceris*); *Mellinus*	—	All

Pedal 1	Prey held in middle legs (sometimes with aid of forelegs or hindlegs) and carried forward, usually in flight.	*Sphecius* (on ground); most Psenini (Pemphredoninae), Crabronini (Crabroninae), and Philanthini (Philanthinae); all Bembicini, Gorytini, and Stizini (Bembecinae)	—	—
Pedal 2	Prey held in hindlegs and carried forward in flight.	Some *Oxybelus*	—	—
Abdominal 1	Prey carried in flight held on sting, sometimes with support of hindlegs.	Some *Oxybelus* and *Sericophorus*	—	—
Abdominal 2	Prey carried in flight held in special devices.	*Clypeadon*; some *Cerceris* (with aid of mandibles)	—	—

Sources: Evans 1962, Krombein 1981, Radović & Sušić 1997.

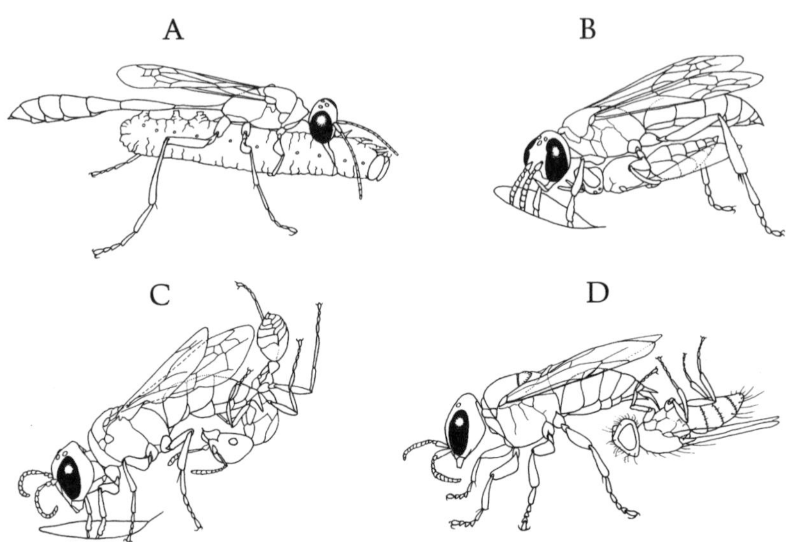

Figure 3-7. Prey-carriage mechanisms (redrawn by Catherine Seibert from the cited sources). (A) *Sphex* carrying caterpillar with legs and mandibles (Evans 1963). (B) Beewolf (*Philanthus*) holding bee with middle legs as she enters her nest (Evans and O'Neill 1988). (C) *Clypeadon* carrying ant on "ant clamp" at tip of abdomen (Evans and West-Eberhard 1970). (D) *Oxybelus* carrying a fly impaled on her sting (Peckham et al. 1973, Peckham and Hook 1980).

some *Oxybelus* and *Sericophorus* species transport prey impaled on their sting (Fig 3-7D), which may be minutely barbed to aid in securing the prey (Evans 1962, Peckham 1985, McCorquodale 1988a, Radović and Sušić 1997). Even more elaborate structures are found in *Clypeadon*, which use a special "ant clamp," a modification of the terminal abdominal segment, to grasp their ant prey between their forelegs and middle legs (Fig. 3-7C). Similarly, some *Cerceris* carry beetles in their mandibles, with the aid of a "buprestid clamp," a modification of the fifth abdominal sternite that holds the prey steady while the wasp is in flight (Krombein 1981).

Prey Storage

Many wasps that stock multiple prey items in each cell store prey in the nest burrow during a sequence of foraging trips, rather than move each one into a cell after each hunting trip (Evans 1957a, Evans and O'Neill 1988). Storing prey allows females to maximize the number of foraging trips during the short daily interval when abiotic conditions are conducive to hunting and when prey are available. Females of *Aphilanthops subfrigidus*, for example, are confined to hunting during the several hours each day in which their only prey, winged alates of the ant *Formica subpolita*, are available in mating swarms (O'Neill 1990). Between hunting trips they spend an average of less than 2 minutes within the nest.

Homing and Orientation

Consider the plight of a female *Bembix* returning to her nest with prey and having to find it among closely packed thousands of other nests on a featureless dune surface. The problem of relocating a nest is compounded when females conceal nest entrances to hide them from their enemies (Chapter 7). It is astounding to watch a female return to a nesting area and unerringly proceed directly to her concealed nest. After capturing prey up to 6–10 m from its nest, a female *Priocnemis minorata* walks backward "over a rough bed of leaves and sticks" in a straight line to its nest (Yoshimoto 1954). *Ammophila* maintain this ability even when observers do their best to disorient them by placing obstacles in the paths or placing the wasps in a dark container and releasing them at a new location (Thorpe 1950). Tinbergen's classic work on *Philanthus triangulum* demonstrated that the problem of relocating the nest has been solved (at least partially) by the ability of females to learn the image of the landscape surrounding the nest and to match the memorized image to the configuration of local landmarks when the female returns to the area.

Before Tinbergen's work, the navigational skills of female wasps were remarked on by early wasp biologists, including J. Fabre (1915), who displaced *Cerceris* females 3 km from their nests, thereby demonstrating that at least some of them could find their way back. Since then, solitary nest-provisioning wasps have continued to be popular subjects for studying homing behavior. Research on *Philanthus triangulum* was followed by work with *Ammophila* (Baerends 1941, Thorpe 1950), *Argogorytes* (Schöne et al. 1993, 1994), *Bembix* (van Iersel 1952; Tsuneki 1956; Chmurzynksi 1964, 1967; van Iersel and van den Assem 1964; Tengö et al. 1990; Schöne and Tengö 1991), *Cerceris* (Zeil 1993a,b; Zeil and Kelber 1991; Zeil et al. 1996),

Odynerus (Tsuneki 1961, Zeil and Kelber 1991), *Sceliphron* (Ferguson and Hunt 1989), and several pompilids (Tsuneki 1950). In combination, these studies suggest that there is no simple rule of thumb followed by all wasps homing to nests. For example, *Bembix rostrata* focus on the total surface area of landmarks rather than on features such as their height and width, the key properties learned by *Philanthus* (van Iersal 1952, van Iersel and van den Assem 1964). Furthermore, severe disruption of the local landscape does not prevent female *Microbembex monodonta* from finding their nests, suggesting that obvious landmarks near the nest are not used for homing (Evans 1966c). Even orientation flights vary widely among species: for example) the flights of *Gorytes canaliculatus* are "loops and figure eights;" *Glenostictia scitula* flights are "low and circling" (Evans 1966c); *Sphex argentatus* orients in a series of linear flights extending like spokes outward from the nest (Tsuneki 1963); *Astata unicolor* a orients in complex array of meanderings both on foot and on the wing (Peckham and Peckham 1898, Evans 1957a); and *Cerceris rybyensis* flies in a sequence of ever-widening and ascending arcs on one side of the nest (Zeil 1993a). Recent evidence suggests that, during orientation flights, female *Cerceris* learn the configuration of local landmarks and determine the absolute distance between the landmarks and the nest (Zeil 1993a,b).

The study of homing behavior is not an isolated exercise in sensory physiology but is related to the broader issue of the evolution of nesting strategies and social behavior in the Hymenoptera. The ability of females to orient to their nests was a necessary prerequisite for the evolution of the construction of nests before hunting and the provisioning of multiple prey items for each cell and multiple cells in each nest. Although parasitoids may have no need for homing, once wasps began caching prey and moving off a short distance to dig and nest, they would have needed to relocate the prey after the nest was built and to relocate the nest after the prey was retrieved. It is these rudimentary homing abilities, perhaps, that were built on in species that began moving greater distances in search of nest sites and prey.

Comparative Survey of Oviposition Sites

Nest-provisioning aculeates lay their eggs, usually one in each nest cell, in one of two locations: on the wall of the cell or on the prey.

Oviposition within the empty cell before provisioning has evolved independently in several taxa of wasp (Evans and West-Eberhard 1970). All eumenines lay the egg in the empty cell, either loosely on the floor of the cell or suspended from the ceiling of the cell by a narrow thread (e.g., *Ancistrocerus*) (Fig. 3-8J). A similar oviposition strategy is considered to be the derived evolutionary condition in the Sphecidae, where it has arisen perhaps five to eight times independently in the subfamily Bembicinae alone (Evans 1966c). In this subfamily, the egg may be found lying on the cell floor, leaning against the back of the cell wall, glued erect to the substrate, or glued erect to a small pedestal made of sand grains (Fig. 3-8F, G). Other bembicines, as well as all other Sphecidae and all Pompilidae lay the egg in a specific location on one prey item in the cell (Fig. 3-8) or simply attach it to the mass of prey. As do parasitoids (Chapter 2), predatory wasps often place the egg near a vulnerable location on prey (a membrane), where the weak, newly hatched wasp larva can break through the exoskeleton of the prey. The egg of the Sphecidae may even be attached to the prey so that the hatching larva has its head free to reach other prey items in the cell (Evans and West-Eberhard 1970).

Feeding by Adults

Most, if not all wasps, feed on plant juices in the form of nectar from flowers or extrafloral nectaries. Many that feed on flowers have a short tongue, and so are restricted to plant species with a short corolla, but many Bembicini have elongate tongues that allow feeding on deep flowers (Evans 1966c). *Exeirus* feed on the sap of trees that oozes from feeding wounds created by their cicada prey, and many solitary wasps feed on homopteran honeydew. Host feeding is also common among female predatory solitary wasps, who may feed at wounds or eat the body tissue of prey (A. Steiner 1982). Wasps of the genera *Anoplius* (Shimizu 1992), *Oxybelus* (Snoddy 1968), *Pemphredon* (Danks 1970), and *Diodontus* (J. Powell 1963) use their mandibles to crush or puncture their prey to release body juices; feeding by *Diodontus* becomes more frequent with increasing temperature (Lin 1978b). Other wasps, notably some *Bembix* and *Prionyx*, pierce and suck juices from their prey with their proboscis (A. Steiner 1982). *Oxybelus emarginatus* exploit a nonarthropod protein source during

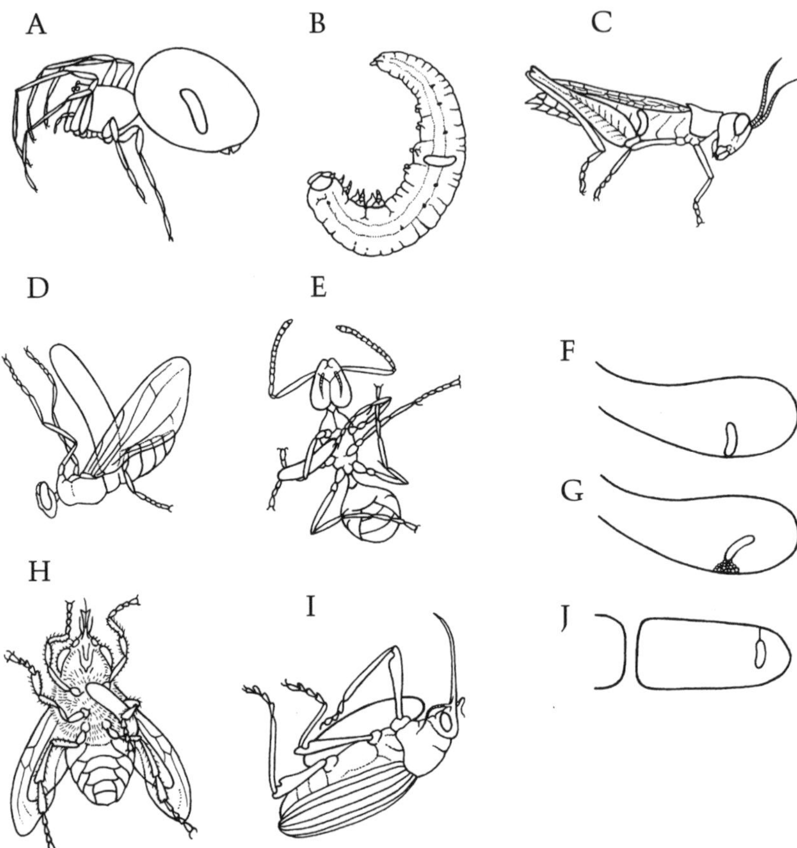

Figure 3-8. Oviposition sites of selected species of Sphecidae and Pompilidae. (A) *Episyron quinquenotatus* egg on *Araneus* spider. (B) *Ammophila infesta* egg on noctuid caterpillar. (C) *Prionyx* egg on adult grasshopper. (D) *Bembix niponica* egg on blow fly. (E) *Tracheliodes quinquenotatus* egg on worker ant. (F) *Microbembex monodonta* egg in empty cell. (G) *Bembecinus neglectus* egg on pedestal in empty cell. (H) *Philanthus triangulum* egg on honey bee. (I) *Cerceris halone* egg on weevil. (J) *Ancistrocerus antilope* egg hanging from silken thread in empty cell. (A–J redrawn by Catherine Seibert from Evans and West-Eberhard 1970, where original sources can be found.)

their forays for flies around livestock, licking fresh blood from wounds inflicted by biting flies (Snoddy 1968).

Impact on Prey Populations

The nest-provisioning aculeates probably have less potential as biological control agents than their parasitoid relatives because they are usually less specialized in their prey preferences and have nest site preferences that may limit their activities in habitats where pests are abundant (Chapter 6). One exception may be greenhouses, where trap-nesting species could perhaps be raised as aphid predators (Corbet and Backhouse 1975). One nest provisioner, *Liris subtesselatus*, was imported into Hawaii in 1922, but it has had no substantial effect on populations of its cricket prey (Funasaki et al. 1988).

Nevertheless, a consideration of the prey preferences of some endemic nest-provisioning solitary wasps suggests that they could play important roles in the natural regulation of pest populations. On cropland and pastures, possible examples of beneficial wasps (and their pest prey) include *Palmodes* (LaRivers 1945), *Prionyx* (Evans 1958c, O'Neill 1995), *Sphex* (Haskell 1955), and *Tachysphex* (Newton 1956) (grasshoppers); *Podalonia* (O'Brien and Kurczewski 1982a) (cutworms); *Passaloecus* (Corbet and Backhouse 1975) (aphids and thrips); *Odynerus dilectus* (G. Bohart et al. 1982, Schaber 1985) (alfalfa weevils); and *Bembix, Rubrica,* and *Stictia* (Evans 1966c, Nazarova and Baratov 1982) (horse flies and deer flies). The beneficial activities of the latter wasps in hunting tabanid flies near livestock have earned them such common names as cowfly tigers, horse guards, and insecto policia (R. Bohart and Menke 1976, Philippi and Eberhard 1986). In forests, beneficial wasps include *Ancistrocerus* and *Euodynerus* as predators of spruce budworm (Jennings and Houseweart 1984, Evans 1987) and *Zethus spinipes* as predators of pine tip moths (*Rhyacionia frustrana*) (Lashomb and Steinhauer 1975).

A few studies have attempted to assess the impact of solitary wasps on pest populations. J. Collins and D. Jennings (1987b) estimated that predation by three species of eumenines at five sites resulted in reductions in spruce budworm populations by just 5% to 9%. I. LaRivers (1945) estimated that about 30,000 *Palmodes laeviventris* females killed about half a million Mormon crickets

(*Anabrus simplex*) within an area of 0.65 km^2 during an outbreak of this pest in Nevada, but LaRivers did not assess the proportion of the population killed. With less hard evidence, R. Newton (1956) claimed that three species of *Tachysphex* were responsible for a drastic reduction in populations of the grasshopper *Oedaleonotus enigma* during a severe infestation in Idaho.

Other wasps prey on arthropods considered to be beneficial to humans. *Philanthus triangulum* in Europe and Egypt (Simonthomas and Simonthomas 1977, 1980) and *Palarus latifrons* in southern Africa (Clauss 1985) can cause losses of worker honey bees in commercial apiaries. In the Dakhla Oasis in Egypt, where *Philanthus* made beekeeping nearly impossible, local authorities offered a bounty for the capture of adult *P. triangulum*, but even the capture of 24,000 wasps one year had no marked effect on their abundance during the following year. Less well documented is the negative effect of species of such genera as *Bembix, Cerceris, Palarus, Philanthus,* and *Trachypus* on pollinators of crops and wildflowers and on hymenopteran and dipteran enemies of pest insects.

Conclusion

The review presented here barely scratched the surface of the vast amount of information available on wasp diets. Our understanding of the foraging strategies of nest-provisioning solitary wasps could be greatly aided with more information on variation in diets in time and space, search and attack methods, and the location of hunting grounds relative to nests. Although further descriptive studies of foraging by solitary wasps are always welcome and useful, selected species of wasp could also be used to test specific models of foraging behavior. For example, central-place foraging models seem particularly suited to studying foraging strategies of nest-provisioning wasps. Central-place foraging models (Stephens and Krebs 1986) predict which sizes of prey items should be taken at different distances from a nest, taking into account the energetic values of different-sized prey and the cost of carrying those prey different distances back to a nest. In addition, interactions between females of species with facultative prey theft (Chapter 4) have all the hallmarks of producer-scrounger games, a set of game theory models that attempt to explain the coexistence of two types of foraging tactic in a population: foragers that find their own prey (producers) and those that steal food from producers (scroungers) (Barnard and Sibley 1981, Giraldeau et al. 1994).

In principle, we should be able to determine how shifts in prey preferences map onto the evolutionary history of wasps. Such information would allow us to ask questions concerning the direction, frequency, and magnitude of evolutionary transitions between different prey types and modes of foraging. However, our ability to test precise hypotheses will depend on the availability of phylogenetic analyses that resolve the evolutionary relationships among subfamilies, tribes, and genera of solitary wasps. The available phylogenies provide a good start for those interested in macroevolutionary studies of wasp behavior, but we should eagerly await further phylogenies, particularly those based on molecular analyses, that could provide powerful tools for testing phylogenetic hypotheses of wasp foraging behavior using standard techniques (Martins 1996).

When compared with parasitoid wasps, nest-provisioning predators have more complex foraging strategies because of their added need for carrying prey between hunting grounds and the nest and memorizing and relocating the position of the nest during foraging trips. In contrast, cleptoparasitic species, discussed in the next chapter, pare down their own foraging strategies by taking advantage of the hard work of other wasps.

4/ Foraging Behavior of Cleptoparasites

In the broad sense, cleptoparasitism refers to the theft of food from another individual, but solitary wasps exhibit two quite different forms of this feeding strategy: theft of individual food items and brood parasitism. Animals that steal food may adopt theft as either a part-time or more or less full-time foraging mode. Among birds, frigate birds and parasitic jaegers are full-time cleptoparasites that make their living by forcibly stealing fish from birds such as terns. On the other hand, many nest-provisioning solitary wasps are part-time prey thieves (or facultative cleptoparasites) that normally forage for their own food, stealing only when opportunity arises (Field 1992b, 1994).

A more elaborate and specialized form of cleptoparasitism is sometimes termed brood parasitism. Female brood parasitic wasps insert their eggs into a host's nest cell, either by entering the cell or by depositing their young on a prey item destined to be placed in a nest cell (Field 1992b). Ornithologists use the term *brood parasitism* to refer to the activities of birds such as cuckoos or cowbirds that lay eggs in other birds' nests (Wittenberger 1981). Entomologists refer to laying eggs in another wasp's nest as either as brood parasitism (Field 1992b) or cleptoparasitism (Evans and West-Eberhard 1970). I follow Field (1992b) and use brood parasitism when referring to wasps that insert their young into the nests of other wasps. The adult or larva of the brood parasitic solitary wasp may kill and eat host wasp egg or young larva, but the brood parasitic larvae gain most of their sustenance from provisions provided by the host. Adult females of the many obligate brood parasites in the Chrysididae, Pompilidae, Sphecidae, and Sapygidae deposit their eggs within the nests of other aculeate species, thus taking advantage of both the food and the nest provided by the host species.

Intraspecific Prey Theft

Normal foraging can be costly; a foraging female spends much time and energy traveling to foraging areas, searching for prey, subduing prey, and hauling it to a nest site. Stealing prey from a conspecific that has already brought prey to a nest site would substantially reduce this cost. Thus, female sphecids, pompilids, and eumenines that typically capture their own prey may also practice a bit of cleptoparasitism on the side (Field 1992b, 1994).

Prey may be stolen from a female from inside or outside of her nest and at almost any time after it has been captured and before it has been consumed. They may be stolen when left unguarded or may be taken directly from a prey-carrying female (Field 1992b). The prey of pompilids that cache prey while they are excavating a nest is sometimes stolen while the female is busy digging (Endo 1981). The prey of sphecids that dig the nest before foraging may be stolen while it is left briefly unguarded at the burrow entrance before the female pulls it in (Field 1989a,b, 1992c). Even within the confines of a nest, prey are not safe from marauding neighbors, who may enter and make off with individual prey. When robbing paralyzed ants from nests, *Clypeadon laticinctus* females usually enter only open nest burrows (Alexander 1986). *Ammophila sabulosa* females, however, can detect closed nest entrances and remove the soil plug to get at the caterpillars within (Field 1989a,b). Some *C. laticinctus* provision entire nests with paralyzed ants stolen from other nests (Alexander 1986), and as many as 23% of *A. sabulosa* prey may be stolen from inside nests (Field 1989a,b).

Prey are not always free for the plucking from unwary neighbors. Fights that erupt between females may involve grappling, biting, kicking, and (attempted) stinging (Field 1992b). For example, a *Stictia heros* female may crash into a prey-laden female with an audible impact and initiate a wrestling match that may be joined by nearby females and last for up to a minute before the victor makes off with the spoils (Sheehan 1984).

Brood Parasitism

CHRYSIDIDAE

Parasitoids in the subfamily Chrysidinae feed on a host only after it has eaten its provisions and become a prepupa (Chapter 2). However, the widely used English vernacular name for chrysidids, cuckoo wasps, refers to a somewhat different way of life in this group. The larva of true cuckoo wasps usually kills and eats the host's egg or young larva and then derives most of its sustenance from the host's provisions. Thus, brood parasitic chrysidines lay their eggs only in nests containing eggs or newly hatched host larvae.

Case Study: *Argochrysis armilla*

Argochrysis armilla (Fig. 4-1) is a cleptoparasite of *Ammophila dysmica*, which nest in aggregations and provision their shallow single-celled nests with one or two caterpillars (Rosenheim 1987a,b, 1988, 1989). During nest excavation, which takes from 20 minutes to 3.9 hours, an *Ammophila* female provides visual cues to *Argochrysis* females seeking a potential host nest. Responses to lures (i.e., dead *Ammophila* females made to simulate digging movements on the end of a fishing line) indicate that *Argochrysis* females can detect a digging host within about 50 cm. However, while the digging *Ammophila* female is present, the *Argochrysis* female cannot gain

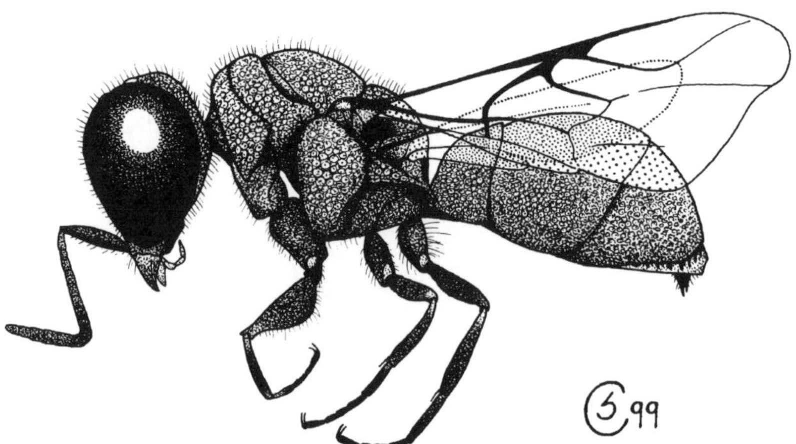

Figure 4-1. *Argochrysis armilla* female (drawing by Catherine Seibert).

entrance to the nest. Therefore, the cleptoparasitic female remains nearby, watching the digging sphecid and perching motionless, presumably to avoid detection. Avoiding detection is critical because if a host female perceives an *Argochrysis* nearby, she may chase it away or abandon the nest.

When the *Ammophila* finishes digging and plugs the nest entrance with soil and pebbles, she leaves on a foraging trip that takes from 12 to 42 minutes. During her absence, the *Argochrysis* female may try to dig through the plug, but few females reach the cell using this method. Rather, the best chance for the *Argochrysis* comes only after the *Ammophila* has returned with prey, removed the plug, and entered the nest. At this time, the cleptoparasite follows the host into the cell and quickly glues an egg to the cell wall. Subsequent feeding by the *Argochrysis* larva deprives the *Ammophila* larva of its food, so that it is rare for both the host and cleptoparasite to complete development. Over a 4-year period, 25% of 275 *Ammophila dysmica* nests were attacked by *Argochrysis armilla* (Rosenheim 1987b).

The ability of *Argochrysis armilla* females to exploit host nests is enhanced by their ability to learn and remember the locations of more than one active nest. After visiting a nest and conducting surveillance of one digging *Ammophila* female, an *Argochrysis* female may leave to visit up to three other nests along a "trapline" as long as 42 m (Rosenheim 1987a). When leaving a nest, an *Argochrysis* female circled the host nest in flight up to 12 times, much in the manner of a nesting female wasp undertaking an orientation flight.

Comparative Survey of Chrysidid Brood Parasites

The evolutionary shift to brood parasitism probably occurred during the radiation of the Chrysidinae. All chrysidid brood parasites are in the subfamily Chrysidinae, and all attack nest-provisioning wasps. Unlike the parasitoids of the subfamily Chrysidinae, chrysidine brood parasites do not attack bees or masarine wasps, possibly because they have not been able to make the switch to feeding on pollen. However, the range of hosts attacked suggests that chrysidid brood parasites are able to eat a wide variety of arthropods, including spiders, cockroaches, aphids, and grasshoppers (Table 4-1). As noted in Chapter 3, there are some host affinities at the subfamily level within the Chrysididae. At the generic level, however, there is no well-defined pattern of either host type or lifestyle (i.e., parasitoids versus brood parasites). Some species (e.g., *Chrysis fuscipennis,*

Table 4-1 Host records and oviposition sites of selected species of brood parasitic Chrysididae, subfamily Chrysidinae

Species	Host families,[1] subfamilies, and genera	Host prey	Oviposition site	Notable observations	References
Chrysidini					
Argochrysis armilla	S, Sphecinae: *Ammophila*	Caterpillars	In partially stocked cell	1–6 larvae per host cell	Rosenheim 1987a,b
Chrysis carinata	S, Larrinae: *Trypoxylon*	Spiders	In partially stocked cell	Larva eats host egg first	Krombein 1967
C. coerulans	V, Eumeninae: *Ancistrocerus*, *Euodynerus*, *Symmorphus*	Caterpillars	In partially stocked cell	Larva eats host egg first	Krombein 1967
C. purpurata	S, Larrinae: *Trypoxylon*	Spiders	In partially stocked cell	Larva eats host egg first	Danks 1970
Neochrysis alabamensis	S, Sphecinae: *Podium*	Cockroaches	In host cell	Larva eats host egg	Krombein 1967
Praestochrysis lusca	S, Sphecinae: *Sceliphron*	Spiders	In nearly completed cell	Larva fights host larva to the death	Bordage 1913 (in Kimsey & Bohart 1990)
Elampini					
Elampus viridicyaneus	S, Bembecinae: *Hoplisoides*, Pemphredoninae: *Mimumesa*	Leafhoppers, treehoppers	—	—	Krombein et al. 1979b, Rosenheim & Grace 1987
Hedychridium fletcheri	S, Larrinae: *Tachysphex*	Grasshoppers	On prey in provisioned cell	Larva presumably eats host larva first	Kurczewski 1967b

H. solierellae	S, Larrinae: *Solierella*	Lygaeid bugs	In provisioned cell	Larva may also feed on young host larva	Carrillo & Caltagirone 1970
Hedychrum intermedium	S, Philanthinae: *Philanthus*	Bees	On prey	Female may oviposit on prey as it is being brought into the nest	Simonthomas & Simonthomas 1972
Omalus aeneus	S, Pemphredoninae: *Passaloecus, Pemphredon, Stigmus*	Aphids	—	—	Krombein 1967, Fricke 1992a
Pseudolopyga carrilloi	S, Larrinae: *Solierella*	Lygaeid bugs	In free-living lygaeid prey of *Solierella*	Egg hatches only after prey brought into nest by host; larva kills host larva before feeding on prey	Carrillo & Caltagirone 1970
Pseudomalus auratus	S, Pemphredoninae: *Passaloecus, Pemphredon, Rhopalum*; Larrinae: *Trypoxylon*	Aphids	In partially completed cell	Single larva may attack more than one cell	Danks 1970, Kimsey & Bohart 1990

Note: The host is the species that provisioned the nest.
[1]S, Sphecidae; V, Vespidae.

Stilbum cyanurum) attack more than one family of hosts, and some chrysidine genera contain both parasitoids and brood parasites (e.g., *Chrysis, Praestochrysis*) and attack a variety of host types.

Although most brood parasitic chrysidines lay their eggs on prey within host nests (Table 4-1), several oviposit on or within the host's prey before it is taken into the nest. *Hedychrum intermedium*, for example, sometimes lays its eggs on the honey bee prey as it is being carried into a nest by a *Philanthus triangulum* female. Female *Pseudolopyga carrilloi* get their larvae into a host cell via an even more indirect route. A female locates a living first- or second-instar nymph of the genus *Nysius* (Homoptera: Lygaeidae) and lays an egg within its abdomen. The chrysidid hatches after about 9 days but does not emerge as a second-instar larva until the *Nysius* nymph is paralyzed and brought into a nest by a hunting female *Solierella blaisdelli* or *S. peckhami*. In the nest, the *P. carrilloi* larva feeds on the other *Nysius* prey of the *Solierella* female. If the lygaeid is not taken by a female *Solierella*, the chrysidid larva dies without further development (Carrillo and Caltagirone 1970).

POMPILIDAE

Cleptoparasitic chrysidids are all obligate brood parasites. Within the Pompilidae and Sphecidae, however, some species are obligate brood parasites and others are facultative brood parasites that sometimes provision their own nests and sometimes lay eggs in the nests of conspecific females (Field 1992b).

Case Study: *Evagetes mohave*

H. Evans and colleagues (1953) observed *Evagetes mohave* females that are obligate brood parasites upon another pompilid, *Anoplius apiculatus*, along Blackjack Creek in northeastern Kansas. Female *A. apiculatus* prey upon wolf spiders (*Arctosa littoralis*) that they deposit in shallow, single-celled nests. Female *E. mohave* search for recently completed nests or nests being provisioned by a female *A. apiculatus*. They enter the nest, by digging if necessary, and then enter the provisioned cell containing a single *Arctosa littoralis* on which the host species has deposited an egg. The *Evagetes* female quickly dispatches the host's egg, sometimes eating it, and lays her own smaller egg on the anterior dorsum of the spider's abdomen. If the spider has recovered from the temporary paralysis inflicted by the *A. apiculatus* female, the *Evagetes* female stings it herself before ovipositing. The egg hatches within 2 days and the larva, free from competition from the host's young, devours the spider within a week before spinning a cocoon.

Comparative Survey of Pompilid Brood Parasites

Obligate brood parasitic behavior has been recorded in just three genera of Pompilidae: *Evagetes*, *Ceropales*, and *Irenangelus*, all of which attack other Pompilidae. Female *Ceropales* take a different approach to brood parasitism than *Evagetes* (Olberg 1959, Evans and West-Eberhard 1970, Iwata 1976). Female *C. maculata* seek out females of *Anoplius* or *Pompilus* that are carrying spider prey and follow them, awaiting a chance to strike (Fig. 4-2). *Ceropales* occasionally fight the host female, but more often they mount a sneak attack when the host female temporarily abandons her burden to scout for potential nest locations. When this opportunity arises, the *Ceropales* female quickly jumps on the prey. She inserts the wedge-shaped tip of her abdomen into a respiratory slit (book lungs) on the spider's abdomen, oviposits, and departs. The egg, now hidden within the spider, is then deposited in the nest by the *Anoplius* or *Pompilus* female. The *Ceropales* egg hatches before that of the host, and the first-instar larva emerges to kill the other wasp's egg before feeding

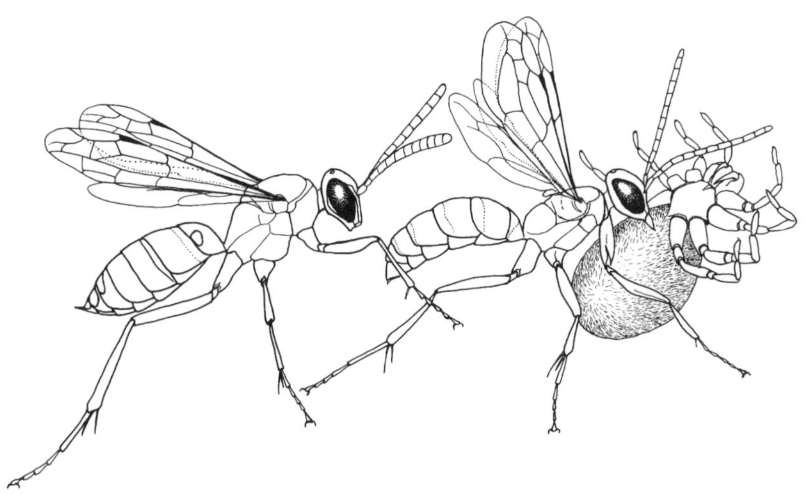

Figure 4-2. A female (left) of the cleptoparasitic pompilid *Ceropales* following a female of *Pompilus* (redrawn by Catherine Seibert from Olberg 1959).

on the spider. *Ceropales* are known to attack pompilids of the genera *Ageniella* (Krombein 1955), *Auplopus* (Weaving 1994), *Dichragenia* (F. Gess 1981), *Elaphrosyron* (Evans and Matthews 1973a), *Phanagenia, Pompilus,* and *Priocnemis* (Krombein et al. 1979b).

Although detailed life histories of *Irenangelus* are lacking, it is clear that females of this genus are also brood parasitic; *I. eberhardi* attacks *Auplopus semialatus* (Wcislo et al. 1988), and *I. luzonensis* is a cleptoparasite of *Tachypompilus analis* and *Auplopus nyemitawa* (Williams 1919a). Limited evidence suggests that *Irenangelus* deposit their egg on the spider after it has been placed in the nest (Wcislo et al. 1988) and that the cleptoparasite larva kills the host egg (F. Williams 1919c).

Obligate cleptoparasitism in the Pompilidae apparently arose via at least two separate evolutionary pathways. Phylogenetic analyses strongly suggest that Ceropalinae (which lack tarsal combs found in ground-nesting Pompilidae) evolved cleptoparasitism directly from (non-nest-building) parasitoid ancestors that also lacked the combs (Day 1988, Shimizu 1994). The phylogeny of the Pompilinae is not completely resolved. However, *Evagetes* bear tarsal combs on their forelegs that they undoubtedly inherited from ancestors that prepared ground nests after hunting (Evans 1953, Shimizu 1994). *Batozonellus annulatus* females may enter nests of conspecifics, kill the egg, and lay their own eggs (Tsuneki 1968). Presumably, a similar facultative brood parasitism was present in the ancestral species of Pompilinae that gave rise to the obligate brood parasites.

SPHECIDAE

Females of only a few genera of sphecid wasps are obligate or facultative brood parasites. The best-documented sphecid brood parasites are in the genera *Stizoides, Nysson,* and *Epinysson* (Evans 1966c, Bohart and Menke 1976), but there is indirect evidence that *Acanthostethus* also includes brood parasites (Evans and Matthews 1971). All four genera are in the subfamily Bembicinae.

Case Study: *Stizoides renicinctus*

Stizoides renicinctus is an obligate brood parasite that practices brood parasitism convergent in many ways with the behavior of *Evagetes* (Evans 1966c). Female *S. renicinctus* attack at least three species of sphecine wasps: *Prionyx atratus* (F. Williams 1914) and *P. thomae* (Rau and Rau 1918), which provision nests with acridid grasshoppers, and *Palmodes laeviventris*, a predator of Mormon crickets, *Anabrus simplex* (LaRivers 1945). As do *Evagetes* females, a *S. renicinctus* female searches for a closed host nest and burrows into it, where it destroys (but does not eat) the host wasp's egg. After laying an egg of its own, the *Stizoides* female returns to the soil surface and plugs the entrance and main burrow with debris. LaRivers found that the *Stizoides* females avoid competition with sarcophagid parasitoids that also attack the host (see Chapter 6) by not ovipositing on prey harboring fly maggots.

Case Study: *Ammophila sabulosa*

Ammophila sabulosa are facultative brood parasites. Females are predators of caterpillars; they provision each of their shallow, single-celled nests with one to five prey items, depositing their egg on the first prey item placed in the nest (Field 1989a,b). After provisioning, a female plugs the top of the burrow with sand and pebbles, but another female *A. sabulosa* may remove the plug and transfer the prey to its own nest, destroying the first female's egg. In 7% of nests, however, other females acted as classic brood parasites. Brood parasitic *A. sabulosa* remove all of the caterpillars from the nest, plucking the egg from the last prey item and sometimes eating it. The prey are then returned to the same nest, along with an egg laid by the brood parasite. In *A. sabulosa*, brood parasitism is part of an opportunistic strategy whose occurrence is conditional on the availability of fully provisioned conspecific nests and perhaps on thermal conditions that define the times of day at which searching is possible (Field 1989a, 1992c). Brood parasitism appears to be a cheap and easy route to producing offspring: it takes a female *A. sabulosa* about 10 hours to construct and provision a nest but only about 30 minutes to switch eggs in an already provisioned nest. However, the success of a brood parasite is far from assured, because over 80% of brood-parasitized nests were later parasitized by another *A. sabulosa* female (Field 1989a).

Comparative Survey of Sphecid Brood Parasites

Stizoides renicinctus attacks members of a different sphecid subfamily (Evans 1966c). In contrast, brood parasites of the genera *Nysson* and *Epinysson* attack members of their own subfamily (Bembicinae), although of a different tribe (Gorytini). The hosts of these two genera, including species of *Argogorytes*, *Dienoplus*, *Gorytes*,

Hoplisoides, *Lestiphorus*, and *Oryttus* provision their ground nests with treehoppers, leafhoppers, and spittlebugs to which the brood parasites gain access by digging through temporary closures of the nest entrances (Evans 1966c, Bohart and Menke 1976, F. Gess 1981).

Nysson females lay an egg cryptically on one of the many prey items in the cell (Fig. 4-3), before the host's egg has been laid, thus having no opportunity themselves to destroy the host's offspring-to-be. However, the brood parasite's egg hatches first, and the *Nysson* larva itself destroys the host egg with its long thin mandibles. Thus, the independent evolution of two lineages of brood parasites in the Bembicinae has produced genera that differ in when the brood parasites' eggs are laid and in whether it is the mother or the first-instar larva that destroys the hosts' eggs. All of the bembicine brood parasites are obligate brood parasites. Within both the bees and solitary Sphecidae, obligate brood parasitism appears to be more common in temperate regions than in the tropics, perhaps because the greater seasonality of host populations in temperature zones provides a more predictable resource (Wcislo 1981).

Facultative (intraspecific) brood parasitism has been reported in species of two genera of Sphecinae (*Ammophila* and *Podalonia*) and in one genus of Larrinae (*Trypoxylon*) (Field 1992b). In some species, such as *Ammophila aberti* (F. Parker et al. 1980), brood parasitism is similar in form to that of *A. sabulosa.*

SAPYGIDAE

The family Sapygidae contains only about 80 described species (Goulet and Huber 1993), including 17 in North America (Krombein et al. 1979b). As does Chrysidinae, Sapygidae contains a mix of ectoparasitoids and cleptoparasites, and most species attack bees. Although all chrysidines that attack bees are ectoparasitoids of mature larvae and pupae, many of the sapygids that attack bees are cleptoparasites that feed on pollen. Thus, feeding habits of sapygids have more in common with cleptoparasitic bees of such genera as *Sphecodes* of the Halictidae and *Epeolus* and *Nomada* of the Anthophoridae (Goulet and Huber 1993) than with the foraging behaviors of other solitary wasps.

Case Study: *Sapyga pumila*

From 1970 to 1977 in southern Idaho, P. Torchio (1972a,b, 1979) studied cleptoparasitism by *Sapyga pumila* in nests of the leafcutter

Figure 4-3. Egg (smaller one on prey at bottom of picture) of *Nysson* on prey stored in nest of *Hoplisoides placidus*, whose larger egg is near the top (from Evans 1963).

bee *Megachile rotundata*. *Megachile rotundata* is a solitary bee native to Europe, but in the mid-1900s it was accidentally introduced into North America. It turned out to be a lucky accident, because it became an important pollinator of alfalfa seed crops, largely because females will nest in holes drilled into boards placed adjacent to alfalfa fields. However, the efficacy of this species as a pollinator is threatened by a high level of nest cell parasitism by *S. pumila*, which destroyed from 2% to 74% of over 20,000 *M. rotundata* nest cells examined during an 8-year period (Torchio 1972a). Unlike *M. rotundata, S. pumila* is native to North America, where it evolved attacking native bees of the genera *Ashmeadiella, Anthocopa, Dianthidium, Megachile, Osmia,* and *Heriades* (Krombein et al. 1979b).

Within their nests, *M. rotundata* females construct a linear sequence of cells, each containing a mass of alfalfa nectar and pollen on which an egg is laid. Adjacent cells are separated by leaf fragments that are ineffective as protection against *S. pumila* invasion. When a female *S. pumila* (Fig. 4-4) enters an active *Megachile* nest and encounters a recently provisioned cell, she turns about, braces herself against the nest walls, and pierces a small hole through the leaf barrier with a pair of serrated stylets on her ovipositor. She then inserts her ovipositor into the hole, depositing one to three eggs in the host cell, which contains a pollen mass and the bee's egg. Female wasps deposit a single egg 83% of the time and then leave quickly in order to avoid the female bee, who may still be provisioning the same nest (Torchio 1972a). If the wasp encounters the returning bee, she is attacked and always loses several antennal segments to the bee's mandibles before escaping.

The wasp eggs hatch within 1 or 2 days, depending on the temperature in the nest. Such a short incubation period is typical of many cleptoparasites whose own eggs must hatch before that of their hosts. Upon hatching, the wasp larva wanders about the host cell. When it encounters a host egg, the larva punctures it with its sharp mandibles and (usually) eats the contents, before feeding on the nectar-pollen mass for the rest of its larval period. After feeding on only one-third to two-thirds of the provision, it spins a cocoon in the host cell and overwinters as a prepupa.

Matters are more complicated when the young wasp larva encounters one or more *S. pumila* eggs or larva, placed there by its own mother or another female wasp. When a conspecific egg is found, it

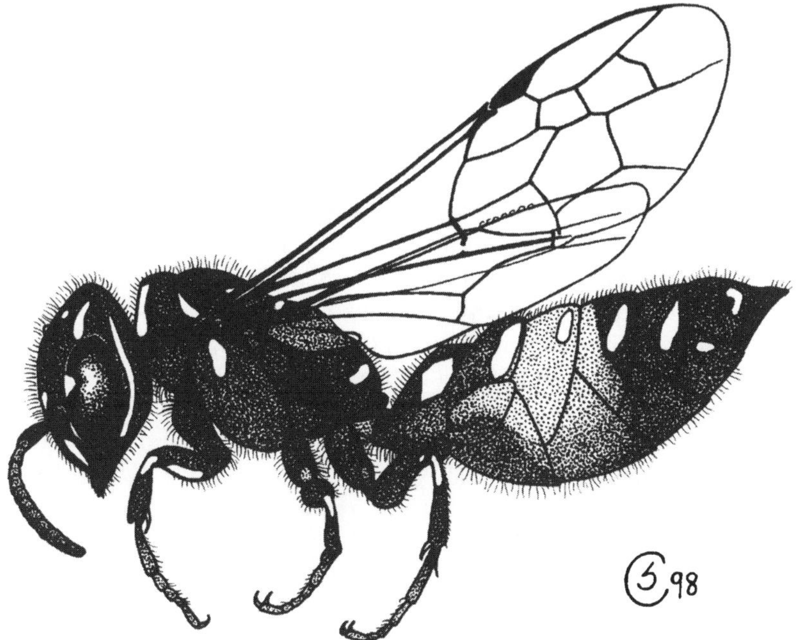

Figure 4-4. *Sapyga pumila* female (drawing by Catherine Seibert).

is destroyed, but (curiously) not eaten. If two *S. pumila* larvae are present, a skirmish ensues in which one or both of the contestants are killed. When both are killed before the host egg is found, the bee can complete its development unmolested. Wasp larva may also die if the mother accidentally places the egg between leaf fragments in the cell closure, so that the larva cannot reach the host cell.

Comparative Survey of Sapygid Host Records

Not much is known about the behavior of other Sapygidae, although a number of host records have been reported. Among the species studied to date, *Fedtschenkia anthracina* is the only sapygid that may have a host other than bees. Females of this species have been seen entering nests of the eumenid wasp *Pterocheilus trichogaster* in the western United States (Bohart and Schuster 1972), but it is by no means certain whether it is a cleptoparasite or an ectoparasitoid, if either (i.e., it has not been reared from nests nor has it been seen feeding).

The hosts of all North American species of Sapygidae, all of which are in the genera *Sapyga* and *Eusapyga*, are bees of the family Megachilidae (Krombein et al. 1979a). For example, among the cleptoparasites, *S. centrata* has been found in nests provisioned by *Osmia pumila* and *O. bucephala* (Krombein 1967). As do those of *S. pumila*, first-instar larvae of *S. centrata* consume the host's egg before moving on to feed on provisions. Among the ectoparasitoids, *Eusapyga rubripes* is hosted by bees of the genus *Dianthidium* (H. Hicks 1927), and *S. confluenta* attacks several species of *Osmia* (Krombein et al. 1979b). Similarly, in Europe *Sapyga quinquepunctata* attacks megachilids of the genera *Heriades*, *Chalicodoma*, and *Osmia* (Iwata 1976). However, other families of bees are also used. For example, *Polochrum*, a genus not present in North America, parasitizes carpenter bees (Anthophoridae: *Xylocopa*) in Brazil (Hurd and Moure 1961) and Europe (Iwata 1976).

Conclusion

As do some humans, some solitary wasps have taken a sneaky shortcut to success. Cleptoparasites may be facultative thieves that steal just a few prey items to supplement their own foraging, or they may be brood parasites, appropriating both prey and nest cell. Obligate brood parasitism has evolved multiple times independently in the aculeate wasps: twice each in the Pompilidae and Sphecidae, perhaps multiple times in the Chrysididae, and at least once in the Sapygidae. Its absence in the solitary Eumeninae seems odd, particularly because usurpation of nests (see Chapter 6) and facultative prey theft have been reported (Field 1992b). Given that obligate brood parasitism has also evolved in bees, the habit is curiously absent in the Masarinae, the pollen wasps that are the subjects of the next chapter.

5/ Pollen Foraging and Pollination

Among the solitary wasps, pollen feeding is rare, occurring only in the Sapygidae and in the vespid subfamily Masarinae. Most sapygids are pollen cleptoparasites in bee nests (Chapter 4), and all masarine wasps provision nests with pollen and nectar mixtures (S. Gess 1996). Because the Sapygidae and Masarinae probably constitute no more than 1% of all aculeate wasps, they seem out of place in this generally carnivorous group. However, from a strict phylogenetic perspective, pollen feeding is not really an oddity in the aculeate wasp lineage. Bees, after all, are really just exceedingly hairy wasps that happen to have a different common name. Although the generally accepted common name for masarines is pollen wasps (Houston 1984, S. Gess 1996), their habits would perhaps not seem so unusual if we called them "vespoid bees" (Malyshev 1968).

Foraging Behavior of Pollen Wasps

In his revisional study of the pollen wasps, O. Richards (1962:35) stated that "very little has been recorded as to what the mother wasp brings in [to the larvae] or how she carries it." However, much progress has been made since 1962, due in particular to the efforts of the team of Sarah and Fred Gess working in areas of South Africa that harbor 7 of the 18 described genera of Masarinae. By comparison, only one genus (*Pseudomasaris*) occurs in North America.

Case Study: *Pseudomasaris edwardsii*

Pseudomasaris edwardsii females are relatively stout wasps, approximately 1 cm long, with a black-and-yellow–striped abdomen and a large knob at the apex of each antenna (Fig. 5-1). Most of our knowledge of this species, which occurs throughout much of the western United States and northern Mexico, comes from work done in Utah with a greenhouse population (Torchio 1970). H. Hicks (1929), K. Cooper (1952), K. Cooper and J. Bequaert (1950), and O. Richards (1963) provided shorter reports of its nesting and foraging biology.

Pseudomasaris edwardsii females forage for pollen and nectar that they use to provision cells in multicellular nests constructed of moistened soil attached to rocks or to plant stems (Chapter 6). In Utah, females collected pollen and nectar only from *Phacelia leuco-*

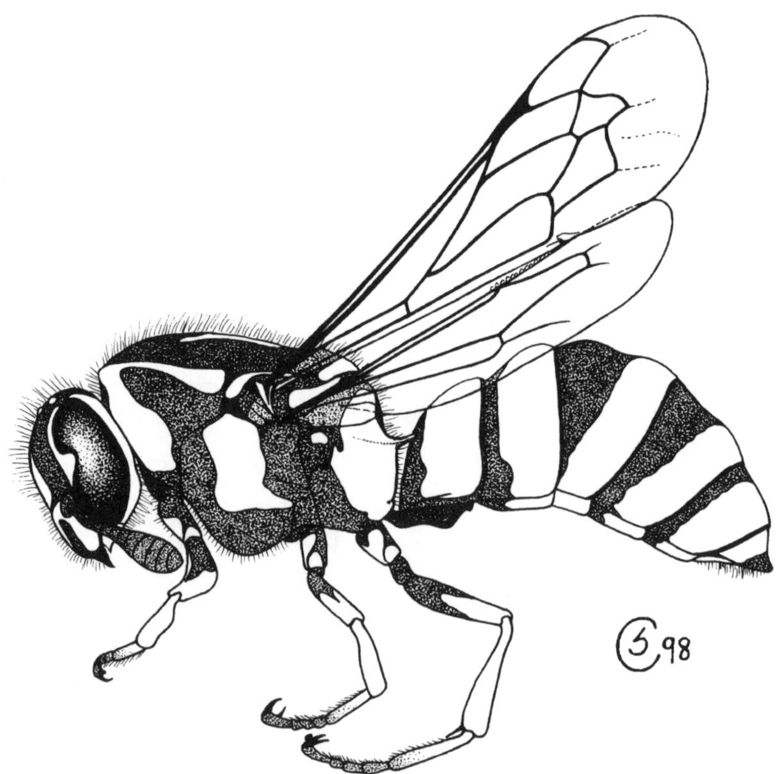

Figure 5-1. *Pseudomasaris edwardsii* female (drawing by Catherine Seibert).

phylla (Hydrophyllaceae) (Torchio 1970), and, even when given access to many types of flowers in a greenhouse, females foraged only on *P. leucophylla* and *P. tanacetifolia*. Elsewhere, however, *P. edwardsii* females are known to visit Asteraceae, Boraginaceae, Lamiaceae, Onagraceae, Tamaraceae, and other Hydrophyllaceae (Cooper and Bequaert 1950, Richards 1963). These records either represent geographic variation in provisioning habits or visits by wasps seeking nectar for self-maintenance.

A *P. edwardsii* female uses two different techniques to collect pollen from *Phacelia* (Torchio 1970). If perched on a *Phacelia* blossom, a female grasps several of the protruding stamens with her forelegs and removes pollen from the anthers with her mouthparts. She swallows the pollen, temporarily storing it in her crop, where it mixes with *Phacelia* nectar. To remove pollen from stamens *without* landing on the flower, a female hovers above it and bounces her extended legs on the stamens before grasping them and feeding directly on pollen on the anthers. The hovering-bouncing technique is more common when there is relatively little pollen on each anther, suggesting that the bouncing loosens pollen. Presumably because it is more energetically costly to hover at lower air temperatures, the females use the hovering method less often when it is cooler.

Typically, a female makes eight foraging trips to provision a single cell (carrying the pollen in her crop), in the end producing a "tacky, homogeneous mass of *Phacelia* pollen bound with *Phacelia* nectar and shaped into a solid [about 12.5 × 3.5 mm] cylinder" (Torchio 1970). Most of the cylinder is covered with "cone-shaped papillalike projections," and the entire mass fills most of the cell volume. The *P. edwardsii* egg, which was laid before the cell was provisioned, hatches in about 2 days, and the larva either burrows into the pollen mass or feeds on its surface.

COMPARATIVE SURVEY OF POLLEN WASP FORAGING BEHAVIOR

Although each species of pollen wasp may visit flowers of a number of families and species for feeding themselves nectar, they appear to be more specialized when it comes to foraging for pollen to provision nests (S. Gess 1996). When pollen found in nests is compared with pollens available in local flowers, the wasps are usually found to provision with the pollen of a single genus (or even a single species). Table 5-1 lists some of the best known host associations for

Table 5-1 Pollen source for selected species of Vespidae, subfamily Masarinae

Species	Source of pollen used in provisions[1]		References
	Family	Genera	
Australia			
Paragia decipiens	Myrtaceae	*Eucalyptus*	Naumann & Cardale 1987
P. tricolor	Myrtaceae	*Eucalyptus*	Houston 1984
P. vespiformis	Mimosaceae	*Acacia*	Houston 1986
Rolandia angulata	Goodeniaceae	*Goodenia*	F. Gess et al. 1995
R. maculata	Fabaceae	*Jacksonia*	Houston 1995
Europe			
Celonites abbreviatus	Lamiaceae	—	Bellmann 1983 (in S. Gess 1996), Müller 1996
Ceramius tuberculifer	Lamiaceae	*Teucrium*	Mauss 1996
North America			
Pseudomasaris edwardsii	Hydrophyllaceae	*Phacelia*	Cooper & Bequaert 1950, Richards 1963, Torchio 1970
P. maculiphrons[2]	Hydrophyllaceae	*Phacelia, Eriodyction*	Richards 1963, F. Parker 1967
P. marginalis	Hydrophyllaceae	*Phacelia*	Dorr & Neff 1982
P. phaceliae[2]	Hydrophyllaceae	*Phacelia*	Cooper & Bequaert 1950, Torchio 1970
P. vespoides[2]	Scrophulariaceae	*Penstemon*	Hicks 1927, Cooper & Bequaert 1950, Torchio 1970, Tepedino 1979
P. zonalis[2]	Hydrophyllaceae	*Phacelia*	Cooper & Bequaert 1950
South Africa			
Celonites clypeatus	Scrophulariaceae	*Aptosimum*	S. Gess 1996
C. latitarsis	Campanulaceae	*Wahlenbergia*	F. Gess & Gess 1992
C. peliostomi	Scrophulariaceae	*Aptosimum, Peliostomum*	S. Gess 1996
C. wahlenbergiae	Campanulaceae Aizoaceae	*Wahlenbergia* *Coelanthum*	F. Gess & Gess 1992
Ceramius bicolor	Aizoaceae	*Psilocaulon*	F. Gess & Gess 1986
C. capicola	Aizoaceae	*Drosoanthemum*	F. Gess & Gess 1980
C. clypeatus	Papilionaceae	*Aspalathus*	F. Gess & Gess 1986, 1990

Table 5-1 *Continued*

Species	Source of pollen used in provisions[1]		References
	Family	Genera	
C. lichtensteinii	Aizoaceae	*Ruschia*	F. Gess & Gess 1980, 1988c
C. linearis	Aizoaceae	*Drosoanthemum*	F. Gess & Gess 1980
C. nigripennis	Asteraceae	*Dimorphotheca*	F. Gess & Gess 1986
C. rex	Asteraceae	*Berheya*	F. Gess & Gess 1988c
C. socius	Aizoaceae	*Psilocaulon*	F. Gess & Gess 1986, 1988c, 1990
Jugurtia confusa	Aizoaceae	*Drosoanthemum*	F. Gess & Gess 1980
Masarina familiaris	Papilionaceae	*Aspalathus*	F. Gess & Gess 1988b
Quartinia vagepunctata	Asteraceae	*Cotula, Relhania, Leysera*	S. Gess 1996, F. Gess & Gess 1992

Note: See also S. Gess (1996) for summaries of plants that are visited though not necessarily used for pollen provision.

[1]Unless otherwise indicated, records determined from matching pollen provisions to pollen of local to plants or to plants visited by females. [2]Inferred from flower visitation records.

masarines, although the information should be interpreted with some caution. Specialization on a particular family or genus appears to be the rule (S. Gess 1996), but perhaps larger sample sizes over a larger geographic range would broaden the known preferences for some species. In addition, some of the records in Table 5-1 are based on the flowers most often visited by females, which may have been gathering only nectar rather than pollen. The flowers used for provisioning by species in a given geographic region tend to come from a limited range of families (Table 5-1; S. Gess 1996). In North America, for example, *Pseudomasaris* seem to favor Hydrophyllaceae and Scophulariaceae. Southern African species, however, seem to prefer Asteraceae and Aizoaceae, and some Australian pollen wasps forage on *Eucalyptus* (Myrtaceae).

In general, although there has been some convergent evolution in the foraging habits of bees and masarines, the manner in which masarines collect, carry, and process pollen and nectar differs from that of many bees. One major difference is that pollen wasps lack the branched body hairs that trap pollen on many pollen-collecting bees (Goulet and Huber 1993). At least three general forms of pollen-collecting behavior have evolved in the Masarinae: direct feeding from anthers, use of legs, and use of body hairs.

Ceramius clypeatus and *Masarina familiaris* display the simplest form of pollen feeding, consuming pollen directly from anthers with their mouthparts (S. Gess and Gess 1989a). Both feed on flowers of *Aspalathus*, whose protruding anthers are readily accessible to a female perched outside a flower. *Pseudomasaris edwardsii* females also feed directly from the anthers, and, although they may use their legs to loosen the pollen first, they do not use their legs to direct pollen into their mouths (Torchio 1970).

Ceramius braunsi and some *Quartinoides* species feed on composite flowers, landing on the center of the blossom and alternately rotating their front legs beneath them, loosening the pollen and drawing it toward their mouth (S. Gess and Gess 1989a). Unlike these species, which use their legs to collect pollen from readily accessible anthers, female *Trimeria buyssoni* must remove pollen from small tubular flowers containing anthers that are difficult to reach. Female *T. buyssoni* insert their forelegs deep into the corolla of a flower and collect pollen using specialized hooked hairs on the foretarsi that remove the pollen from the anthers and deposit it in an elongate concavity on the first tarsal segment.

Many flowers are structured so that a visiting pollinator receives a load of pollen in a location that promotes deposition of the pollen onto the stigma of the next flower visited. Although unable to control where the pollen lands, some masarines are able to move much of the pollen adhering to their bodies to their mouths after they leave a flower. The dorsum of the thorax of a *Pseudomasaris vespoides* female becomes covered with pollen grains while she is visiting *Penstemon*. The pollen grains are later transferred to the mouthparts with the aid of combing movements of the forelegs (Torchio 1974). Similarly, when a female *Ceramius tuberculifer* visits a *Teucrium* blossom, the front and top of her head becomes coated with pollen. After visiting several flowers, the female uses her forelegs to brush pollen to her mouth, with the use of dense brushes of short, thick hairs on the forelegs (Mauss 1996). The dense brushes of hairs on the foretarsi of most female masarines (Richards 1962) may also aid pollen handling in those species that rake the pollen directly from the flower into the mouth.

Females of other masarine species direct pollen onto particular areas of their bodies that are covered with hairs specialized for collecting pollen. While visiting a flower, a female *Celonites abbreviatus* rubs her face rapidly back and forth over the anthers, and large amounts of pollen adhere to the many knobbed hairs (curiously, both males and females sport these hairs). Upon leaving the flower, the wasp immediately transfers the pollen to its mouth with its forelegs (Müller 1996). Female *Rolandia maculata* also have specialized pollen-collecting hairs on the underside of the thorax, where they pick up pollen when a foraging "female first probes deep into the nectary, then backs up slightly to hunch over the anthers" (Houston 1995).

Although pollen wasps have also been referred to as honey wasps (Schwarz 1929), this is a misnomer and is more appropriately applied to nectar-storing social vespids of such genera as *Brachygastra* (Richards 1978). Nectar becomes honey only after it has been concentrated by evaporating off water (as honey bees do). Masarines do not make honey. Rather, they blend unconcentrated nectar with pollen in their crops and pack the mixture into nest cells. Whereas most bees carry pollen on their hindlegs or on the underside of their abdomen, only colletid bees of the subfamily Hylaeinae and masarines carry pollen mixed with nectar in their crops (Goulet and Huber 1993).

Because the size of a single pollen and nectar load brought into a nest is constrained by the size of the female's crop, she has to make more than a single foraging trip to fully provision a cell for a single offspring (Torchio 1970, Houston 1984). The final mass of pollen deposited in a cell by different masarines varies in shape, surface texture, consistency, color, and the extent to which it fills the cell (S. Gess 1996). The pollen "loaf" in a cell may be an "indiscrete" mass or it may take a fairly constant form in each species. *Paragia tricolor* females, for example, create a cylindrical loaf that is rounded and blunt on one end and tapered on the other (Houston 1984). The surface of a *P. tricolor* pollen loaf is covered with "folds and annulations," the latter presumably representing the boundaries of individual pollen loads (S. Gess 1996). A pollen loaf may also have a central spine of pollen (e.g., *Pseudomasaris phaceliae*) or may be covered with papillae that could represent separate deposits of pollen regurgitated by the female (e.g., *Pseudomasaris edwardsii* [Torchio 1970] and *P. vespoides* [H. Hicks 1927]).

Variation in the consistency of the provision mass apparently derives from the relative proportions of pollen and nectar deposited in the cell. The provision of many *Ceramius* species is a firm, dry loaf (F. Gess and Gess 1980, 1986, 1990). The provision of species such as *Celonites capicola, Jugurtia confusa* (F. Gess and Gess 1980), and *Masarina familiaris* (F. Gess and Gess 1988b), however, is very wet and sticky, sometimes adhering to the cell walls in the nest (e.g., *Quartinia vagepunctata* [F. Gess and Gess 1992]. The color of the pollen mass derives from the color of the pollen collected: for example, the pollen mass of *Rolandia angulata* is white (F. Gess et al. 1995); that of *Celonites wahlenbergiae* is olive-green (F. Gess and Gess 1992); and that of *C. abbreviatus* is orange (Bellman 1984, in S. Gess 1996). The size of the loaf relative to the cell size also varies, from loaves that partially fill the cell (e.g., many species of *Ceramius* [S. Gess 1996]) to those that completely fill it (e.g., *Quartinia vagepunctata* [F. Gess and Gess 1992]).

With the exception of a very few genera of solitary wasps (e.g., *Parnopes* of the Chrysididae, *Steniolia* and *Zyzzyx* of the Bembecinae, and *Synagris* of the Eumeninae), most aculeate wasps have a short tongue that restricts them to feeding on nectar easily accessible in flowers with a short corolla. However, as do bees, many masarines have a relatively long tongue that allows them to feed within deep flowers. Even in the "short-tongued" species, the tongue may be 20% to 30% of the total body length, and a substantial number of the "long-tongued" species have females with a tongue exceeding 5 mm in length and over 50% of body length (S. Gess 1996). Most impressive on a relative scale are the small *Quartinoides*, whose 4-mm-long females have a tongue that averages ≥5 mm. The tongue of *Quartinoides capensis* is so long that it doubles upon itself four times when retracted beneath the head (Richards 1962). S. Gess (1996) provides detailed discussion of how masarine mouthpart size is correlated with the structure of the flowers the masarines pollinate.

Although adult males and females of all masarine species undoubtedly feed on nectar, there is evidence that adults of both sexes also feed on pollen. V. Mauss (1996), for example, found pollen not only in the crops of females, where it is stored before regurgitation in the nest, but throughout the length of the adult's digestive tract, as far back as the rectum. The pollen and nectar feeding habits of adults may explain why the known range of flowers visited by pollen wasps often greatly exceeds the range of species known to be used for feeding offspring.

The Role of Solitary Wasps in Plant Pollination

To some flowering plants, many Hymenoptera are effectively winged genitalia that can be "bribed" or "tricked" into moving pollen from blossom to blossom. Most of the important hymenopteran pollinators of angiosperms are bees (Proctor et al. 1996). The impact of solitary wasps on the flowers they visit ranges from inefficient pollen transfer by those species that are essentially nectar robbers to relationships in which plants are critically dependent on wasps for pollen transfer.

POTENTIAL POLLINATION BY NECTAR FORAGERS

Although most solitary wasps are not pollen feeders, adults forage for nectar and may inadvertently carry pollen between flowers. The efficiency of pollen transfer, however, depends on the wasps' behaviors during and after the visit to the flower, as well as on the wasps' morphology and relative size. A wasp can cross-pollinate flowers only if it picks up a sufficient amount of pollen on one flower and then deposits it on the appropriate female structure on the next flower visited. Many solitary wasps probably fail on this account because they are relatively nonhairy (so cannot carry much pollen) and probably visit few flowers during a day. Some wasps also approach the flower from below, often cutting into the blossom, where they rob nectar without picking up pollen (Rau and Rau 1918, Haeseler 1980). Solitary wasps are also poor pollinators if the next flower visited is likely to be of a different species or if the second visit occurs much later. Although most nonmasarine solitary wasps probably have little value as pollinators (Proctor et al. 1996), some have great value. For example, although flowers of the orchid *Epipactus palustris* are visited by a large number of bees, flies, and ants, it is the less common visitor *Eumenes pedunculatus* that is its most important pollinator (Nilsson 1978).

The range of effects that nectar-foraging wasps have on flowers can be seen among the wasps that visit *Kallstroemia grandiflora* (Zygophyllaceae) in southwestern U.S. deserts. *Kallstroemia grandiflora* has circular flowers containing 10 pollen-bearing stamens surrounding a central stigma. The stigma can be cross-pollinated only when it first becomes receptive, but it can be self-pollinated later. The solitary wasps visiting *Kallstroemia* blossoms for nectar vary widely in behavior and body size (Cazier and Linsley 1974). Two of the wasps, *Bembix u-scripta* and *Myzinum navajo*, cannot act as pollinators because they do not pick up pollen while sipping nectar. *Bembix u-scripta* perch beneath blossoms and feed on nectar by inserting their tongue between the sepals to reach the nectary. *Myzinum navajo* approach nectaries from within the blossom, but they are so small that they move between and beneath the anthers. Although these species are nectar robbers, a third wasp, *Campsoscolia octomaculata*, is probably a fairly effective pollinator. When *C. octomaculata* visits *Kallstroemia*, it moves about in the anthers and is large and hairy enough to pick up large amounts of pollen on its underside (for potential later cross-pollination) and to bend the anthers so that they come into contact with the stigma (effecting self-pollination). Thus, although the behavior, size, and morphology of nonmasarine wasps usually limit their value to flowers, they may not always be poor pollinators.

POLLEN WASPS AS POLLINATORS

In general, few plants seem totally dependent on masarines for pollination. Although there are correlations between the distribution of some plants and the masarines that visit them, the wasps often have narrower distributions, suggesting that the plants have other pollinators (S. Gess and Gess 1994, S. Gess 1996). With one minor exception, masarines appear to have little importance in the pollination of cultivated crops. Because honey bees (*Apis mellifera*) are inefficient pollinators of rooibos tea (*Aspalathus linearis*), growers in South Africa depend on the masarines *Ceramius clypeatus* and *Masarina familiaris* (as well as on several solitary bee species) for seed production (S. Gess and Gess 1994).

Pollen wasps, although absent or relatively rare in many parts of the world, may occasionally play an important role as members of pollination guilds in plant communities (S. Gess 1996). For example, the structure and pollinating mechanism of some *Penstemon* in western North America are particularly suited to the foraging habits and body size of *Pseudomasaris vespoides* (Torchio 1974). When a female *P. vespoides* crawls into a tubular corolla of a *Penstemon cyananthus* or *P. sepalulus* to reach nectar, she is forced into a position that causes the anthers to deposit a small load of pollen on the anterior dorsum of her thorax. Later, upon pushing her way into another *Penstemon*, the female exerts pressure that initiates a sequence of mechanical events that causes the pollen-receiving surface of the stigma to touch the pollen-laden area of her thorax. The color of the flowers of *Penstemon* species pollinated by *Pseudomasaris* differs from that of *Penstemon* pollinated by other animals. Both *P. cyananthus* and *P. sepalulus* have violet flowers, whereas those *Penstemon* pollinated by bees are blue and those pollinated by hummingbirds are red (Torchio 1974). In general, flowers of southern African species that depend on masarines for pollination tend to be light-colored (never red or purple, rarely blue), sweetly scented, diurnally open flowers that produce relatively nonviscous nectar in concealed locations (S. Gess 1992, 1996).

DECEPTIVE POLLINATION

In most plant-bee pollinator relationships, of course, female insects are manipulated into moving pollen from flower to flower, lured by rewards of nectar and pollen (Proctor et al. 1996). Both parties benefit from the relationship, the plants securing aid in gamete transfer and the bees obtaining food. Although this is the most common arrangement, some species of orchids have adopted a different and one-sided approach. Some species, referred to as food-flower mimics have conspicuous, fragrant flowers that attract hymenopteran pollinators but provide no reward to visitors. For example, the Madagascaran orchid *Cymbidiella flabellata* has yellow flowers with red and ultraviolet markings and "delicate vanilla-like perfume" (Nilsson et al. 1986). Although the flowers produce no nectar, they are apparently pollinated by the wasp *Sceliphron fuscum*, which shares its marsh habitat with this flower.

Many orchids are sexual mimics that coopt male Hymenoptera for their own purposes by taking advantage of the eagerness of males to find mates. Males duped by the flowers gain nothing and, in fact, waste time and energy that could be better spent pursuing real females. In some parts of Europe, Africa, and Australia, one might be lucky enough to experience the spectacle of a male wasp excitedly attempting to copulate with an orchid. Close inspection would reveal that the orchid lacks the nectaries and abundant pollen normal for flowers pollinated by female bees. A typical sequence, illustrated by the behavior of male *Argogorytes mystaceus* toward the orchid *Ophrys insectifera*, might proceed as follows (Ågren and Borg-Karlson 1984, Borg-Karlson 1990). A male, flying in an agitated manner, approaches the flower from downwind and lands on the labellum, a modified petal that provides a landing platform (Fig. 5-2). The male is so highly aroused that he exposes his genitalia even before alighting, although males do not actually ejaculate on the flower, unlike males of *Lissopimpla semipunctata* (Ichneumonidae), which are thoroughly duped by the orchid *Cryptostylis leptochila* (Coleman 1928). As the male *A. mystaceus* attempts to copulate with the labellum, a pollinium (pollen package) attaches to his head. If the male is later fooled by another *O. insectifera*, pollen grains are transferred to the stigma of the second flower.

The types of traits that orchids mimic give insight into how male wasps perceive their own females. The flowers are able to lead the male to the pollen-bearing part of the flower using several cues that act in sequence at decreasing distances on different sensory modalities of the wasp. From a considerable distance, males orient to volatile chemicals released from the flower, chemicals that resemble those in the Dufour's glands of female *A. mystaceus* (Ågren and Borg-Karlson 1984). As a male nears a flower, he responds to visual cues. To a male wasp, whose eyes have poor image resolution (compared with humans), the labellum probably simulates the female wasp's body in size and shape. The labellum is also dark in color like the female and reflects ultraviolet light in the same range of wavelengths as those reflected from the females' wings. The deception continues after the male lands, as he probably receives further positive feedback from fine hairs on the labellum that closely match those on the dorsum of a female *A. mystaceus'* abdomen in length, density, and pattern. Finally, the height at which some plant species present their phony females is optimal for attraction of males (Stoutamire 1983, Borg-Karlson 1990, Handel and Peakall 1993).

Even though the orchids succeed well enough to achieve pollination, the extent of their mimicry suggests that deception has selected for improved discrimination by males. Males of *Zaspilothynnus trilobata* quickly locate *Drakaea glyptodon* flowers that appear in their habitat, but they usually cut short their visit and so complete the full sequence of behaviors leading to pollen attachment and transfer only 16% of the time (Peakall 1990). Nevertheless, the interactions could

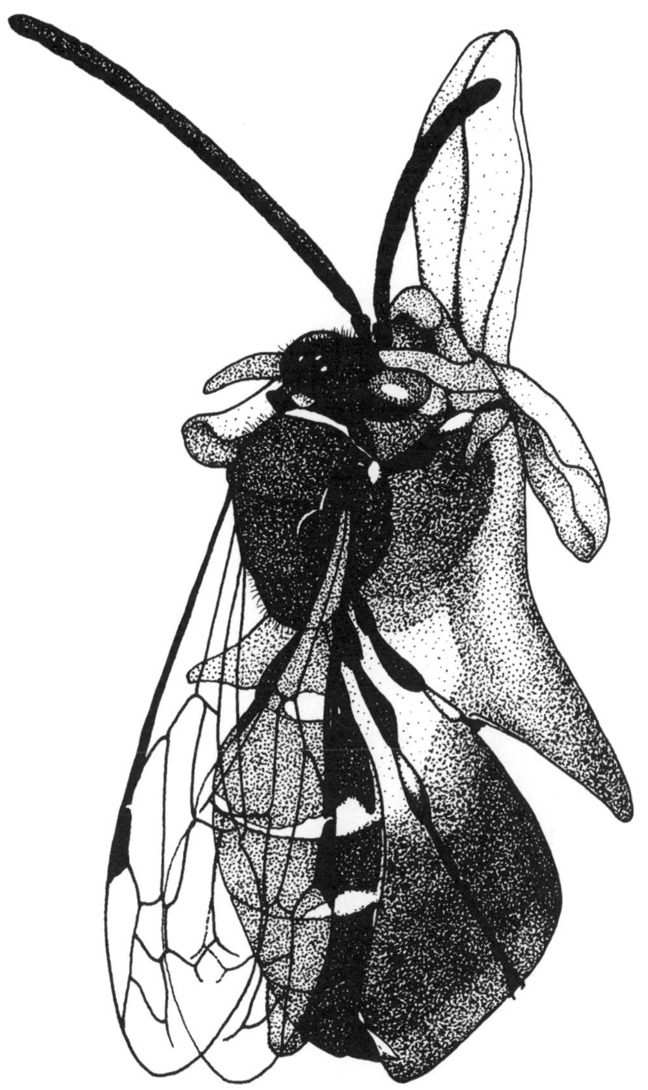

Figure 5-2. *Argogorytes mystaceus* male on flower of *Ophrys insectifera* (redrawn by Catherine Seibert from Proctor et al. 1996).

be costly to males, who may be attracted to the flower from a great distance. Males that have recently visited a pseudofemale apparently recognize their mistake and go through a refractory period during which they do not visit nearby flowers.

Deceptive pollination is a common form of parasitism by orchids, species of which are known to recruit males from four families of solitary aculeate wasps (Table 5-2) as well as four families of bees, two families of beetles, and an ichneumonid wasp (Coleman 1928, Borg-Karlson 1990, K. Steiner et al. 1994). Over 70 species of orchids in Australia use male thynnines to transfer pollen (Stoutamire 1983, Handel and Peakall 1993). The relationships between orchids and

Table 5-2 Species of solitary aculeate wasps known to pollinate sexually deceptive orchids

Wasp species	Flower species	Reference
Pompilidae: Pepsinae		
Hemipepsis hilaris	*Disa bivalvata*	K. Steiner et al. 1994
Scoliidae: Scoliinae		
Campsoscolia ciliata	*Ophrys speculum*	Ågren et al. 1984
C. tasmaniensis	*Calochilus campestris*	Fordham 1946
Campsoscolia sp.	*Calochilus holtzei*	Jones & Gray 1974
Sphecidae: Bembecinae		
Argogorytes fargei	*Ophrys insectifera*	Kullenberg 1973
A. mystaceus	*Ophrys insectifera*	Kullenberg 1973
Sphecidae: Sphecinae		
Podalonia canescens	*Disa atricapilla*	K. Steiner et al. 1994
Tiphiidae: Thynninae		
Neozeloboria sp.	*Chiloglottis reflexa*	Handel & Peakall 1993
Tachynomyia sp.	*Caladenia multiclavia, C. macrostylus*	Stoutamire 1983
Thynnoides bidens	*Caladenia lobata, C. barbarossa*	Stoutamire 1983
Thynnoides spp.	*Caladenia dilatata*	Stoutamire 1983
Zaspilothynnus trilobatus	*Drakaea glyptodon*	Peakall 1990
Zaspilothynnus sp.	*Caladenia huegelii*	Stoutamire 1983

wasps tend to be very specific. The scoliid *Campsoscolia ciliata* is the only known pollinator of the European *Ophrys speculum*, and two species of *Argogorytes* suffice for *Ophrys insectifera* (Borg-Karlson 1990). The chief pollinator of the southern African *Disa bivalvata* is *Hemipepsis hilaris*, although pollen is occasionally transferred by *Hemipepsis capensis* and a hopliinid beetle (*Peritrichia* spp.) (K. Steiner et al. 1994). That each orchid specializes on a narrow set of wasp species is reflected in the orchids' morphology and biochemistry. L. Ågren et al. (1984) compared the hairs on the abdomen of female wasps with those on the labellum of the flower. The high density, even distribution, short length, and microstructure of the hairs on *Ophrys insectifera* mimic those on the dorsum of the abdomen of female *Argogorytes*. By comparison, the dense fringe of long hairs around a central bare area on *Ophrys speculum* closely imitates that on females of its pollinator, *Campsoscolia ciliata*.

Whereas these European orchids facilitate pollen transfer by inducing a male to "copulate" in a position that causes him to pick up a pollen package, some Australian orchids take advantage of the unique mating strategies of male thynnines (described in Chapter 9). Before normal mating, a male thynnine carries off the female in flight. However, when a male lands on a sexually deceptive orchid, the results are quite different. The female-mimicking labellum, which the male grips tightly, is attached to the plant via a hingelike structure, thereby causing the flying male to pivot in an arc and slam against the anthers, where he picks up a pollen package (Fig. 5-3).

Conclusion

Specialized pollen feeding is common in the Aculeata, having evolved independently at least three times: once in the Apoidea by the bees and twice in the Vespoidea by the Sapygidae and Masarinae. Recently, K. Krombein and B. Norden (1997) presented evidence that the sphecid wasp *Krombeinictus nordenae* also feeds pollen to its young. They found pollen within the feces of larvae and on the mandibles of an adult female, but they found no prey fragments within nests.

The masarine wasps represent an independent "experiment" in the evolution of pollen provisioning, a pathway that has been more thoroughly explored by the bees with their greater taxonomic, morphological, and behavioral diversity. Among the bees, some species of the

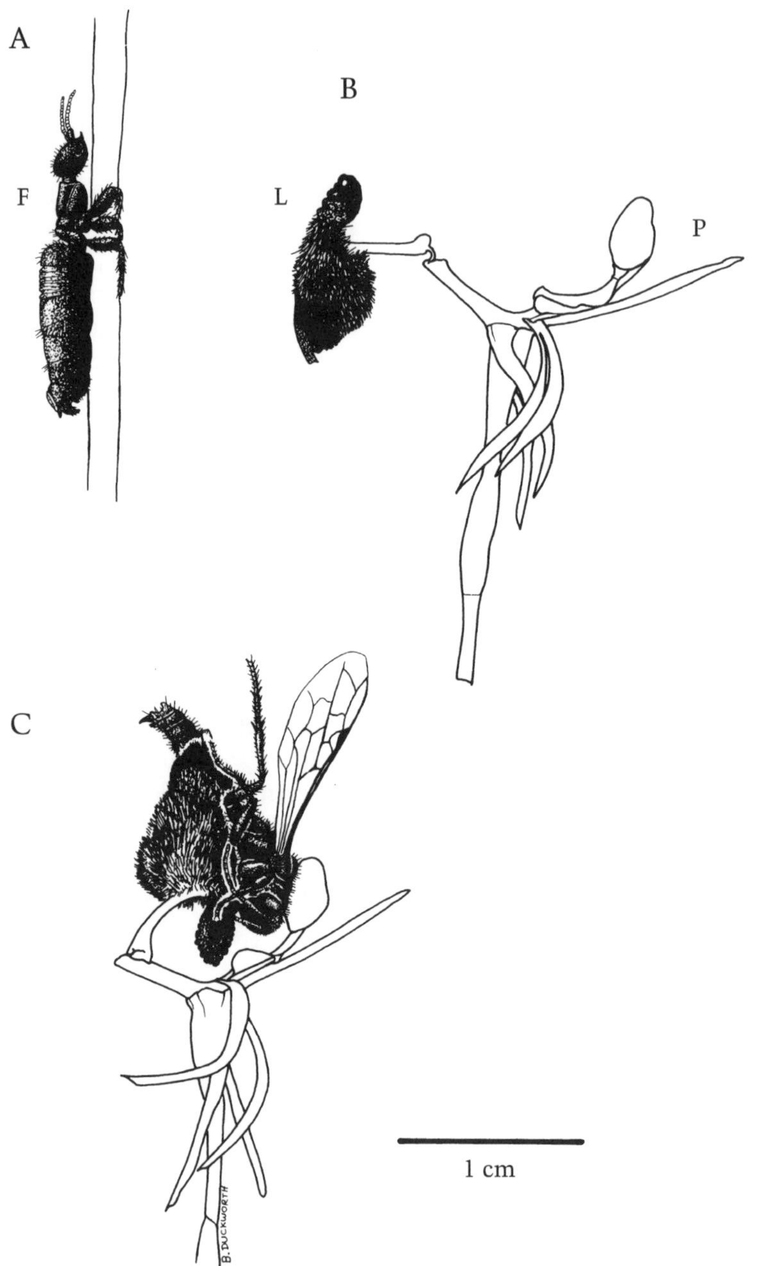

Figure 5-3. (A) *Zaspilothynnus trilobatus* female (F) in premating posture. (B) Position of pseudofemale labellum (L) and pollen-bearing column (P) of orchid *Drakea glyptodon*. (C) Male *Zaspilothynnus trilobatus* tipped against the column after attempting to fly off with the pseudofemale (from Peakall 1990).

relatively primitive family Colletidae share much in common with masarines, having relatively sparse body hairs, a habit of carrying pollen mixed with nectar in the crop, relatively narrow flower preferences at the species level, and a greater taxonomic diversity in the Southern Hemisphere (Krombein et al. 1979b, Michener 1979, Goulet and Huber 1993, S. Gess 1996).

6/ Nesting Behavior

Many insect larvae live in refuges of some sort. Some live in naturally occurring nooks and crannies or cavities formed as a side effect of feeding inside wood, fruits, or leaves. Other larval insects create their own refuges. Tiger beetles dig short burrows; bagworms build retreats of plant materials bound with silk; and some caddisflies construct cases of plant debris, pebbles, and snail shells. All of these refuges are found or created by the larvae themselves. In other insect species, however, it is the adults that build the refuges used by the young; examples include certain crickets, carrion beetles, scarabs, and many Hymenoptera (E. Wilson 1971). Although not all insect species that care for offspring after oviposition build nests (Tallamy and Schaeffer 1997), the building of nests is certainly a hallmark of a high level of parental care.

Among the nonsocial Hymenoptera, nest building is common in bees and in wasps of the families Sphecidae, Pompilidae, and Vespidae. Several authors have provided general classifications of the types of nests made by solitary wasps; some more detailed (F. Gess 1981), some less so (Evans and West-Eberhard 1970, Iwata 1976). I favor a simple scheme involving four basic (but interrelated) types of nests and nesters: (1) nests excavated in soil (the ground-nesters), (2) nests excavated in plant tissue (the pith- or wood-burrowing wasps), (3) nests built by modifying preexisting cavities (the cavity-nesters), and (4) free-standing nests (including the mud-nesters). In this chapter, I discuss variation in nest-site selection, nest-building behaviors, nest structure, the spatial dispersion of nests, and interactions among nesting females. The role of nests in defense is covered in Chapter 7, and the evolution of nesting behaviors in Chapter 10.

Ground-Nesting Wasps

Case Study: *Bembix pallidipicta*

Bembix pallidipicta (Sphecidae) is a large, black wasp with pale stripes that preys on flies of at least 11 families. The wasp occurs widely in the United States and tends to nest in extremely large and dense aggregations, making it convenient for studies of nesting behavior. The first substantial descriptions of the behavior of *B. pallidipicta* (as *B. pruinosa*) were provided by H. Evans (1957b, 1966c), who worked at 10 sites from New York to Utah; Rubink (1978, 1982) followed with an in-depth study of nest-site selection in New Mexico.

More is known about the specific variables affecting nest-site selection in this species than for any other fossorial wasp, with the possible exception of *Sphex ichneumoneus* (Brockmann 1979). Evans's geographically wide-ranging studies made it clear that *B. pallidipicta* consistently nests in expanses of open sand with little vegetation. Rubink took these general observations a step further to investigate specific habitat characteristics influencing nest-site selection. In comparing 22 nest sites (separated by as much as 4 km) with similar adjacent habitats, he found that nesting areas had (1) higher soil surface temperatures in the late morning (~6°C higher on average) and (2) flatter slopes (20% vs. 36% on average) that tended to face east to northeast (rather than north). Dropping to a lower spatial scale, he found that nest density within a large aggregation was correlated with (1) morning but not afternoon soil surface temperatures (areas of high nest density were ~2°–3°C warmer in the morning than areas of low density) and (2) soil particle size (areas of high nest density had less-variable particle sizes). Although soil moisture was not an important characteristic in nest-site selection, *B. pallidipicta* females apparently adjust nest depth in response to subsurface conditions, digging deeper in drier parts of the dunes. Two other factors apparently influence nest-site selection at a smaller spatial scale. First, using nearest-neighbor analysis, Rubink showed that females are probably sensitive to overcrowding: each tries to maintain some minimum distance between her own nest and her neighbors' nests. Second, it appears that the sand surface at the point of nest initiation must have some minimum cohesiveness to provide structural stability to the upper reaches of the nest tunnel (i.e., females like their sand crusty on top).

Soil surface cohesiveness and other site characteristics are probably evaluated during the exploratory stages of nesting, when a female makes a series of shallow excavations, several centimeters apart, as she walks backward over a short distance. When she begins digging the actual nest, the female stands head downward, with her body at about a 45° angle to the sand surface, and uses her front legs in synchrony to rake sand back beneath her body. Because sand can be thrown as far as 20 cm, an elongate mound forms as the female alternates time digging within the burrow with clearing sand from the entrance.

A nest is initiated when the *B. pallidipicta* female digs a short, obliquely descending entranceway, after which she excavates a horizontal "preliminary tunnel", about 10 mm high, 15–30 mm wide, and 20–51 cm long (Fig. 6-1A) (Evans 1957b). Though just 2–4 cm beneath the soil surface, the tunnel (which takes from 40 to 60 minutes to complete) seems fairly stable, perhaps because the female has chosen a site with adequate surface cohesion during her initial explorations. Upon completing the preliminary tunnel, the female lands on the accumulated mound of sand and, while facing away from the nest, levels the mound and covers the nest entrance with raking movements of her forelegs. Some females follow up by digging a 3–10 cm deep false, or accessory, burrow just next to the true nest entrance (Fig. 6-1B) (the hypothesized function of these burrows, which do not house the young of the wasp, will be discussed in Chapter 7).

After reentering the preliminary tunnel (a unique feature of this species) and closing it from the inside, the female digs a 23–64 cm long "true burrow" that descends obliquely to a depth of 16–54 cm. She then excavates a horizontal 11–33 cm long brood cell (Fig. 6-1C), the only cell in the nest. As the soil is cleared from the burrow, it is deposited in the preliminary tunnel (rather than outside the burrow entrance, as is typical for other fossorial wasps). After laying an egg at the far end of the empty cell, the female closes the entrance of the cell with soil dug from a short "spur" at the end of the main burrow (Fig. 6-1D). Finally, when the female emerges from her completed nest to begin provisioning, she does so through a new entrance at the

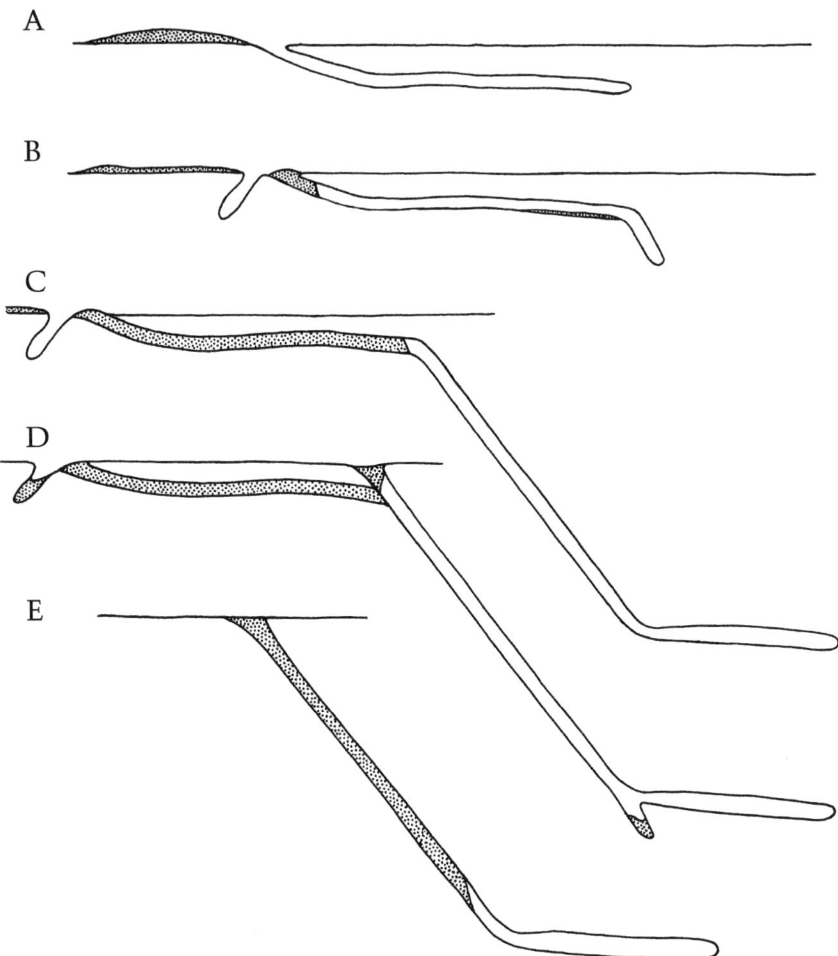

Figure 6-1. Lateral profiles of the nest of *Bembix pruinosa* (*B. pallidipicta*) in various stages of construction (egg, larvae, and prey not shown). (A) Initial form of preliminary tunnel with mound at entrance. (B) Preliminary tunnel with leveled mound and accessory burrow. (C) Filled preliminary tunnel, with oblique main burrow and elongate horizontal cell. (D) Spur at end of main burrow, which is now plugged with a temporary closure. (E) Nest at stage of final closure. (A–E redrawn from Evans 1957b.)

top of the main burrow; the preliminary burrow is not used again. While on provisioning trips, the female almost always seals the nest entrance with a temporary plug of sand.

Upon the completion of provisioning, which is progressive (Chapters 3 and 10), the female performs two final duties. First, she completely fills the main burrow with sand (Fig. 6-1E), which comes from the burrow walls and (mostly) from outside of the nest. Second, the female levels the soil around the burrow to such an extent as "to render the burrow entrance quite invisible" (Evans 1957b:164).

COMPARATIVE SURVEY OF GROUND-NESTING SOLITARY WASPS

The nest of *Bembix pallidipicta* is "one of the most remarkable structures of any digger wasp" (Evans 1957b). But, as we will see, the nesting behavior of the fossorial wasps is so variable that no species can be considered typical. Ground nests are the most common type of nest among the solitary wasps (Table 6-1). The seven major subfamilies of Sphecidae all include ground-nesters, and ground nesting is the only nesting habit in the large subfamilies Bembicinae and Philanthinae. Most nest-provisioning Pompilidae are also ground-nesters, as are a large number of Vespidae of both the Eumeninae and Masarinae.

Clearly, most fossorial wasps have specific nest-site preferences. Take, for example, the North American *Bembix*, which variously prefer "large tracts of open, loose, more or less blowing sand" (*B. occidentalis, B. pallidipicta*), "smaller tracts [of sand] where there is less blowing" (*B. amoena, B. sayi*), "firm sand, soft earth, or fine gravel" (*B. americana*), "firm, coarse sandy gravel" (*B. belfragei, B. hinei*), or the "hard-packed soil" of coastal salt flats (*B. cinerea*) (Evans 1957b). Preferences vary in a similar manner in *Philanthus*: *P. albopilosus* nests in dune blowouts; *P. psyche* in the sparse vegetation at the edge of blowouts; *P. pulcher* and *P. crabroniformis* in moderately hard-packed alluvial soils; and *P. inversus* in vertical soil banks (Evans and O'Neill 1988). Although many ground-nesting species prefer relatively dry and friable soils, others, such as *Astata occidentalis* (Evans 1957a), *Spilomena subterranea* (McCorquodale and Naumann 1988), and *Cerceris antipodes* (McCorquodale 1989c), nest in hard-packed soils. In contrast are the *Alysson*, which nest in damp, shaded soils (Evans 1966c, O'Brien and Kurczewski 1982b), and the *Anoplius* that prefer sites along the edges of ponds and streams, even digging in muddy ground on occasion (Evans 1949, Roble 1985, Shimizu

Table 6-1 Nest types of selected genera of solitary Sphecidae, Pompilidae, and Vespidae

Subfamily	Genera in which female modifies preexisting cavity	Genera in which female excavates cavity in			Genera in which female builds mud nest
		Soil	Plant stems	Rotten wood	
Sphecidae					
Sphecinae	*Chalybion, Isodontia, Podium*	*Ammophila, Chlorion, Palmodes, Podalonia, Prionyx, Sphex, Stangeella*	—	—	*Sceliphron, Trigonopsis*
Pemphredoninae	*Passaloecus, Pemphredon, Psenulus, Spilomena, Stigmus*	*Diodontus, Mimesa, Mimumesa, Nesomimesa, Pluto, Pulverro*	*Carinostigmus, Passaloecus, Pemphredon, Spilomena, Stigmus*	*Mimumesa, Passaloecus, Pemphredon, Psen, Spilomena*	—
Astatinae	—	*Astata, Dinetus, Diploplectron*	—	—	—
Larrinae	*Nitela, Pison, Pisonopsis, Solierella, Trypoxylon*	*Miscophus, Palarus, Plenoculus, Tachysphex, Tachytes*	—	—	*Pison, Trypoxylon*
Crabroninae	*Crossocerus, Tracheliodes*	*Belomicrus, Crabro, Crossocerus, Ectemnius, Enchemicrum, Lestica, Lindenius, Moniaecera, Oxybelus, Rhopalum*	*Crossocerus, Dasyproctus, Ectemnius, Rhopalum*	*Ectemnius, Lestica*	—

Table 6-1 *Continued*

Subfamily	Genera in which female modifies preexisting cavity	Genera in which female excavates cavity in			Genera in which female builds mud nest
		Soil	Plant stems	Rotten wood	
Bembicinae	—	*Alysson, Bembecinus, Bembix, Bicyrtes, Exeirus, Gorytes, Hoplisoides, Mellinus, Microbembex, Rubrica, Sphecius, Steniolia, Stictia, Stictiella, Stizus, Zyzzyx*	—	—	—
Philanthinae	—	*Aphilanthops, Cerceris, Clypeadon, Eucerceris, Philanthus, Trachypus*	—	—	—
Pompilidae					
Pepsinae	*Auplopus, Dipogon, Fabriogenia*	*Ageniella, Cryptocheilus, Dichragenia, Pepsis, Priocnemis*	—	—	*Auplopus, Macromerella, Macromeris, Phanagenia*

Pompilinae	*Agenioideus*	*Agenioideus, Anoplius, Batozonellus, Episyron, Poecilopompilus, Pompilus, Tachypompilus*	—	*Anoplius*	*Anoplius*
Vespidae					
Eumeninae	*Ancistrocerus, Euodynerus, Monobia, Montezumia, Odynerus, Stenodynerus, Symmorphus, Tricarinodynerus, Zethus*	*Ancistrocerus, Anterhynchium, Euodynerus, Odynerus, Montezumia, Paralastor, Pterocheilus*	*Raphiglossa*	—	*Abispa, Ancistrocerus, Eumenes, Euodynerus, Montezumia, Orancistrocerus, Paraleptomenes, Synagris, Xenorhynchium*
Masarinae	*Celonites*	*Celonites, Ceramius, Jugurtia, Masarina, Paragia, Quartinia, Rolandia, Trimeria*	—	—	*Celonites, Gayella, Pseudomasaris*

Sources: Information primarily from F. Williams 1919c; Krombein 1967; Danks 1970; Evans & Matthews 1973a; Bohart & Menke 1976; Iwata 1976; Evans et al. 1980c; F. Gess 1981; Cowan 1991; Shimizu 1992, 1994; F. Gess & Gess 1991; S. Gess 1996.
Note: Some genera occur in more than one category.

1992). The species mentioned above rarely nest outside of their preferred soil types, although a few other species seem less finicky. *Pompilus scelestis*, for example, can be found on beaches and inland dunes as well as in grasslands and in forests (Gwynne 1979). Nevertheless, suitable soil substrate (near sources of prey and nectar) is probably a major limiting factor in the distribution of many species of ground-nesters.

The digging behavior of ground-nesters has been divided into four (somewhat overlapping) categories: rakers, pullers, pushers, and carriers (Olberg 1959, Evans and West-Eberhard 1970, F. Gess 1981). Rakers use their front legs to rake loose soil backward beneath their body, usually after the soil has first been broken up with the mandibles. During raking, soil is thrown backward beneath the wasp, where a small a mound often forms. Pompilids and some sphecids rake with alternate use of the left and right forelegs; many other sphecids, including *Bembix* and other Bembicines, use their forelegs synchronously. Most fossorial sphecids are rakers, but the trait appears to be rare in the solitary Vespidae (Evans and West-Eberhard 1970, S. Gess 1996).

Pullers also loosen soil with their mandibles and front legs. However, rather than raking it away, they pick the soil up in a small pellet between the mouthparts and forelegs, carry it backward, and drop it a short distance away. The result is a series of pellets, usually left around the entrance (e.g., *Tachytes mergus*) (Krombein and Kurczewski 1963). Most rakers and pullers dig an oblique burrow, whereas pushers are commonly found among those species that dig a vertical or nearly vertical burrow. After loosening the soil, pushers walk backward up the burrow, nudging soil along with the legs and the tip of the abdomen. Usually, the result is circular tumulus, with a hole near the middle. Some *Cerceris* species push the plug of soil out of the nest and discard it just outside the entrance as a sausage-shaped cylinder of sand (F. Gess 1981). Finally, carriers (e.g., many *Ammophila*) carry soil, often in flight, depositing it far from the nest entrance. In contrast to the nests of pullers, no conspicuous ring of soil pellets accumulates around the entrance of the nests of carriers.

In ground-nesters, morphological evolution has gone hand in hand with behavioral evolution to increase the efficiency of loosening and dispersing soil. Rakers have rows of long, inwardly curved rake spines (forming a pectin, or tarsal comb) on the front tarsus that aid in raking soil. In species of Sphecidae that dig in the pith of twigs with their mandibles (e.g., many Pemphredoninae; see the next section), the row of spines is rudimentary, whereas in *Larra* (parasitoids that use the ready-made burrows of their hosts) the rake is absent (Bohart and Menke 1976). In pushers, the movement of soil out of the burrow is made more efficient by the presence of broadened pygidial plates (on the upper surface of the tip of the abdomen) or by serrations on the hind tibia in pompilids of the subfamily Pepsinae. Finally, several groups of carriers have independently evolved fringes of long hairs (psammophores) to hold soil pellets between the head and thorax (*Belomicrus* of the Sphecidae, *Rolandia* of the Masarinae) or between the mandibles and palpi (*Pterocheilus* of the Eumeninae) (Evans and West-Eberhard 1970, S. Gess 1996).

Both the behavior and morphology associated with digging can vary within a genus. *Oxybelus* includes rakers, pushers, species that both push and rake, and carriers that occasionally push and rake (Peckham et al. 1973). In a study of seven South African *Ammophila* species, A. Weaving (1989) found that, for excavating and closing, all species were rakers, but for soil disposal one was a puller and six were carriers. Species of *Bembecinus* that are rakers have well-developed rake spines, whereas those that are carriers lack the spines (F. Gess 1981). Variation in rake spines can also be seen in male *Bembecinus*: male *B. quinquespinosus*, which dig while searching for preemergent females (Chapter 10), have rows of short rake spines on the front legs, but male *B. nanus*, which do not dig for females, have much shorter leg spines (O'Neill and Evans 1983a).

Besides the digging techniques and tools mentioned above, fossorial wasps have further soil-loosening techniques in their repertoire. Some pompilid, eumenine, and masarine wasps that dig in hard-packed soils first carry water to the digging site, regurgitate the fluid, and work it in with their mandibles to soften the soil (F. Gess and Gess 1974, F. Gess and Gess 1975, F. Gess 1981, Cowan 1991, S. Gess 1996). The mud may then be discarded as a pellet or used to build a turret at the nest entrance (see below). Another distinctive method of loosening soil involves the use of vibrations produced in the flight muscles and transmitted to the soil via the mandibles in the manner of a high-frequency jackhammer (Spangler 1973, F. Gess 1981).

The placement, number, spacing, depth, and form of brood cells within nests are highly variable (Fig. 6-2). In the Pompilidae (e.g., *Anoplius* and *Pepsis*) and Sphecidae (e.g., *Prionyx* and *Podalonia*) that

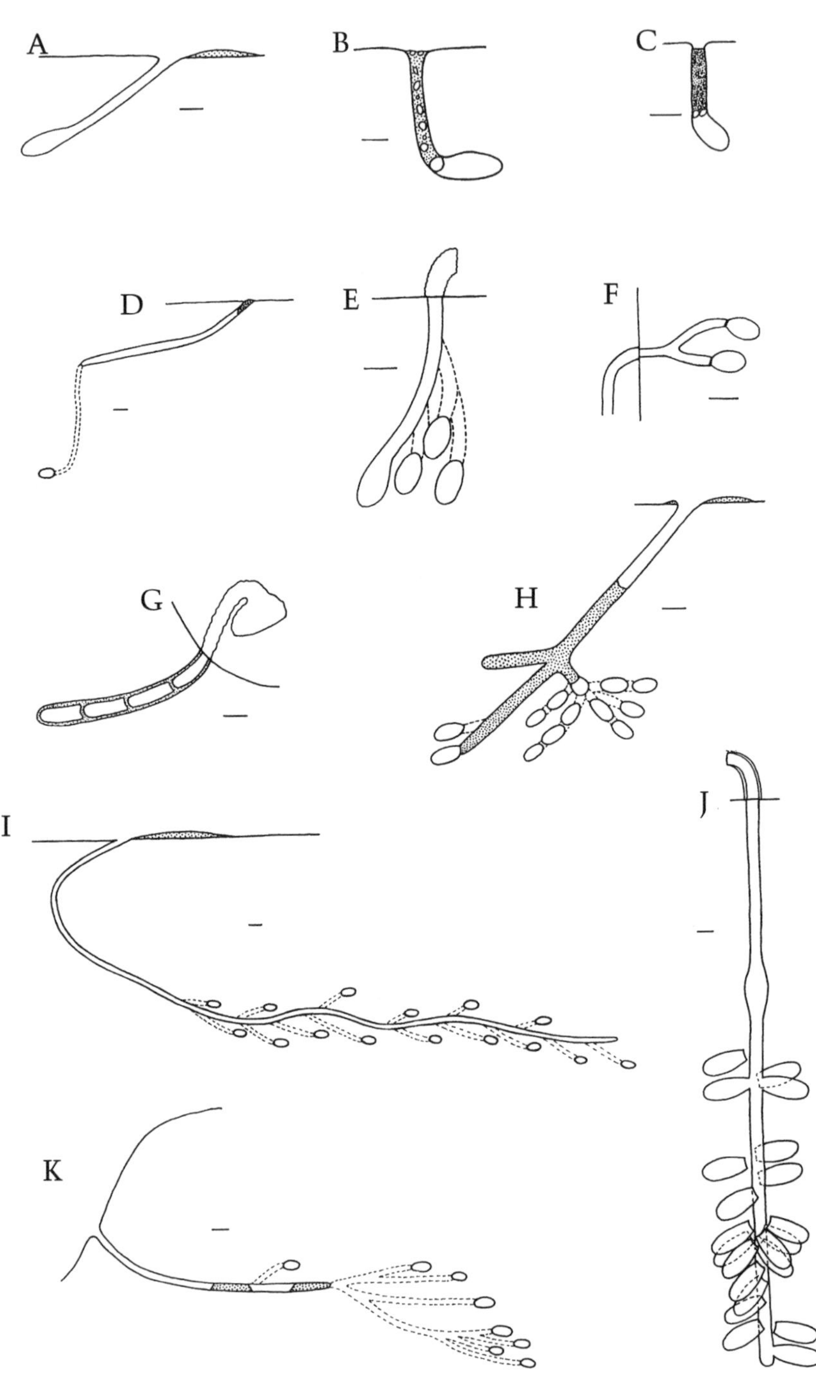
A
B
C
D
E
F
G
H
I
J
K

quickly build single-celled nests after hunting, the burrow may be very short. In these nests, the single nest cell may be no more than an expansion of the terminus of the main burrow and may lie just several centimeters deep (Fig. 6-2A–D). The cells of species that provision single-celled nests built before hunting may be quite variable in depth: 2–5 cm for *Ammophila harti* (Hager and Kurczewski 1986), 3–11 cm for *Steniolia obliqua* (Evans 1966c), 6–10 cm for *Rubrica surinamensis* (Evans 1966c), 9–17 cm for *Bembecinus nanus* (Evans and O'Neill 1986), 8–25 cm for *Microbembex monodonta*, 13–30 cm for *Stictia carolina* (Evans 1966c), and 27–75 cm for *Ammophila ferrugineipes* (Weaving 1989). The structural simplicity of single-celled nests does not mean that these genera are primitive sphecids. Many Bembicinae dig single-celled nests, but this subfamily is considered relatively evolutionarily derived within the Sphecidae (Bohart and Menke 1976), and Evans (1966c) considered single-celled nests a derived feature within this group.

Multicellular ground nests are common within Pompilidae and Sphecidae, are almost the rule among fossorial Eumeninae, and are the only kinds of ground nests in the Masarinae (Iwata 1976, S. Gess 1996) (Fig. 6-2E–K). Most Pompilidae that build nests before hunting have multicellular nests, up to 12 cells for some *Priocnemis* (Yoshimoto 1954). Most multicellular nests of the Sphecidae have fewer than 10 cells, but some high maximum values include 16 for *Crabro*

Facing page

Figure 6-2. The nest structures of a variety of ground-nesting solitary wasps (brood, prey, and eggs are not shown). Dashed lines indicate the estimated form of burrows leading to cells; stippling indicates burrows filled with temporary or final closures; short horizontal line near each nest is a 1-cm scale (redrawn from the cited sources). (A) *Episyron quinquenotatus* (Evans 1970a). (B) *Ammophila azteca,* with final closure (Evans 1965). (C) *Pterocheilus texanus,* with final closure (Evans 1956). (D) *Philanthus psyche,* with temporary closure (Evans and O'Neill 1988). (E) *Euodynerus annulatus,* with turret (Evans 1956). (F) *Stenodynerus microstictus,* with turret extending to left from vertical clay bank (Evans 1956). (G) *Paralastor* sp., with turret (A. Smith 1978). (H) *Astata occidentalis,* with soil from final closure (Evans 1957a). (I) *Philanthus crabroniformis* (Evans and O'Neill 1988). (J) *Ceramius lichtensteinii,* with turret and bulb in main burrow that female uses for turning around (S. Gess 1996). (K) *Cerceris fumipennis,* in steep bank with inner closures (Evans 1971).

advenus, 17 for *Stictiella formosa*, 18 for *Nesomimesa hawaiiensis*, 20 for *Sericophorus victoriensis*, and 24 for *Cerceris sabulosa* (Iwata 1976). In the Masarinae, *Ceramius lichtensteinii*, builds nests with up to 21 cells (S. Gess 1996). Certain genera, such as *Oxybelus* (Peckham et al. 1973) and *Tachysphex* (Kurczewski 1987b), contain some species that make unicellular nests and others that provision multicellular nests.

The cells in multicellular nests can bear various spatial relationships to the main burrow, to side burrows, and to one another. Cells may be arrayed in a radial fashion around the end of the burrow as are *Nesomimesa hawaiiensis* nests (F. Williams 1919d), in sequence within different side burrows as are *Astata occidentalis* nests (Fig. 6-2H), or at various distances along the main burrow as are *Philanthus crabroniformis* nests (Fig. 6-2I).

The cells within a nest may also be provisioned in varying temporal sequences relative to the nest entrance. *Sphex argentatus* and *Alysson melleus* build cells in a regressive pattern, beginning at the end of the burrow and working their way back toward the entrance (Tsuneki 1963, Evans 1966c). Many species of *Philanthus*, on the other hand, construct successive cells in a progressive pattern, extending the main burrow and adding new cells as time goes on (Evans and O'Neill 1988). (A progressive pattern of cell construction should not to be confused with progressive provisioning of each cell; see Chapter 10.)

The burrows and cells of the fossorial pompilids and sphecids are simple cavities excavated in the soil either at the end of a burrow or off to the side along a secondary shaft. The cells, and often the burrows, of some eumenines and masarines, however, are lined with material that increases their structural integrity (Cowan 1991, S. Gess 1996). Special cell linings are particularly common among fossorial masarines (S. Gess 1996). The source of materials for cell lining varies. Various species of *Paragia*, *Ceramius*, and *Jugurtia* line the cell with mud made from water mixed with soil collected within the burrow, whereas *Celonites latitarsus* mix nectar with earth quarried from outside the burrow. *Quartinia vagepunctata* line the cells with sand bonded with silk (produced from glands in the head, unlike Pemphredoninae, which produce it in the abdomen). Finally, *Paragia tricolor* lines its cells with some unidentified substance that forms a polished and waterproof coating (Houston 1984).

Although soil mounds or tumuli are simply by-products of excavation, other structures associated with the nest entrance are made deliberately, as in the accessory burrows of *Bembix pallidipicta* (Chapter 7). Wasps in several lineages build external mud tubes (turrets) at the nest entrance (Fig. 6-2E, F, G, J) (Evans 1956, 1970a; A. Smith 1978; F. Gess 1981, F. Gess and Gess 1988b, S. Gess 1996). Turrets are often curved and of even diameter along their length, but they may also fan out at the end as do those of *Paravespa* (F. Gess and Gess 1988b) and *Paralastor* (A. Smith 1978) (Fig. 6-2G). Turrets are usually made by successively applying rings of individual mud pellets to form a tube, but the turret of *Quartinia vagepunctata* is made of a composite of sand and silk produced by the wasp. Turrets probably had an independent evolutionary origin in the Sphecidae, Pompilidae, and Vespidae and probably appeared early in the evolution of the Vespidae (S. Gess 1996). Despite their elaborate structure and widespread occurrence, the function of turrets is unknown. It has been suggested that turrets serve as orientation markers or as antiparasitoid devices, but I am aware of no evidence to support these hypotheses (Miotk 1979, S. Gess 1996).

An open nest can be invaded by various natural enemies (Chapter 7). Thus, many (though not all) wasps block the nest burrow in some manner before they depart to forage (a temporary closure) or leave for good when the nest is finished (a final or permanent closure). The nest may be blocked either at the level of the cell (an inner closure) or in the main burrow and at the entrance (an outer closure). Different taxa use these in different combinations. Within the subfamily Bembicinae, some species typically construct no closures at any time (e.g., *Sphecius speciosus*), whereas others maintain both inner and outer closures (e.g., most *Bembix* and all *Stictia*). Species such as *Hoplisoides nebulosus* have hastily assembled final closures, whereas others, such as many *Bembix*, carefully fill the main burrow and then spend much time leveling the soil around the entrance (Evans 1966c, Evans and Matthews 1973b). Only four of the nine *Oxybelus* species reviewed by D. Peckham and colleagues (1973) constructed temporary closures even though species of this genus appear to have great problems with cleptoparasites' entering their nests. Most *Philanthus* species use temporary closures while hunting, but the existence of a final closure seems to depend on the number of nests made by a female. The females of species that build multiple nests during their lives close off each nest when finished. However, the females of species that have a single large nest apparently continue to expand the nest until they die (Evans and O'Neill 1988).

The Sphecinae, usually considered to be fairly primitive sphecids (Bohart and Menke 1976), usually have no inner closure, but outer final closures are common (Brockmann 1985b). For the outer closure, soil discarded during digging may be scraped back into the hole, but females may also dig new soil for the closure. Alternatively, females may pick up stones or lumps of soil and place them in the hole. Once soil is in place, the female compacts it by pressing her clypeus and open mandibles against the substrate, or she pounds the substrate with her head. In both cases, the behavior is often accompanied by loud buzzing that may serve to break up soil clumps and aid in compaction. Many sphecines hold small stones while pressing the soil in the nest hole, and some *Ammophila* hold a stone while pounding the soil, then later discard it. Females of some of the species that use stones for pressing or pounding soil spend a lot of time selecting a stone of the proper size. Some early investigators considered this to be an example of an insect's improvising a tool and making intelligent use of it (to paraphrase Peckham and Peckham 1898), but the actions do not require insightful thinking (Evans 1959, Brockmann 1985b).

Many masarines and eumenines are known to have fairly sophisticated inner closures (F. Gess and Gess 1988b, S. Gess 1996). After enclosing a fully provisioned cell in its mud case, ground-nesting species of *Ceramius* (Fig. 6-2J) and *Jugurtia* fill the secondary shaft leading to the cell with soil and then seal the juncture to the main burrow with mud, which they then smooth over. Final closures have been noted for several eumenines that block the upper portion of the burrow with loose soil and then, sometimes, tamp it down with their mandibles (Evans 1956, Evans and Matthews 1974, Grissell 1975).

The nest-site selection decisions of a female wasp are critical to her offsprings' ability to thermoregulate and control water balance. Because the nest cells of many ground-nesting species lay deep within the ground, where temperatures fluctuate less than air temperatures, their eggs, larvae, and pupae develop in a relatively stable thermal environment. Evans (1957b) and L. Kimsey and colleagues (1981) noted that, even when temperatures were 45° to 52°C at the soil surface, temperatures at the level of nest cells of *Bembix* were only 27° to 31°C. Similarly, for *Bembix pallidipicta*, W. Rubink (1978) found that temperatures at cell depth remained stable at about 24°–29°C, while soil surface temperature ranged from about 17° to 58°C. Temperature (Fig. 6-3) and relative humidity an 8-cm depth in a nest of *Cerceris arenaria* exhibit little fluctuation relative to conditions at the soil surface and in the air above the nest (Willmer 1982); a similar buffering was observed within nests of the *Passaloecus* species that nests in beetle emergence tunnels in wood (Corbet and Backhouse 1975).

A. Weaving (1989) examined nest temperatures throughout seasons by recording soil temperatures at depths typical for nest cells of *Ammophila insignis*. This species nests in vertical soil banks in South Africa, where soil surface temperatures reach as high as 64°C in the summer and where temperatures at 2.5 to 5.0 cm below the surface range from 7° to 40°C. During the hottest months, November through April, females nest primarily in shaded areas of south-facing banks, but they switch to nesting on the more insolated north-facing banks in May. In switching, they avoid potentially stressful surface temperatures while digging and position their

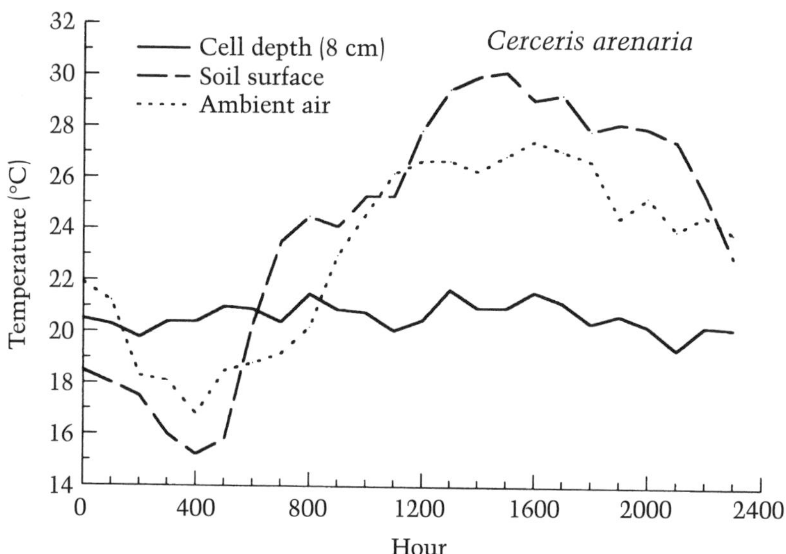

Figure 6-3. Daily changes in temperature within and above a *Cerceris arenaria* nest (redrawn from Willmer 1982).

nest cells where temperatures remain in the range of about 20°–30°C. Another species, *Ammophila ferrugineipes*, apparently adjusts cell depth in response to the amount of sunlight striking the soil surface. Cells ranged in depth from 30 to 51 cm in nests shaded by a tree, but they were 55–75 cm deep in nests dug in fully insolated soil.

Much of the buffering of temperature and humidity in the nests of ground-nesting species may be due to the presence of moisture at cell depth (Willmer 1982). Several sphecids dig deeper nest cells in drier soil than in moister soil, presumably to reach moist (but not saturated) soil at greater depths (Rubink 1978, Hager and Kurczewski 1986, Kurczewski and Wochaldo 1998). Saturation of the soil for long periods may be harmful. For example, Howard Evans and I had to cut short our study of *Bembecinus quinquespinosus* when weeks of heavy rains saturated the soil, apparently killing most of the pupae in their cocoons. On the other hand, in the tropics, *Stictia signata* apparently survive when nesting in sand banks that are consistently flooded during the wet seasons (Evans 1966c). The ability of prepupae and pupae of ground-nesters to deal with variations in humidity and free water may also be linked to cocoon structure. The unique structures of the cocoons of Bembicinae, in particular, may allow them to survive long periods of saturation or drought (Evans 1966c). At least one species of *Bembix*, *B. coonundura*, is believed to survive in cocoons in nests over long dry periods, waiting for rain that will soften the soil and fill lakes that will supply damselfly prey (Evans and Matthews 1973c).

Pith- and Wood-Burrowing Wasps

Case Study: *Pemphredon lethifer*

Pemphredon lethifer (Sphecidae), a Holarctic species, was studied in England by H. Danks (1970), in France by H. Janvier (1960), in Japan by K. Tsuneki (1952) and R. Ohgushi (1954), and in the United States by P. Rau (1948), K. Krombein (1958, 1964), and M. McIntosh (1996). Females of *P. lethifer* are tiny wasps with a relatively large head. Their mandibles bear sets of stout teeth that they use to burrow nests in the soft pith of the twigs of various species of shrub, including *Brassica* (mustard), *Foeniculum* (fennel), *Hibiscus, Rhus* (sumac), *Rosa* (roses), *Rubus* (bramble, raspberry), and *Sambucus* (elderberry). In experiments conducted by McIntosh (1996), female *P. lethifer* showed a preference for elderberry (77% of nests) over mustard (20%) and fennel (3%), perhaps because elderberry has sturdier woody stems. She found no correlation between female size and the diameter of twigs used, but females of all sizes tended to prefer larger stems, which allowed them to build more extensive nests.

Within narrow stems, females excavate a single unbranched tunnel in which they provision a linear sequence of cells with aphids. In wider stems, females prepare a branched system of 3–5 mm tunnels (Fig. 6-4) that may house up to 33 cells, each typically 8–9 mm long

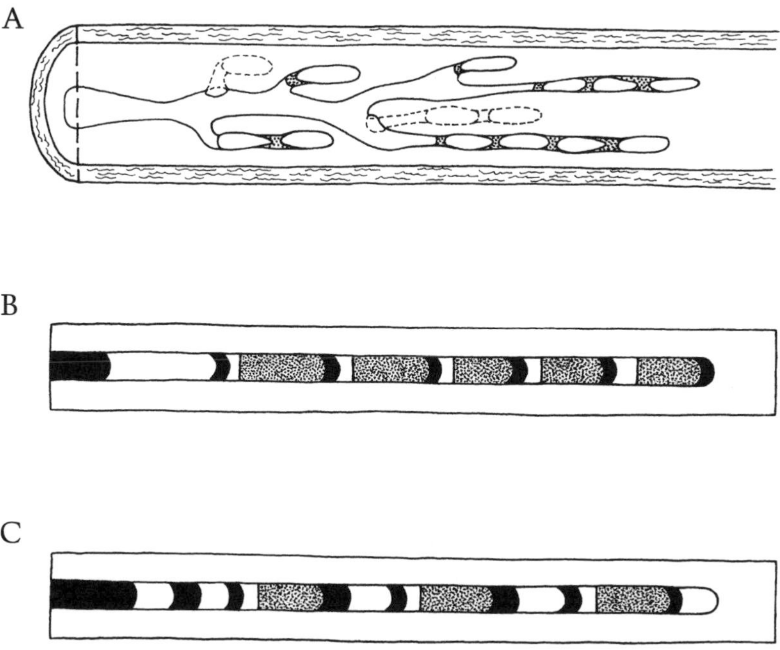

Figure 6-4. (A) Cross section of nest of *Pemphredon lethifer* excavated in the pith of a stem; the opening on the left leads to a branched system of tunnels containing cells separated by partitions (stippling) (redrawn from McIntosh 1996). (B) Cross section of a trap nest containing a diagrammatic representation of a nest with alternating cells and partitions (stippling indicates approximate size of provision mass, white the empty space, and black the partitions and closures) (redrawn from Krombein 1967). (C) Cross-section of a trap nest containing a diagrammatic representation of a nest with intercalary cells interspersed among provisioned cells (redrawn from Krombein 1967).

(Danks 1970). The most variable aspect of the nests, however, is the thickness of partitions between cells, which are made of pith particles scraped from the side of the burrow. Partitions are sometimes absent, but when present they are up to 27 mm thick. The plug at the entrance of the nest is also highly variable, but even when it is absent the nest may be protected by the adult female, who often sits in the vestibule guarding the nest against intruders. McIntosh (1996) observed female *P. lethifer* fighting with each other over a nest and attempting to extract *Ceratina* bees that had entered their nests.

COMPARATIVE SURVEY OF SOLITARY WASPS THAT BURROW IN PLANT TISSUE

With a few exceptions in the Pompilidae and Eumeninae (Gess 1981, Shimizu 1992), wasps that excavate nests in plant tissue are restricted to certain genera of the Pemphredoninae and Crabroninae (Table 6-1). Although most pith-burrowers probably enter stems only after the stems have been broken, *Dasyproctus* species enter stems by burrowing through the side wall of intact stems, including those of gladiolas, sorghum, and grasses (Bowden 1964, F. Gess 1980b). Species that burrow in plants have a ready-made source of raw materials in the form of pith and sawdust, but they face a potential problem not experienced by other types of nest provisioners: the material they nest in could be eaten by large herbivores. To avoid this problem, many species nest in stems that are covered with thorns and prickles (e.g., roses, raspberries, and brambles), in plants toxic to herbivores, or in plants that grow within patches of thorny shrubs (F. Gess 1980b).

Although several genera contain species that nest in rotten wood, these wasps have been less well-studied than pith-burrowers (probably because it is a bit bothersome to drag rotten logs into a lab). Wood burrowers of the genera *Crossocerus*, *Ectemnius*, and *Mimumesa* excavate branched galleries in rotting stumps and logs, constructing partitions of sawdust and beetle frass (Davidson and Landis 1938, Michener 1971, Kurczewski and Miller 1986, Rosenheim and Grace 1987). Nest burrows may initially follow abandoned burrows of beetles, then strike off into rotten tissue following the grain of the wood.

Some of the genera with species that burrow in plant tissue also include species that nest in soil (e.g., *Ectemnius*, *Mimumesa*), in preexisting cavities *(Pemphredon)*, or in both *(Crossocerus)*. Thus, there appears to be considerable evolutionary flexibility in nesting behavior in these groups. Evolutionary transitions to pith-burrowing behavior have apparently involved parallel morphological changes: mandibles with extra teeth for tearing pith and a narrow, concave pygidial plate for scooping pith (Evans and West-Eberhard 1970, Bohart and Menke 1976, Krombein 1984). Krombein (1984) noted that *Stigmus* that nest in pith have tridentate mandibles, whereas one species that nests in preexisting cavities has only two teeth on each mandible. Similarly, females of those *Crossocerus*, *Ectemnius*, and *Lestica* species that nest in rotten wood have a narrow, scooplike pygidial plate, whereas ground-nesters of these genera have broad, flat plates for tamping sand (Bohart and Menke 1976).

Cavity-Nesting Wasps

Case Study: *Passaloecus cuspidatus*

Some species of *Passaloecus* (Sphecidae) share with their relatives in the genus *Pemphredon* the habit of digging nests in the pith of plant stems. Others, such as *Passaloecus cuspidatus*, nest in preexisting cavities in wood after the emergence of wood-feeding insects (Fye 1965; Krombein 1967; Vincent 1978; Fricke 1991, 1992a,b, 1993, 1995). Cavity-nesters have proved to be excellent subjects for studies of nesting biology because they readily nest in "trap nests," holes drilled in blocks of wood and placed in appropriate habitats.

When given access to trap nests ranging in diameter from 1.6 to 6.4 mm, *P. cuspidatus* prefer those of 2.8–4.8 mm diameter. Upon locating a potential nest hole, females first clear debris from the tunnel, carrying it to the entrance with their mandibles and dropping it. Before provisioning begins, the female visits resin flows on conifers to pick up droplets the size of her head. She spreads the resin in a ring around the nest entrance (perhaps as a barrier to marauding ants). She then stocks a linear sequence of cells with aphids, forming 0.1–5.0 mm thick partitions of pine resin between cells. The cells themselves range in length from 5 to 23 mm, those provisioned for daughters tending to be larger than cells used for sons. When the tunnel is nearly full of cells, the female plugs the entrance with a large blob of resin (0.25–4.0 mm thick), but she usually leaves one to three empty vestibular cells between the plug and the outermost cell. Basing his calculations on the typical volume of resin closures and the average amount of resin gathered on each trip, J. Fricke (1995) estimated that females typically require 8 trips to plug a tunnel with a 3.2-mm diameter and 18 trips to close a 4.8mm diameter nest. When the resin plug is complete, the female makes an additional 6–12 trips for wood and bark fragments that she places on the sticky surface of the closure. Probably because *P. cuspidatus* use resin for nesting, their nests tend to be found around pines rather than hardwoods; even when nesting near hardwoods they go elsewhere to collect pine resin.

COMPARATIVE SURVEY OF CAVITY-NESTING SOLITARY WASPS

The literature on species that nest in preexisting cavities is extensive, because of the invention of the trap-nest method. Trap nests are fairly inexpensive to construct in large numbers and can be made of readily available materials; besides drilled wood blocks, one can also use bundles of paper straws or hollow reeds. K. Krombein provided (1967) a huge compendium of information on trap-nesting wasps and bees based on his studies, mostly in the eastern United States. Other important studies of communities of cavity-nesters include those of R. Fye (1965) and H. Danks (1970). These studies (and many others) reveal that many Sphecidae and Eumeninae, as well as species in several genera of Pompilidae and Masarinae, nest in cavities (Table 6-1). The natural cavities used by solitary wasps for nesting include not only abandoned emergence tunnels of wood-boring insects but also the abandoned mud nests and old ground burrows of wasps and bees, as well as empty snail shells and plant galls, rolled leaves, tiger beetle burrows, and gaps between stones. *Trypoxylon* have been recorded nesting in keyholes, folded newspapers, an old hummingbird nest, and a pipe stem (Bohart and Menke 1976).

Although cavity-nesters take over ready-made holes, they differ in several ways from parasitoids that deposit hosts in preexisting niches (e.g., ampulicine Sphecidae). First, cavity-nesters find and prepare the niche before foraging. Second, nonpompilid cavity-nesters place multiple prey in each cell. Third, cavity-nesters substantially modify the cavity with the addition of partitions and plugs using a variety of materials (Krombein 1967, Iwata 1976). The cavity-nesting eumenines, as well as many sphecids (e.g., *Solierella*, and *Trypoxylon*), make the barriers of mud or agglutinated sand. Pemphredonines construct theirs of such stuff as resin and masticated plant materials, including wood fibers. *Dipogon* may use a composite variously made of mud, plant materials, insect parts, spider silk, and caterpillar frass, but the most visually striking plugs are those of *Isodontia* species, which stuff large wads of grass stems in the nest hole. Other modifications may also be made to nests. *Tricarinodynerus guernii*, for example, adds mud turrets to the nest entrance (Weaving 1994), and both *Celonites wahlenbergii* (S. Gess 1996) and *Auplopus caerulescens* (Krombein 1967) build complete mud cells within preexisting cavities.

The nests of many species, such as *Passaloecus cuspidatus*, consist of a series of provisioned cells interrupted only by partitions and finished with one or more empty vestibular cells (Fig. 6-4B). Other species intersperse provisioned cells with empty cells called intercalary cells, adding one or more empty vestibular cells at the end (Fig. 6-4C). The number of cells in these species' nests can be quite variable because it depends on the length of the burrow (which in the trap-nesting studies is determined by the investigator). Ten or more cells per nest are not uncommon (Iwata 1976), although not all species build discrete cells within nests. In Krombein's (1967) study, two species, *Solierella affinis* and *Tracheliodes amu*, did not divide the nest into discrete brood cells, and several *Isodontia* species made a single large brood cell shared by all of the larvae in the nest.

Mud-Nesting Wasps

Case Study: *Sceliphron caementarium*

Sceliphron caementarium (Sphecidae), the black-and-yellow mud-dauber, is familiar throughout North America because it builds nests on human-made structures. The *S. caementarium* female is a slender wasp, about 2.5 cm long, with a mostly black body, bright yellow legs, and what seems like an impossibly thin petiole connecting the posterior half of the abdomen to the rest of the body. The narrative of the nesting biology of *S. caementarium* offered here is a composite of accounts given by G. Peckham and E. Peckham (1898), P. Rau and N. Rau (1918), M. Muma and W. Jeffers (1945), G. Shafer (1949), and C. Ferguson and J. Hunt (1989).

Sceliphron caementarium nests are constructed in sheltered places, including human habitations (under the eaves of roofs or in barns); their natural habitat is protected places on cliff faces, on vines, or in the hollows of trees. Shelter is required because nests struck by rain tend to "fall to pieces" (Peckham and Peckham 1898).

A nest built by a female *S. caementarium* over the course of a season may include up to 25 cells in a cluster, each made entirely of mud (Fig. 6-5A). The process of building the first cell begins when a female flies to a source of mud and gathers up a globule of mud with her mandibles. When she returns to her nest site, she places the mud pellet firmly on the substrate and spreads it out into a crescent shape that tapers to a point. The next pellet is placed opposite the first and is drawn toward it until they meet to form a semicircle or arch. As subsequent loads are added serially, the arches formed by each pair eventually become oriented perpendicular to the substrate. From 30 to 40 loads are required to complete a cell, but a cell can be completed in less than one hour. A completed cell is somewhat longer than the female and more or less cylindrical, with its long axis

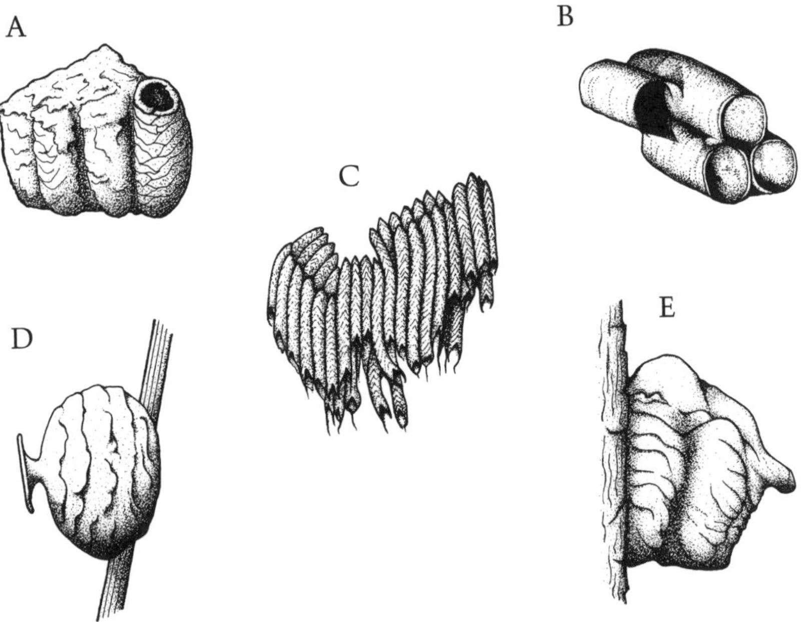

Figure 6-5. Mud nests (redrawn from the cited sources). (A) *Sceliphron caementrarium* (Shafer 1949). (B) *Pseudomasaris edwardsii* (Torchio 1970). (C) *Trypoxylon politum* (Brockmann 1988). (D) *Eumenes colona* (Iwata 1976). (E) *Auplopus nyemitawa* (F. Williams 1919c).

usually oriented vertically. After the first cell has been completed, most subsequent cells are adjoined to it in the same orientation.

Usually, 6–15 prey spiders are placed in each cell, although Peckham and Peckham (1898) cited instances of up to 40 spiders in each cell. If a female does not finish stocking before evening, she adds a temporary closure, a thin curtain of mud to block parasites. After the final prey is placed in the cell, the female lays an egg on one of the last prey and then permanently closes the cell with a thick plug of mud. The outside of a cell cluster is eventually plastered with more mud, so that contours of individual mud loads are obscured.

COMPARATIVE SURVEY OF MUD-NESTING SOLITARY WASPS

The building of free mud nests probably evolved independently in at least five different lineages of solitary aculeate wasps (Table 6-1): the subtribe Sceliphrina of the Sphecinae, the tribe Trypoxylonini of the Larrinae, the tribe Ageniellini of the Pepsinae, the Eumeninae, and the Masarinae. Future phylogenetic analyses may lead to the conclusion of further independent origins of mud nesting, particularly in the Vespidae. For example, free mud nests occur in three relatively unrelated genera of masarines, although most masarines build mud cells within ground nests (S. Gess 1996). In addition, some relatives of wasps that build free mud nests use mud for cells, partitions, plugs, and turrets as parts of other kinds of nests. The cavity-nesting *Trypoxylon* and Eumeninae build mud partitions, and most ground-nesting Masarinae build mud cells and mud turrets. A full analysis of the evolution of mud-nesting will thus require consideration of the various combinations of mud cells, partitions, and turrets within excavated burrows, preexisting cavities, and free-standing nests.

The independent evolution of mud nesting in different taxa is reflected in the various forms taken by mud nests, which may be made up of both mud cells and a separately applied covering over a group of a dozen or more individual cells (Iwata 1976). Cells are often cylindrical, as in the case of *Sceliphron, Pseudomasaris edwardsii* (Fig. 6-5B), and *Auplopus nyemitawa*, but groups of cells many take on a more irregular appearance, particularly when a separate covering is added to the cluster (Fig. 6-5E). The "pipe-organ" mud-daubers of the genus *Trypoxylon* build a linear sequence of cells within a tube of mud made of a series of "interlocking half-arches" of mud; the entrances of cells face downward (Brockmann 1988) (Fig. 6-5C). Cells may also be more or less spherical in many Pompilidae (F. Williams

1919c) and Eumeninae. Members of the genus *Eumenes*, the so-called potter wasps, build jug- or pot-shaped cells, complete with a funnel-shaped "spout" (Fig. 6-5D; see also F. Williams 1919c and Olberg 1959). *Eumenes* pots are attached to plant stems or are plastered to tree trunks and have openings smaller than the diameter of the wasps. Once the pot has been completed and is ready to provision, the female first inserts the tip of her abdomen into the nest, where she lays an egg in the empty cell. Then, after finding caterpillar prey, she has to stuff them through the small aperture, which is eventually plugged with mud (sometimes after the funnel has been chewed off) (F. Williams 1919c).

Plant materials are sometimes incorporated into mud nests, as in the use of a layer of lichen on nests of *Auplopus* (F. Williams 1919c). Moreover, some Eumeninae (in the genera *Calligaster* and *Zethus*) build their free-standing nests completely out of pieces of leaf, macerated leaf, or bits of moss. *Calligaster cyanoptera*, a communal species, builds its nest solely of leaf fragments excised from large leaves (F. Williams 1919c).

Nest-site selection among mud-nesting wasps may be driven by several factors, including the need for a water or mud source nearby. *Trypoxylon* and *Sceliphron* require a source of mud, whereas eumenines and pompilid mud-nesters first collect water and then make their own mud. In addition, because mud nests are susceptible to damage from rain, many species build nests beneath protective overhangs or under leaves. *Trypoxylon politum* choose smooth vertical surfaces, near water (for mud) and woods (for prey), and protected from rain (Cross et al. 1975). *Sceliphron laetum* nest under the shelter of rock overhangs or in hollow logs, but always near a mud source (A. Smith 1975). *Trigonopsis cameronii* place the nests on rocks at the edges or in the middle of streams (Eberhard 1974). In the Philippines, Williams (1919c) found a variety of mud-nesting Sphecidae, Pompilidae, and Eumeninae building nests on the underside of leaves, within hollows of bamboo stalks and tree trunks, or within the silken retreats of jumping spiders (their prey). Although *Auplopus nyemitawa* build in the open, they finish nests (Fig. 6-5E) by covering them first with a "varnish" of waterproof "plant gum" and finally a layer of masticated lichens (Williams 1919c). Similarly, *Xenorhynchium nitidulum* coat their barrel-shaped mud cells with gummy substances that they collect from fig and acacia trees and that strengthen the fragile mud cells (West-Eberhard 1987).

Nest Dispersion

The nests of most Pompilidae and many Sphecidae and Vespidae are widely dispersed, probably because their prey and potential nest sites are not concentrated in any one area. Detailed information on these species is often obtained only after years of chance observations by a number of biologists or by a single hard-working individual with the foresight to accumulate records slowly. Fortunately for those of us with less patience, some solitary wasps nest in aggregations, making it easier to find and observe a large number of females. Some examples of extremely high nest densities within aggregations are truly impressive. H. Evans (1966c) found one aggregation of *Alysson melleus* that included about 300 nests in a 0.75 × 1.5 m patch of soil (>250 wasps/m^2) and another of about 200 female *Bembix cinerea* whose nests were just 1–3 cm apart at the center of the aggregation (Evans 1957a). Mud-nesters, such as *Trypoxylon* and *Sceliphron*, also build their nests in tightly bunched clusters (Cross et al. 1975, Brockmann 1988, Hunt 1993).

The strong tendency for ground-nesters to nest near one another can be seen in some species by tracking the ontogeny of nest aggregations. Although nesting females sometimes maintain some minimum distance to their nearest neighbors (Rubink 1982), nest aggregations often maintain a tight cohesion that cannot be explained solely with reference to habitat characteristics that limit females to nesting in a small portion of the environment. During a season, nesting areas do not necessarily fill an area like stars gradually appearing on a clear night. Rather, they often grow from a central core, with new nests being initiated near those already present, as for example in *Bembix americana, Stictia carolina* (Evans 1957b, 1966c), and *Philanthus bicinctus* (Gwynne 1980). The entire aggregation of *Bembecinus quinquespinosus* nests shifts to nearby spots within available nesting habitat (Evans et al. 1986) (Fig. 6-6).

At some sites, different species nest together in more or less defined aggregations (Evans 1957b, 1966c; Evans and O'Neill 1988). One well-studied example comes from Evans's (1970a) work at one site in Grand Teton National Park in Wyoming. Here, *Bembix americana* and *Oxybelus uniglumis* nests intermingled in sand along the Snake River, whereas *Ammophila azteca* and *Stenodynerus papagorum* nests were in compact soil along a nearby dirt road and *Steniolia obliqua* and *Diodontus argentina* nests occupied partially shaded,

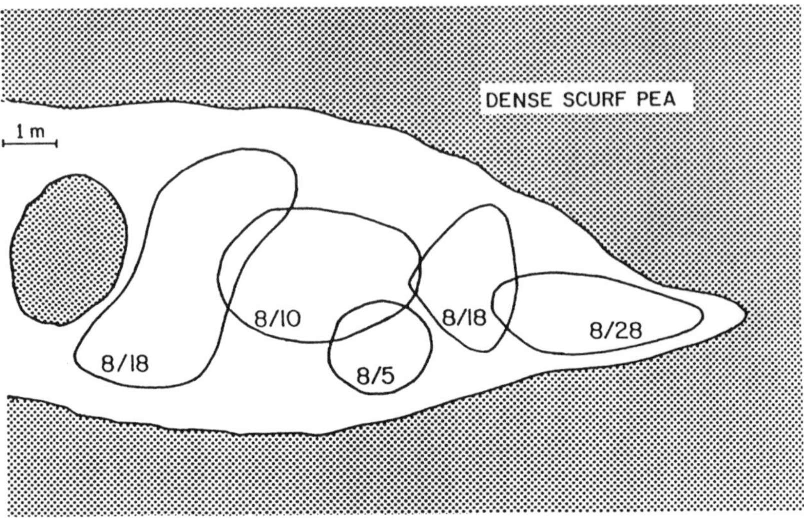

Figure 6-6. Changes over a 24-day period in the distribution of nests within a nesting aggregation of *Bembecinus quinquespinosus*. Circled areas indicate portions of an area of bare sand occupied by active nests on dates indicated (from Evans et al. 1986).

vegetated soils. Among the other ground-nesters at the site were *Episyron quinquenotatus, Hoplisoides spilographus, Nitelopterus evansi,* and *Podalonia communis* (Fig. 6-7).

Not surprisingly, because of strict nest-site preferences, aggregations often occur in the same place from year to year if not disturbed by human activities. For example, an aggregation of *Philanthus bicinctus* first found near the south gate of Yellowstone National Park in 1957 by K. Armitage (1965) was still active in 1987 (Evans and O'Neill 1988). Many other examples of short-term stability can be found in the wasp literature (e.g. Evans 1957b, 1966c; Evans and O'Neill 1988; Brockmann 1985a). Localized nesting aggregations of Bembicinae may also persist for years, often moving "more or less *in toto*" if "there is a change in physical conditions" (Evans 1966c). One dramatic example of the mass movement of aggregations was observed in *Bembecinus quinquespinosus*, which also provides a rare example of the males' activities affecting female nest distribution (Evans et al. 1986). Males of this species search for newly emerged virgin females in the previous year's nesting area (Chapter 8). Each

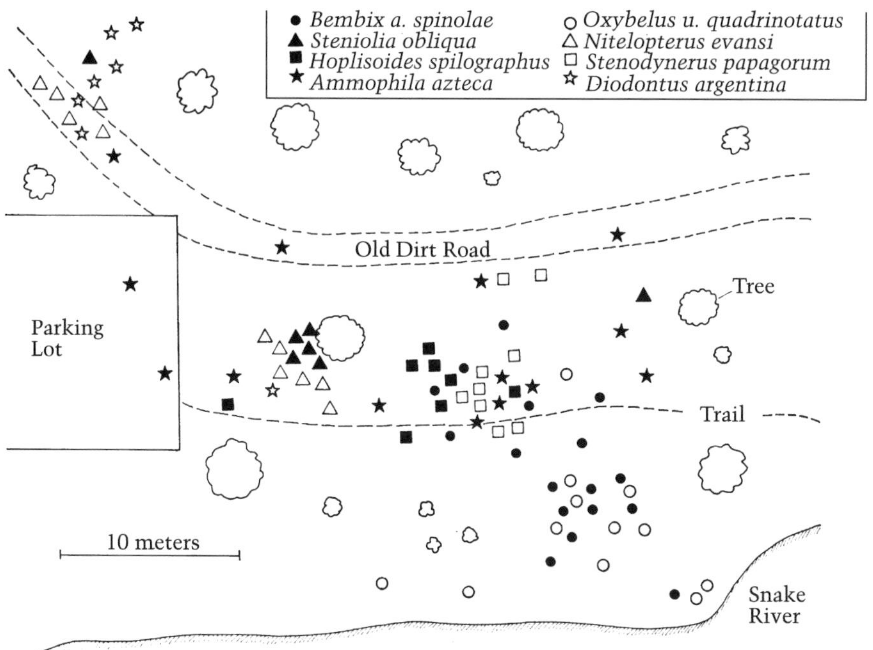

Figure 6-7. Map of an area in Jackson Hole, Wyoming (July 1964), showing location of nests of eight species of ground-nesting wasps (from Evans 1970a).

year over a 4-year study, females that attempted to nest in the emergence area were so harassed by libidinous males that they eventually gave up and joined aggregations elsewhere, up to 150m from their emergence site. Interestingly, most females move to the same new aggregation, with the result that the density of emerging adults in the new location is again high in the following summer (necessitating yet another move).

Nest Usurpation and Supersedure

In Chapter 4, I noted that females of many species steal prey from one another, resulting in a reduction in foraging costs to themselves. Nests can also be a valuable resource worth usurping if there is a shortage of potential nest sites or if part of the time and energy cost of building a nest can be avoided by stealing a nest (Evans 1966c, Field 1992b).

Case Study: *Sphex ichneumoneus*

Female *Sphex ichneumoneus* spend about 100 minutes building a nest with one cell, which they provision with 1–6 katydids over a period of 1–10 days before laying an egg on the last prey item and closing the cell (Brockmann 1985a). During 6 years of study at three sites, H. Brockmann found that a second female entered 5% to 15% of the established nests and began provisioning the same cell. Nest sharing continued until the two females met by chance and fought until the loser was ousted; the victor then finished provisioning and laid an egg. Occasionally, nest-sharing wasps did not meet, so that the first female that made the decision to lay an egg and close the cell was the winner.

Brockmann and colleagues (1979) referred to this as a "winner take all game" and analyzed it using the conceptual tools of evolutionary game theory. The game for the wasps is played basically as follows. A female can make the decision to (1) "found" her own nest, thereby running a small risk of being joined and having her nest usurped or (2) "join" an established nest, thereby forgoing the cost of digging and some of the cost of foraging (but also running a risk of being ejected by the founder after having contributed prey to the cell). If we know the costs and benefits associated with each tactic, we should be able to answer the following question: What frequencies of founding and joining behaviors should we expect to observe in a population? Brockmann and her colleagues argued that the two strategies will coexist in some equilibrium ratio if, on average, the individuals joining and founding have equal expected reproductive success. In principle, the equilibrium frequency of the two behaviors can be reached either with the appropriate mix of obligate founders and joiners or with a mixed strategy in which each individual founds or joins with a specific probability. The existence of a mix of obligate founders and joiners is ruled out by the fact that some females of *S. ichneumoneus* are known to found and join on different occasions.

Obviously, if most females in a population attempted to join, very few nests would be available; joiners would experience a prohibitively high cost of fighting and searching for established nests. On the other hand, if joiners were rare, they would experience a higher than average success in an environment with many established nests. Thus, some intermediate frequency of founding and joining (with founders in the majority) was expected. Brockmann and her colleagues (1979) found that the wasps undertook a mixed strategy, with founders and joiners in one population having equal fitness at the observed rates of joining. Although this balance was found in only one of the two populations examined, the study represents one of the classic examples of the usefulness of game theory models in ethology.

COMPARATIVE SURVEY OF NEST USURPATION AND SUPERSEDURE

Unfortunately, game theory models have not been tested using other solitary wasps because of the daunting task of collecting the requisite data to calculate relative fitnesses. We do know, however, that nest usurpation is common in some groups of solitary wasps (Field 1992b). True usurpation can be distinguished from other situations in which one wasp uses the nest of another female. The first possibility is noncompetitive supersedure, when a cavity-nesting wasp takes over the space remaining in a nest after the original occupant dies or leaves (Krombein 1967). The second possibility occurs when one wasp takes over a mud nest after the brood of the female that built the nest have emerged. For example, *Pachodynerus nasidens* often nests in old mud nests of *Eumenes colona* or *Sceliphron assimile* (Freeman and Jayasingh 1975).

Intraspecific aggressive nest usurpation is common in certain genera of the sphecid subfamilies Sphecinae *(Ammophila, Sphex)*, Larrinae *(Trypoxylon)*, Crabroninae *(Crabro, Lindenius)*, Bembicinae *(Sphecius)*, and Philanthinae *(Cerceris)* (see Field 1992b for citations). Many of these species build deep burrows off of which many cells can be built, so the benefits of usurping (and the costs of being expelled) are potentially high. Even among cavity-nesting species, the cost to the wasp of being usurped can be very high when the usurper takes over and discards the provisions and offspring in cells completed by the previous owner (Field 1992b). Rates of nest usurpation can rival the proportion of broods lost to parasitoids and cleptoparasites (Chapter 7): up to 15% of nests of *Sphex ichneumoneus* (Brockmann et al. 1979), 17% of *Cerceris arenaria* nests (Field 1992b), and 36% of *Zethus miniatus* nests (West-Eberhard 1987) are lost to usurpers of the same species. Wasps may also be victims of usurpation by other species. For example, bees of the genus *Ceratina* sometimes take over *Pemphredon lethifer* nests (McIntosh 1996), and megachilid bees and *Ammophila* usurp nests of *Paravespa mimus* (F. Gess and Gess 1988a).

Because of prey theft and nest usurpation, fighting among female wasps in nest aggregations is common in some species, including

Bembix species and *Microbembex monodonta* (Evans 1966c), *Anoplius viaticus* (Field 1992c), *Cerceris fumipennis* (Mueller et al. 1992), *Pemphredon lethifer* (McIntosh 1996), *Podalonia valida* (A. Steiner 1975, O'Neill and Evans 1999), and *Sphecius speciosus* (Pfennig and Reeve 1989, 1993). Studies of fighting *S. speciosus* have added a new twist to interpretations of interactions among nesting wasps: differential treatment of intruders based on kin relations. For example, female *S. speciosus* are less likely to engage an intruder in a violent fight if the individual is a neighbor that nests within 1 m of her own nest. Perhaps neighbors are more tolerated because they can accidentally enter each others' burrow and it does not pay to escalate aggression when the intruder has come in by mistake. Nonneighbors, on the other hand, are unlikely to accidentally enter a nest distant from their own unless they are attempting to usurp it. D. Pfennig and H. Reeve (1993) also found that neighbors were more closely related genetically than nonneighbors, leading them to hypothesis that tolerance of neighbors (i.e., closer relatives) is a form of kin-selected nepotism. Tolerance of other females, especially sisters, was probably an important preadaptation for the evolution of communal nesting and eusociality in the Hymenoptera (Pfennig and Reeve 1993).

Conclusion

Nest building sets the solitary wasps (and bees) apart from most other nonsocial insects. They show themselves to be sophisticated engineers adept at site selection, mining, transportation and preparation of building materials, and construction of structures to tight specifications. Their skills allow them to construct (1) nests stable enough to remain intact during the adult's tenure at the nest and (2) cells that protect the vulnerable eggs, larvae, and provisions over many months. These are not trivial feats, considering, for instance, that ground nests may be excavated in hard-packed soils that would be challenging for a human to dig or in unstable sands where maintaining an intact burrow and entrance is difficult. Furthermore, nests must not only withstand physical rigors but must also protect the vulnerable young from all sorts of parasites, parasitoids, predators, and cleptoparasites. These natural enemies and the role of nests in providing defenses are subjects of the next chapter.

7/ Natural Enemies and Defensive Strategies

Few groups of terrestrial insects and spiders are immune to attack by solitary wasps. Even prey with formidable defenses are vulnerable: honey bees are hunted by *Philanthus*, wolf spiders by pompilids, and larval tiger beetles by *Methoca*. Nevertheless, despite their foraging prowess, solitary wasps receive similar treatment from their own set of natural enemies. A great deal of our knowledge of these enemies comes from examining nests, where various arthropods feed on wasp larvae or on the rich store of nest provisions. When nests occur in high densities at the same location each year, nesting aggregations provide easily found resources for each generation of natural enemy. Thus, many arthropods have evolved as specialists on the nests of solitary wasps.

A Survey of the Natural Enemies of Solitary Wasps

Information on the natural enemies of wasps is ample but is sometimes tricky to interpret because of the complexities of the trophic relationships centered around nests. Often, there is direct evidence that a species commonly feeds on a wasp or its nest provisions. In other cases, the pupa or emerging adult of a natural enemy is found in nest cells, suggesting that it completed development at the expense of the wasp. Nevertheless, when only a fully developed "natural enemy" is found in a nest, we may have no direct evidence that it fed on the wasp species of interest. There are several alternative scenarios, including the possibilities that the so-called enemy is (1) a harmless scavenger that fed on refuse within a nest (Vesey-Fitzgerald 1940, Evans 1966c, Evans and Matthews 1973c, Evans and West-Eberhard 1970, Disney 1994), (2) a parasitoid (i.e., a hyperparasite) of the wasp's real enemy, (3) an inquiline that took over the nest after the successful emergence of the original resident's brood, or (4) a natural enemy of the inquiline. Despite these possibilities, however, the relationships between solitary wasps and their natural enemies can often be sorted out with careful observations and a general knowledge of host relationships.

CLEPTOPARASITES: BROOD PARASITES

As noted in Chapter 4, *cleptoparasitism* embraces two quite different forms of attack on a host. Some cleptoparasites steal prey, whereas others (brood parasites) insert their young into a host's nest, where they consume most or all of the provisions. Many of the important obligate brood parasites of the nesting solitary wasps are the Chrysididae, Pompilidae, Sphecidae, and Sapygidae discussed in Chapter 4. Here, I will discuss the nonaculeate brood parasites and prey thieves.

Flies

Three families of true flies (Diptera) contain brood parasites of solitary wasps (Table 7-1). *Lepidophora lepidocera* (Bombyliidae) and *Megaselia aletiae* (Phoridae) are brood parasites in nests of cavity-nesting species, but their closest relatives occupy a diversity of other trophic niches. In the Sarcophagidae, the tribe Miltogrammini includes a number of destructive brood parasites whose maggots either kill the host egg directly or feed on provisions so fast that the host larva starves to death (Myers 1927, Reinhard 1929, Endo 1980, Spofford et al. 1986). The maggots of some species may devour all of the prey in one cell, then move to other cells in the same nest (Krombein 1967).

Miltogrammines may have quite broad host ranges. *Senotainia trilineata*, for example, attacks at least 44 species in 27 genera of Sphecidae, Pompilidae, and Eumeninae (Vespidae), thereby feeding on arthropods as diverse as aphids, bees, beetles, caterpillars, flies, katydids, and spiders. Similarly, *Amobia floridensis* has been found in the nests of 7 sphecid and 5 eumenine species, and *Metopia*

Table 7-1 Selected nonaculeate cleptoparasites of nest-provisioning solitary wasps

Natural enemy family	Selected natural enemy genera	Wasp host family[1]	Type of feeding[2]	Selected references
Acarina				
Acaridae	*Lackerbaueria, Pymotes*	S	BP	Krombein 1967, Hook 1987
Coleoptera				
Cicindelidae	*Cicindela*	P	PT	Alm & Kurczewski 1984
Dermestidae	*Trogoderma*	S	BP	Molumby 1995
Meloidae	*Ceroctis*	VE, VM	BP[3]	S. Gess 1996
Diptera				
Bombyliidae	*Lepidophora*	S	BP	Hull 1973, Garcia & Adis 1993
Calliphoridae	Not known	VE	IC	Freeman & Taffe 1974
Phoridae	*Megaselia, Phalocrotophora*	S, VE	BP	Krombein 1967, Coville & Griswold 1984, Disney 1994
Sarcophagidae	*Amobia, Hilarella, Metopia, Phrosinella, Protomiltogramma, Ptychoneura, Senotainia, Sphenometopa, Taxigramma*	S, P, VE	BP	Linsley & MacSwain 1956; Ristich 1956; R. F. Chapman 1959; Evans 1966c, 1987; Itino 1988; Spofford et al. 1986; Wcislo 1986; Evans & Hook 1986a
Sarcophagidae	*Blaesoxipha, Senotainia*	S	IC	Peckham 1991
Tachinidae	*Exorista, Leskellia, Nemorilla, Stomatomyia*	VE	IC	Krombein 1967, O'Neill & Evans 1999

Table 7-1 *Continued*

Natural enemy family	Selected natural enemy genera	Wasp host family[1]	Type of feeding[2]	Selected references
Hymenoptera				
Braconidae	*Macrocentrus, Meteorus*	S	IC	Krombein 1967, Bohart & Menke 1976
Bethylidae	*Goniozus*	VE	IC	Krombein 1967
Formicidae	*Crematogaster, Pheidole, Solenopsis*	S	PT	Eberhard 1974, Peckham & Hook 1980, F. Gess & Gess 1982, Brockman 1985a, Peckham 1985, Rosenheim 1990a
Gasteruptiidae	*Carinafoenus*	VM	BP	Houston 1984
Ichneumonidae	*Bathyplectes, Campoplex, Glypta, Netelia, Oxyrrhexis*	S, VE	IC	Hicks 1933a, Endo 1981, Fye 1965, Itino 1986
Ichneumonidae	*Messatoporus*	S	BP	Krombein 1967, Coville & Coville 1980

Note: Host records for hymenopterans can also be found in Krombein et al. 1979a,b.

[1]P, Pompilidae; S, Sphecidae; VE, Vespidae (Enmeninae); VM, Vespidae (Masarinae). [2]BP, brood parasite; IC, incidental cleptoparasite; PT, prey thief. [3]May also kill and feed on larva.

argyrocephala attacks at least 18 sphecid species (Krombein et al. 1979b).

Miltogrammines differ in the kinds of nests attacked, the manner in which they find hosts, and the way in which they insert larvae into host cells; miltogrammines produce young by a process termed *larviposition*, so called because eggs hatch into maggots as soon as they are deposited. *Amobia* and *Ptychoneura* attack mud- and hole-nesting wasps. *Amobia* trail the host female back to her nest, wait nearby, and enter to larviposit only when the wasp leaves to forage (Krombein 1967, Itino 1986). *Ptychoneura* larviposit on the adult host female as she enters the burrow, where the larvae then drop off (Day and Smith 1981). *Metopia, Hilarella, Phrosinella,* and *Senotainia* attack ground-nesting wasps. After being attracted visually to an open burrow or after following a wasp to its nest, *Metopia* larviposit within a cell or leave larvae in the burrow, where the larvae find their own way to a cell (Linsley and MacSwain 1956, Endo 1980, Wcislo 1986). *Phrosinella* attack only closed nests; they dig through the closure with expansions of their foretarsi and larviposit in the burrow (Evans 1966c, Spofford et al. 1986). Species of *Senotainia* vary in their method of attack (Endo 1980; Hager and Kurczewski 1985, 1986; McCorquodale 1986; Spofford et al. 1986). Female *S. vigilans* and *S. sauteri* larviposit within the nest burrow, but *S. sauteri* females may also larviposit on unattended prey at the nest entrance. Female *S. trilineata* (Fig. 7-1), known as satellite flies, sit in a nesting area and wait for a wasp carrying prey to pass by; upon spotting a prey-carrying wasp, they follow the wasp "as closely in her wake as if they were tied to her by invisible traces" (Reinhard 1929). At some point during their pursuit, they larviposit on the prey while the wasp is still carrying it (Reinhard 1929, Spofford et al. 1986).

Mites

Several mites are brood parasites of wasps (Krombein 1967, Hook 1987). Immature stages of *Lackerbaueria krombeini* (Acaridae) find nests of *Diodontus* by hitching rides on the adult wasps and then feeding on the paralyzed aphids after killing the wasp's egg. Infestations of the mite may persist in nests reused by successive generations of wasps. The grain itch mite, *Pymotes ventricosus* (Pyemotidae), is an opportunistic brood parasite in nests with prey, but it also feeds directly on the eggs, larvae, or pupae of wasps (and other insects).

Figure 7-1. Miltogrammine female perched in a nesting area (photo by H. E. Evans).

CLEPTOPARASITES: INCIDENTAL BROOD PARASITES

Prey brought into a nest by a wasp may already harbor eggs or larvae of parasitoids. When these prey are placed in a cell, the larvae of the parasitoid may have a developmental head start on the feeding wasp larva and act effectively as brood parasites. Such enemies are referred to as incidental cleptoparasites, because their foraging strategies have not evolved specifically to exploit the wasp. Incidental cleptoparasitism is probably uncommon, but reports occasionally appear (Table 7-1). T. Itino (1986) reported that 4% of *Anterhynchium flavomarginatum* larvae starved when they were outcompeted by faster developing *Campoplex* larvae present in caterpillar prey. *Oxybelus sparideus* courts disaster by provisioning its nests with sarcophagids, including *Senotainia trilineata* (R. Bohart et al. 1966, Peckham 1991).

The fly larvae sometimes emerge from their paralyzed mothers and destroy eggs and nest provisions within *O. sparideus* nest cells.

CLEPTOPARASITES: PREY THIEVES

Brood parasites steal both prey and nest cells, whereas other enemies (including conspecific females, Chapter 3), steal prey without living in the nest. Ants are ubiquitous marauders of wasp nests (Table 7-1) that also steal prey in transit to the nest (Newcomer 1930), but their actual impact is probably underestimated because they leave little evidence of their attacks and thus are not identified unless caught in the act (Cowan 1991). Provisions may also be stolen from nests by birds such as canyon wrens (*Catherpes mexicanus*) (Martin 1971), downy woodpeckers (*Picoides pubescens*) (K. Smith 1986), and Seychelles fodies (*Foudia sechellarum*) (Brooke 1981). House sparrows (*Passer domesticus*) commonly steal katydids directly from *Sphex ichneumoneus* females by provoking a female to drop her prey or by snatching prey when the wasp enters her nest (Brockmann 1980).

PARASITOIDS

Nonaculeate Wasps

Many of the most important parasitoids of the immature stages of solitary wasps are Chrysididae and Mutillidae (Chapter 2), but there are also important nonaculeate hymenopteran species in this category (Table 7-2). Space limitations prevent discussion of all of these parasitoids, but two families, Eulophidae and Ichneumonidae, warrant a closer look.

The larvae and prepupae of stem- and mud-nesting solitary wasps are among the numerous hosts of the genera *Tetrastichus* and *Melittobia* in the family Eulophidae (Clausen 1940, Krombein 1967). *Tetrastichus* are endoparasitoids (Krombein 1967), whereas the better-studied *Melittobia* are ectoparasitoids (Clausen 1940, Dahms 1984). Exploitation of a host by *Melittobia chalybii* begins when a single, fully winged adult female (Fig. 7-2B) enters an active nest (Buckell 1928, Schmieder 1933). To gain access to a host, *Melittobia* can burrow through the walls and partitions of closed mud nests (Dahms 1984, Trexler 1985, Molumby 1995) and then chew or oviposit through the silk coverings of cocoons (Dahms 1984). If the female finds a fully developed wasp larva, she lays a clutch of eggs. The

Table 7-2 Selected nonaculeate parasitoids of the immature stages of solitary wasps

Natural enemy family	Selected genera of enemy	Wasp host family[1]	Selected references
Coleoptera			
Rhipiphoridae	*Macrosiagon*	S, T, Sc, VE	Wolcott 1914, Barber 1915, Snelling 1963, F. Parker & Bohart 1968, Eberhard 1974, Evans 1957b, Freeman 1982, Itino 1986, Hunt 1993, Hook & Evans 1991
Diptera			
Bombyliidae	*Anthrax, Chrysanthrax, Exoprosopa, Ligyra, Toxophora*	C, T, S, P, VE	Clausen 1928; Marston 1964; Evans 1957b, 1966c; Krombein 1967; Hull 1973; Coville & Coville 1980; Evans & O'Neill 1988; Garcia & Adis 1993
Hymenoptera			
Braconidae	*Meteorus*	S	Matthews 1968
Chalcididae	*Brachymeria*	S	Freeman 1981a, Coville & Griswold 1984
Diapriidae	*Ismarus*	D	Chambers 1955, Jervis 1979
Encyrtidae	*Schedioides*	D	Triapitzin 1964
Eulophidae	*Melittobia, Pediobius, Tetrastichus*	B, C, M, S P, VE	Buckell 1928; Schmieder 1933; Krombein 1967; Medler 1967; Tachikawa & Yukinari 1974; Taffe 1978, 1979; Freeman 1981a, 1982;

			Dahms 1984; Kapadia & Mittal 1986; Freeman & Ittyeipe 1976, 1993; Garcia & Adis 1993; Hunt 1993; Molumby 1995
Eurytomidae	*Eurytoma*	S, VE	F. Parker & Bohart 1968, Danks 1970, Krombein et al. 1979a
Ichneumonidae	*Acroricnus, Ischnurgops, Neorhacodes, Osprynchotus Perithous, Pimpla, Photocryptus, Poemenia*	C S, VE	Roubaud 1911, Fye 1965, Krombein 1967, Danks 1970, Eberhard 1974, Itino 1986, F. Williams 1919c
Leucospididae	*Leucospis*	VE	F. Parker & Bohart 1968, Cowan 1986a
Megalyridae	*Megalyra*	S	Naumann 1987
Perilampidae	*Perilampus*	S	Krombein 1964, Garcia & Adis 1993
Platygastridae	*Tetrabaeus*	S	Krombein 1964
Pteromalidae	*Epistenia, Habritys, Habrocytus*	S, VE	Krombein 1958, 1960; F. Parker & Bohart 1966
Torymidae	*Diomorus, Microdontomerus, Monodontomerus*	S, VE	Bechtel & Schlinger 1957; Krombein 1964; F. Parker & Bohart 1966, 1968; Danks 1970
Trigonalidae	*Lycogaster*	VE	Cooper 1953, F. Parker & Bohart 1966

Note: Host records for hymenopterans can also be found in Krombein et al. 1979a,b.

[1]B, Bethylidae; C, Chrysididae; D, Dryinidae; M, Mutillidae; P, Pompilidae; S, Sphecidae; Sc, Scoliidae; T, Tiphiidae; VE, Vespidae (Enmeninae).

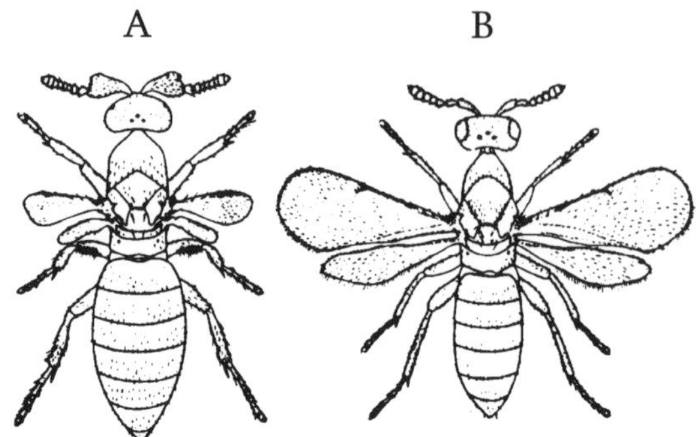

Figure 7-2. *Melittobia chalybii.* (A) Short-winged male. (B) Fully winged female. (A and B from Schmieder 1933.)

resulting *Melittobia* larvae develop into short-winged sons and daughters (Fig. 7-2A) that mate with one another within their natal nest. The daughters, along with their still-living mother, then lay large numbers of eggs that produce a generation of fully winged adults that leave to seek out new host nests. Adult *M. chalybii* females remain within a single nest cell, but *M. australica* females distribute their eggs among several cells when attacking smaller host species (Dahms 1984).

Melittobia exhibit relatively little host specificity. *Melittobia chalybii*, *M. acasta*, and *M. australica* attack a variety of solitary nest-provisioning wasps, solitary bees, and social bees and wasps (Buckell 1928, Clausen 1940, F. Parker and Bohart 1966, Krombein 1967, Freeman and Jayasingh 1975, Dahms 1984). They can also develop as hyperparasitoids within nests already parasitized by chrysidids or mutillids (Krombein 1967). Because of their wide host ranges, *Melittobia* are common pests in laboratory cultures of insects (Evans and West-Eberhard 1970).

The family Ichneumonidae also has some important parasitoids. Even a closed, hard-walled wasp nest can be vulnerable to ichneumonids capable of ovipositing through solid barriers (Askew 1971, Quicke 1997) (Table 7-2). For example, *Acroricnus ambulator* (Itino 1986) and *Photocryptus* sp. (Eberhard 1974) use their ovipositors to pierce crevices in mud nests. Similarly, *Perithous divinator* oviposit through the walls of plant stems housing *Pemphredon lethifer* nests (Danks 1970), and *Pimpla spatulata* bore through the wooden sides of trap nests to reach eumenine larvae (Krombein 1967).

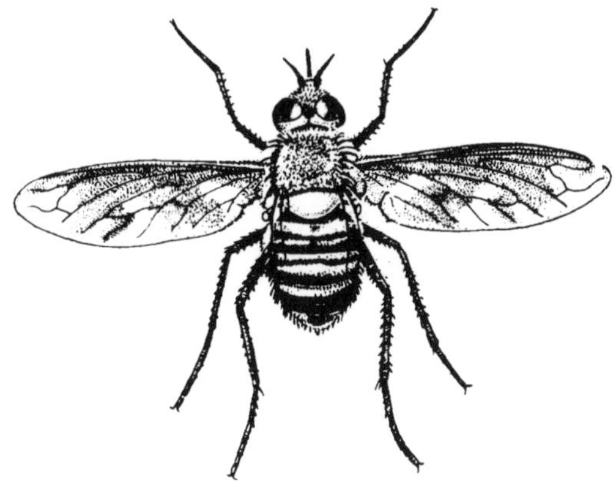

Figure 7-3. *Exoprosopa fascipennis*, a parasitoid of *Tiphia* (from Evans and West-Eberhard 1970).

Flies

Along with the brood-parasitic miltogrammines, several genera of Bombyliidae (bee flies) are perhaps the most common and destructive dipteran enemies in solitary wasp nests (Table 7-2). *Chrysanthrax*, *Ligyra*, and *Rhyanchanthrax* attack Tiphiidae and Scoliidae, whereas *Anthrax*, *Exoprosopa* (Fig. 7-3), and *Toxophora* are parasitoids of the prepupae and pupae of solitary nest-provisioning aculeates. Each species may have a wide host range (Hull 1973); *Anthrax distigmus*, for example, attacks sphecids, eumemines, and bees (Iwata 1933).

Depending on the type of wasps attacked, bombyliids oviposit in one of two ways. Parasitoids of Tiphiidae, such as *Ligyra oenomaus*, scatter large numbers of eggs on the soil surface so that their larvae must search for hosts beneath the soil surface at depths of 1–3 cm (Clausen 1928, 1940; Hull 1973). *Anthrax* and *Exoprosopa* target their eggs more precisely, hovering at the entrances of open nests and tossing eggs into the entrances with a flick of the abdomen. However, females are apparently unable to accurately distinguish active host nests from similar visual stimuli. They have been observed ovipositing into abandoned nests, into glass vials placed in a nesting area to collect eggs, into the eyelets of shoes, and onto dark spots on the substrate or on the wings of a resting moth (Frick 1962, Marston 1964, Evans 1966c). The slender, legless first-instar bee fly larva crawls down the nest burrow with the aid of long bristles on the thorax and abdomen (Hull 1973). The parasitoids of nest-provisioning species forgo feeding until the host pupates, then quickly consume it and pupate within the host's cocoon.

Most fly parasitoids of solitary wasps attack the immature stages, but flies of the genera *Physocephala* and *Zodion*, in the family Conopidae, attack adults of ground-nesting wasps (Evans 1966c, Freeman 1966, Evans and O'Neill 1988). After mounting an adult wasp (usually a male), a conopid female uses the hooked terminus of her abdomen to insert an egg between two abdominal segments. The fly completes development in the wasp's abdomen, killing the host (Evans and West-Eberhard 1970).

Beetles

Female *Macrosiagon* (Rhipiphoridae) scatter eggs on flowers, where their larvae await the appearance of female wasps so that they can obtain rides to nests. In North America, *Macrosiagon* have been found attacking hole-nesting (Krombein 1967), ground-nesting (Barber 1915), and mud-nesting wasps (Hunt 1993). Once inside nests, the beetles develop as parasitoids of larvae (Evans and West-Eberhard 1970). Because the larval wasp may live long enough to spin a cocoon, adult rhipiphorids may be found within the cocoons of the wasp (Barber 1915).

PREDATORS

Immatures within nests of solitary wasps may be fed upon by various predators that either invade the nest and remove larvae or prepupae (Crawford 1982, Coward and Matthews 1995) or feed upon them within the nest. Predators within nests include wasps of the family Gasteruptiidae (Clausen 1940, Askew 1971), beetles of the families Cleridae (Krombein 1967) and Dermestidae (Krombein 1967, Taffe 1979, Itino 1986), and ants (Freeman and Taffe 1974, Freeman and Jayasingh 1975, Endo 1981).

Some of the most important predators of adults are other solitary wasps, namely species of *Bembix*, *Palarus*, and *Philanthus* that specialize on hymenopteran prey (Evans and Matthews 1973c, Evans and O'Neill 1988, Gayubo et al. 1992). Some species of robber flies (Asilidae) that prey primarily on Hymenoptera also include solitary wasps in their diets, even wasps as small as dryinids (Scarborough 1981). At several locations in Montana, five families of solitary wasps made up 39% of the prey of *Megaphorus willistoni*, which exacted an especially heavy toll on *Tiphia nevadana* foraging at honeydew sites (O'Neill and Seibert 1996). Similarly, R. Lavigne and F. Holland (1969) found that 28% of *Diogmites angustipennis* prey were solitary wasps.

However, because robber fly diets are often time- and site-specific (O'Neill 1992a,b), a report that a robber fly takes a large percentage of wasps at one time and one place does not guarantee that it has a consistently high impact on populations.

Ant lions (Tsuneki 1956, Rubink and O'Neill 1980), tiger beetle larvae, and spiders (Obin 1982, Evans and O'Neill 1988) have also been reported as predators of adult solitary wasps, as have horned lizards (*Phrynosoma douglassi*) (O'Neill et al. 1989), toads, nightjars, bats (Bayliss and Brothers 1986), and roadrunners (Punzo 1994a). The most extensive study of vertebrate predators of adults is that of I. LaRivers (1945), who estimated that shrews, mice, and birds took about 10% of a population of *Palmodes laeviventris.*

PARASITES

Certain mites and members of the insect order Strepsiptera are classified as true parasites, rather than as parasitoids because, like mosquitoes, they suck blood without killing their hosts. Mites parasitic on wasps include several genera within the family Saproglyphidae: *Vespacarus*, *Kennethiella*, and *Monobiacarus* (Krombein 1961, 1967; S. Gess 1996). When a female eumenine leaves its natal nest, she may carry *Vespacarus* nymphs that remain on her until she lays an egg. The mites then attach themselves to the wasp larva, but forgo feeding until it spins a cocoon. *Kennethiella trisetosa* hitch rides only on host males, because female larvae eat their mites before spinning a cocoon. Therefore, to reach a new nest, the mites must transfer to a female wasp's genital chamber during copulation (Cooper 1955). The fact that wasps associated with saproglyphids possess special cavities (acarinaria) on their abdomens, whose only known function is to house mites (Cooper 1955, Krombein 1967, S. Gess 1996), suggests that the wasps and mites have some form of symbiotic relationship. As are other symbionts, saproglyphid mites are strongly host-specific, each species attacking a single species of eumenine (Krombein 1961, 1967).

Strepsipteran parasites of solitary wasps are members of two genera of the family Stylopidae: *Pseudoxenos* parasitize ground-nesting sphecids and twig-nesting eumenines (Bohart 1941, Krombein 1967), and *Paragiaxenos* attack masarines (Naumann and Cardale 1987, S. Gess 1996). Stylopids have bizarre life cycles (Bohart 1941, Krombein 1967, Evans and West-Eberhard 1970). The tiny, but active and long-legged, stylopid larva climbs onto a female wasp visiting a flower. Upon reaching a nest, the larva disembarks and burrows into a wasp egg or larva, where it develops slowly and pupates at about the same time as its host. Although stylopids are relatively rare, their presence is easy to diagnose, because the puparium of the stylopid protrudes between two abdominal segments of the adult host. Although strepsipterans do not necessarily kill their hosts, they typically cause pathological changes, including altered coloration, distorted body parts, smaller body size, and a lower propensity to nest and forage (Salt 1931, Bohart 1941). Thus, although parasitized wasps may survive their "infections," their reproductive success is severely curtailed.

Impact of Natural Enemies on Solitary Wasp Populations

One can estimate the impact of natural enemies on nesting wasp populations by examining a random sample of nest cells. This work is most easily done with species of mud-nesters, cavity-nesters that accept trap nests, and ground-nesters that build shallow, easily exhumed nests. If large numbers of nests can be examined, one can obtain precise information on mortality at each developmental stage and can attribute mortality to specific factors. Total developmental mortality in the nest for the 29 species cited in Table 7-3 ranged from 12% to 77%, with 0% to 42% of the mortality caused by natural enemies. The parasitoid wasp *Melittobia* was the primary enemy of over one-third of the species; the other major natural enemies were various bombyliid flies and assorted chrysidid, ichneumonid, and torymid wasps. Mortality *not* due to natural enemies was often ascribed to such factors as mold and failure at hatch, with a substantial residual of unknown causes.

A few cautionary notes are necessary regarding Table 7-3. First, all but two of the wasps studied are mud- or cavity-nesters. The major natural enemies of these species are Hymenoptera (*Brachymeria*, *Melittobia*, *Diomorus*, *Omalus*, *Perithous*, *Photocryptus*, *Poemenia*, and *Ptychoneura*) and Diptera (*Amobia* and *Megaselia*) that are not important enemies of ground-nesting species. Ground-nesters are instead often plagued by mutillid wasps and different genera of chrysidid wasps and miltogrammine and bombyliid flies (Evans 1966c, Evans and O'Neill 1988). Second, in the studies cited, the susceptibility of the nests could have been influenced by variation in the

Table 7-3 Developmental mortality caused by natural enemies of selected species of solitary wasps

Wasp species	Total no. of eggs or cells	Total developmental mortality (%)	Developmental mortality due to natural enemies (%)	Major natural enemies (% mortality)	Reference
Pompilidae					
Dipogon sayi	89	51	0	—	Fye 1965
Episyron arrogans[1]	113	54	39	*Metopia* (18), ants (16)	Endo 1981
Sphecidae					
Ammophila dysmica[1]	275	49	28	*Argochrysis* (23), *Hilarella* (3)	Rosenheim 1987b
Crossocerus capitosus	61	77	23	*Diomorus* (11)	Danks 1970
Passaloecus gracilis[2]	62	53	11	*Perithous* (11)	Danks 1970
P. ithacae	197	50	7	*Omalus* (4), *Poemenia* (3)	Fye 1965
Pemphredon lethifer[3]	1,048	52	20	*Perithous* (11), *Omalus* (7)	Danks 1970
P. schuckardi	74	49	23	*Omalus* (15), *Perithous* (8)	Danks 1970
Penepodium goryanum	111	53	20	*Melittobia* (15)	Garcia & Adis 1993
Psenulus schencki	163	47	4	*Perithous* (1)	Danks 1970
Rhopalum clavipes[2]	398	55	24	*Ptychoneura* (11), *Diomorus* (9)	Danks 1970

Table 7-3 *Continued*

Wasp species	Total no. of eggs or cells	Total developmental mortality (%)	Developmental mortality due to natural enemies (%)	Major natural enemies (% mortality)	Reference
R. coarctatum	471	59	14	*Diomorus* (6), *Ptychoneura* (6)	Danks 1970
Sceliphron asiaticum	2,278	43	39	*Melittobia* (33), *Amobia* (1)	Freeman 1982
S. assimile	2,290	40	31	*Melittobia* (30), ants (<1)	Freeman 1973
S. fistularium	1,966	36	29	*Melittobia* (20), *Amobia* (5)	Freeman 1982
S. laetum	997	38	17	*Melittobia* (8), chrysidids (8)	A. Smith 1975
Trypoxylon attenuatum	129	44	20	*Trichrysis* (13)	Asis et al. 1994
T. palliditarse	2,372	29	19	*Brachymeria* (7), *Photocryptus* (6)	Freeman 1981a
T. politum	2,466	49	36	*Melittobia* (17), *Trogoderma* (7)	Molumby 1995
T. politum	464	44	34	*Melittobia* (19), miltogrammines (10)	Cross et al. 1975
T. rogenhoferi	713	59	42	*Neochrysis* (8), *Lepidophora* (4)	Garcia & Adis 1995

Vespidae (Eumeninae)					
Ancistrocerus catskill	391	25	3	*Chrysis* (2)	Fye 1965
Anterhynchium flavomarginatum	525	53	42	*Amobia* (20), *Megaselia* (15)	Itino 1986
Eumenes colona	1,191	41	30	*Melittobia* (23), *Amobia* (6)	Freeman & Taffe 1974
Euodynerus leucomelas	122	65	6	*Chrysis* (5), *Amobia* (1)	Fye 1965
Orancistrocerus drewseni	395	67	25	*Amobia* (13), *Macrosiagon* (6)	Itino 1986
Pachodynerus nasidens	451	46	34	*Melittobia* (27), *Amobia* (3)	Freeman & Jayasingh 1975
Zeta abdominale	4,748	42	29	*Melittobia* (19), *Amobia* (7)	Taffe 1979
Z. argillacea	1,704	12	6	*Amobia* (3)	Rocha & Raw 1982
Z. canaliculatum	756	43	24	*Melittobia* (18), *Amobia* (3)	Taffe 1979

Note: See Tables 7-1, 7-2, and Appendix II for names of enemies. Species are hole-nesters or mud-nesters, unless otherwise indicated.

[1]Ground-nesting species. [2]Silwood Park site. [3]Silwood Park I site.

study methods used (e.g., use of natural versus artificial nests, selection of microhabitats for sampling, timing of collection, and rearing techniques). Third, there is potential geographic bias in the data because over two-thirds of the species were studied at just a few locations in Canada (by Fye), in England (by Danks), and in Jamaica and Trinidad (by Freeman and his colleagues). Nevertheless, the data in the table demonstrate that a substantial proportion of developmental mortality can be attributed to natural enemies.

Although most studies do not partition mortality among all of the different natural enemies of a wasp, they sometimes report mortality caused by single type of enemy (Table 7-4). Many of the values are low, suggesting that these particular natural enemies had little impact at the population level. However, several authors have suggested that parasitoids can cause major declines (Evans 1957b) or even local extinctions of wasp populations (Simonthomas and Simonthomas 1972). The best evidence that natural enemies regulate wasp populations comes from a few studies showing that mortality caused by *Melittobia* is density-dependent (reviewed in Rosenheim 1990b).

Although the impact of a natural enemy on a wasp population may be low, there may be great variation in its effect on individual nests. Many nests may escape attack, but a few may be hard hit, particularly those of species that place cells close together. The distribution of natural enemies within and between trap nests studied in Florida by Krombein (1967) will serve to illustrate. *Chrysis inaequidens* was reared from only 12% of 183 nests of *Euodynerus foraminatus*, but within affected nests, 25% of the available cells were attacked. Similarly, although *Amobia floridensis* was found in only 3% of the nests, it parasitized 51% of 37 cells in those nests. An even more skewed distribution was found in *Ancistrocerus antilope* nests, where miltogrammines infested 70% of the cells in the 6% of nests attacked (Krombein 1967). Such patterns of mortality indicate that selection pressures due to natural enemies are unevenly distributed across populations and, perhaps, that some females are better than others at defending against natural enemies.

Defensive Strategies: Personal Defense

Besides estimating rates of mortality, another way of inferring that natural enemies have been important in the evolution of solitary

Table 7-4 Percentage of total mortality caused by selected species of natural enemies of solitary wasps

Natural enemy	Host wasp species	Mortality (%)[1]	Reference
Coleoptera: Dermestidae			
Trogoderma ornatum	*Trypoxylon politum*[2]	7	Molumby 1995
Coleoptera: Rhipiphoridae			
Macrosiagon pusillum	*Tiphia pullivora*	28	Clausen 1940
Diptera: Bombyliidae			
Anthrax sp.	*Trypoxylon politum*[2]	3	Molumby 1995
Exoprosopa fascipennis	*Philanthus gibbosus*	>50	Evans & O'Neill 1988
Lepidophora vetusta	*Trypoxylon tenoctitlan*	6	Coville & Coville 1980
Ligyra oenomaus	*Tiphia* sp.	>60	Clausen 1928
Diptera: Phoridae			
Megaselia aletiae	*Ectemnius paucimaculatus*	32	Krombein 1967
Phalacrotophora punctiapex	*Trypoxylon superbum*	8	Coville & Griswold 1984
Diptera: Sarcophagidae			
Hilarella hilarella	*Ammophila dysmica*	3	Rosenheim 1987b
Metopia campestris	*Crabro cribrellifer*	25–42	Wcislo et al. 1985
Metopia sp.	*Ammophila gracilis*	≥11	Gaimari & Martins 1996
Metopia sp.	*Ammophila sabulosa*	5	Field 1992d
Miltogrammini	*Philanthus* (11 species)	8–41	Evans & O'Neill 1988
3 Miltogrammini	*Tachysphex terminatus*	31–58	Spofford et al. 1986
4 Miltogrammini	*Podalonia occidentalis*	75	Evans 1987
Hymenoptera: Chalcididae			
Brachymeria sp.	*Trypoxylon pallidítarse*	7	Freeman 1981a
Hymenoptera: Chrysididae			
Argochrysis armilla	*Ammophila dysmica*	23	Rosenheim 1987b

Table 7-4 *Continued*

Natural enemy	Host wasp species	Mortality (%)[1]	Reference
Caenochrysis mucronata	*Trypoxylon tridentatum*	32	Krombein 1967
C. nigropolita	*Trypoxylon tenocititlan*	3	Coville & Coville 1980
Chrysis sp.	*Trypoxylon palliditarse*	<1	Freeman 1981a
Chrysis sp.	*Trypoxylon rogenhoferi*	13	Garcia & Adis 1995
Hymenoptera: Eulophidae			
Pediobius imbreus	*Goniozus nephantidis*	32–90	Kapadia & Mittal 1986
Hymenoptera: Formicidae			
Various ant species	*Ammophila dysmica*	1	Rosenheim 1987b
	Anterhynchium flavomarginatum	1	Itino 1986
	Eumenes colona	1	Freeman & Taffe 1974
	Episyron arrogans	16	Endo 1981
	Sceliphron asiaticum	1	Freeman 1982
Hymenoptera: Ichneumonidae			
Acroricnus ambulator	*Orancistrocerus drewseni*	1	Itino 1986
Messatoporus sp.	*Trypoxylon tenoctitlan*	5	Coville & Coville 1980
Neorhacodes enslini	*Spilomena* spp.	13	Danks 1970
Photocryptus sp.	*Sceliphron asiaticum*	3	Freeman 1982
Strepsiptera: Stylopidae			
Paragiaxenos decipiens	*Paragia decipiens*	17	Naumann & Cardale 1987
Pseudoxenos hookeri	*Euodynerus foraminatus*	10	Krombein 1967
Pseudoxenos sp.	*Sphex nigellus*	25	Clausen 1940

Note: Values vary in whether they refer to single or multiple generations or single or multiple sites.
[1]Percentage of cells destroyed except for Strepsiptera, where data are the percentage of adults infected. [2]Includes data on brood parasitism but not on other forms of intraspecific parasitism.

wasps is to look for specific morphological and behavioral traits used to defend against enemies. Among the defenses exhibited by adult wasps, some are aimed at personal defense, whereas others are directed at protecting offspring and provisions inside nests. Personal defenses include both passive tactics, such as the use of armor and warning coloration, and active approaches, such as stinging.

All adult wasps have a more or less rigid exoskeleton, but mutillids and chrysidids whose hosts defend their nests with stings and mandibles are exceptionally well armored. Mutillid females are often so hard that it is difficult not only for other insects to bite them but also for entomologists to insert pins through their thoraces (Schmidt and Blum 1977, Deyrup 1988). J. Schmidt and M. Blum (1977) determined that the average force required to crush the thorax of a female *Dasymutilla occidentalis* (27.8 newtons to be exact) was 2.4 times that required to crush a *Scolia dubia* and 11 times that needed to demolish a honey bee. The Chrysididae (Chrysidinae) have a thick, smooth exoskeleton, along with a ventrally concave abdomen that allows them to curl into a tight ball when attacked by their host bees and wasps (Fig. 7-4). However, the amesigine and cleptine Chrysididae that attack defenseless hosts have a thin exoskeleton (Kimsey and Bohart 1990). Tarantula hawks (*Pepsis*) also resist being killed by their powerful prey by having thick, smooth armor (as well as by being quick) (F. Williams 1956, Punzo 1994a). Finally, although immature wasps are soft-bodied, the cocoons of some wasps are hardened by sand grains embedded in silk (Evans 1966c). B. Freeman (1981a) noted that female *Melittobia* have difficulty penetrating cocoons of *Trypoxylon palliditarse.*

Although most people shy away from wasps for fear of being stung, the stings of most solitary wasps cannot penetrate human skin and those that do generally cause little pain. Pain is a somewhat subjective experience, of course, but we can rate the immediate pain caused by stinging Hymenoptera on a scale of 0 to 4, where 0 refers to species whose stings cannot penetrate human skin and 4 to "the greatest-known insect sting pain"; values of 2 or more are considered to have "clear defensive value" (Schmidt 1990, Starr 1985). For example, a sting from the ant *Paraponera clavata* rates a lofty 4 on the immediate pain scale and leaves its victim with intense pain for 3–5 hours. Other social species, such as paper wasps (*Polistes*) and harvester ants (*Pogonomyrmex*), also inflict intense, relatively long-lasting pain.

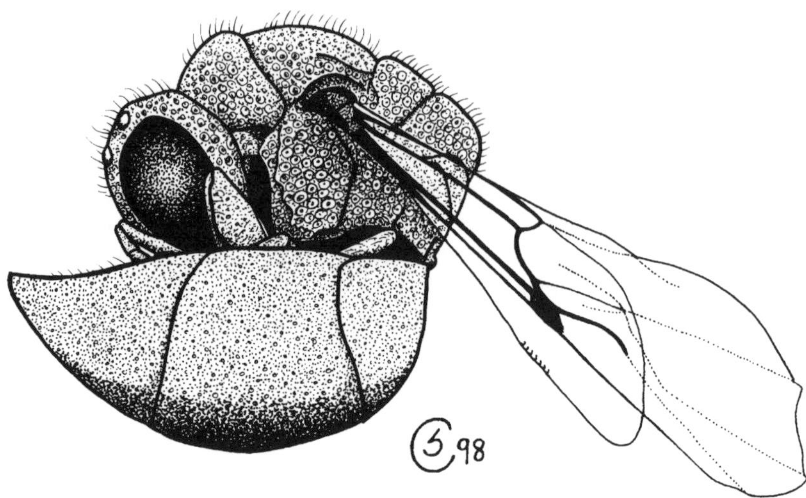

Figure 7-4. A chrysidid wasp in defensive posture (drawing by Catherine Seibert).

The stings and venoms of the social wasps, bees, and ants are usually not used to subdue prey, so they have been free to evolve purely defensive functions. Thus, the venoms of many social species contain substances that cause local tissue damage and severe allergic reactions in vertebrates, effects not taken into account by the immediate pain scale (Schmidt 1990). The stings of solitary species, in contrast, have evolved primarily as offensive weapons for paralyzing prey. Nevertheless, their stings are often painful (Waldbauer and Cowan 1985, Schmidt 1990). Tarantula hawks (*Pepsis formosa*) deliver a sting that earns a 4 on the pain scale, and the velvet ant *Dasymutilla klugii* can inflict a 3. Presumably (and more important), an attacking bird or lizard would have a similarly high-valued pain response when stung on the softer tissues inside the mouth. R. Lane (1957, in Waldbauer and Cowan 1985) noted that, although he personally found the sting of an *Odynerus* ineffective against himself, it was a strong deterrent to feeding by a flycatcher.

A sting may need not be debilitating to be effective, as long as it startles an attacker long enough to facilitate escape. In fact, a "sting" may be effective even when it cannot inject venom. Males of many Mutillidae, Tiphiidae, Sphecidae, and Eumenidae possess "pseudostings," modifications of the genitalia or the abdominal exoskeleton that they use to prick an attacker, using the same probing movements of the abdomen that females use when they sting (Evans and West-Eberhard 1970). The stout, sharp leg spines of male tarantula hawks (*Hemipepsis ustulata*) also jab a would-be attacker when the wasp is attacked (J. Alcock, personal communication). Although stings, pseudostings, and spines are a good way to make a point with a predator (pun intended), there are other ways to startle an attacker. When assailed, both male and female mutillids produce a squeaking sound using a stridulatory apparatus otherwise used only during courtship (Bayliss and Brothers 1996; see Chapter 8). W. Masters (1979) found that stridulating female *Dasymutilla* were more likely to survive an attack by a mouse than were females that had been experimentally "silenced."

Stinging is a last-ditch effort to foil a predator's attack, but the wasps' interests would be better served if predators avoided them altogether. It is generally assumed that the bright color patterns typical of solitary and social wasps serve to warn predators, who have learned from experience that insects with particular color patterns inflict a painful sting (Evans and West-Eberhard 1970). A wasp can benefit from looking like another member of its species that a bird previously attacked, but it can also profit from a lesson learned when the bird attacked a different but similar-looking (and similarly noxious) species. The more striking the color pattern and the greater the number of noxious insects that have it, the greater the probability that a predator will learn to avoid such insects. It is perhaps for this reason that stinging wasps have evolved relatively few standard color schemes, most notably alternating yellow and black bands or some combination of orange and black. The yellow-and-black pattern is typical of many sphecids and solitary and social vespids. Orange-and-black patterns are common among pompilids, mutillids, and sphecids such as *Ammophila* and *Sphex*. Orange insects are commonly poisonous or distasteful and are avoided by experienced predators. For example, lizards learn to avoid orange *Dasymutilla* females (Sexton 1964, Schmidt and Blum 1977).

Groups of poisonous, distasteful, or bad-smelling insects that inhabit the same area and share similar color patterns are known as Müllerian mimics (Wickler 1968). For example, H. Evans (1969c) found a complex of orange Müllerian mimics in South America that included pompilid spider wasps, social vespids, and a sphecid. Where there is Müllerian mimicry, there is often Batesian mimicry, in which palatable, nonnoxious species find safety by resembling species with defensive capabilities. The concept of Batesian mimicry perhaps explains why certain innocuous syrphid flies resemble yellow-and-black wasps (Waldbauer et al. 1977, Waldbauer and LaBerge 1985), while some neotropical spiders look like mutillids (Nentwig 1985), some moths and flies mimic tarantula hawks (Evans and West-Eberhard 1970, Nelson 1986), and some bombyliid flies resemble *Ammophila* (Zalom et al. 1979). The situation can also be reversed, with solitary wasps mimicking nonwasps (though it is not always clear whether the wasps are Batesian or Müllerian mimics). Pompilids of the genus *Iridomimus* mimic ants of the genus *Iridomyrmex* (which produce noxious defensive secretions) (Evans 1970b), and females of the mutillid *Pappognatha myrmiciformis* resemble workers of the ant *Camponotus sericeiventris* (G. Wheeler 1983, Yanega 1994). Some of the most complex situations, apparently involving both Müllerian and Batesian mimicry, arise when different sexes of the same wasp species mimic different models in patterns referred to as dual sex-linked mimicry. Females of the pompilids *Austrochares gastricus* and *Chirodamus longulus* are members of a complex of black-and-red Müllerian mimics, whereas males of the same species are apparently Batesian mimics of social vespids (Fig. 7-5) (Evans 1968, 1969c).

Defensive Strategies: Protecting Offspring

Some enemies of solitary wasps attack nests while the female is absent, whereas others direct themselves toward prey-carrying females in transit to the nest. The diverse locations, times, and modes of attack require varied defensive strategies, some of which conflict with one another. The nests themselves can provide substantial defenses against enemies, but as conspicuous and well-provisioned structures they may also attract attention. However, natural enemies can often be thwarted with adaptations that hide the nest, distract attention elsewhere, and prevent entry after discovery by the attacker.

SPACING OF NESTS

The nests of many ground-nesting and mud-nesting solitary wasps occur in dense aggregations of hundreds or even thousands of nests

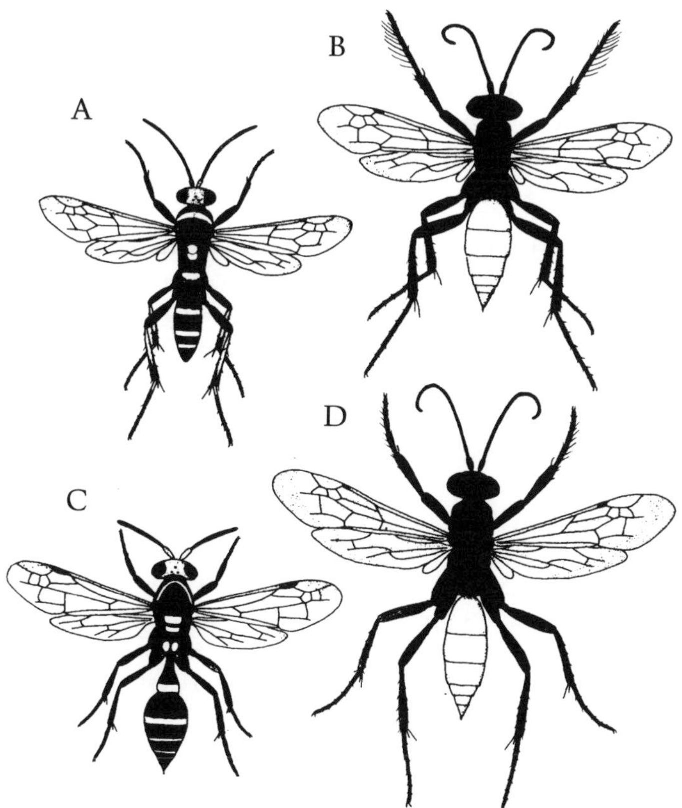

Figure 7-5. Dual sex-linked mimicry in a spider wasp, *Austrochares gastricus.* (A) Male *A. gastricus,* a Batesian mimic of workers of (C) the social wasp *Polybia parvula.* (B) Female *A. gastricus,* part of a complex of Müllerian mimics, which include (D) the spider wasp *Dicranoplius satanus.* (A–D from Evans 1969b.)

spaced several centimeters apart. Several authors have hypothesized that nesting close to conspecifics reduces the risk of parasitism if it lowers the ratio of parasitoids to potential hosts (the so-called dilution effect). In examining different portions of a *Crabro cribrellifer* nesting area, W. Wcislo (1984) found that, as nest density increased, there was a proportional increase in the density of the wasp's major natural enemy, *Metopia campestris.* However, the percentage of nest cells parasitized *decreased* with density. F. Larsson (1986) also found that

Metopia density increased with increases in the density of *Bembix rostrata* nests, but that the ratio of adult *Metopia* per nest decreased with nest density. Although aggregations may provide some measure of protection, it is more common for rates of parasitism to be unchanged or to increase with nest density (Rosenheim 1990b).

NEST STRUCTURES

Nests have evolved as adaptations to protect the wasps' offspring and provisions by hiding the wasps and providing physical barriers to attack while also furnishing a buffered physical environment (Chapter 6). As a ground-nesting wasp prepares her burrow, the mound of soil accumulating outside of the entrance could attract parasitoids and cleptoparasites, such as chrysidids and miltogrammines. Thus, many ground-nesting pompilids and sphecids spend much time either depositing the soil at some distance from the nest (e.g., many *Ammophila*) or leveling the mound so that the surface in the vicinity of the entrance is indistinguishable from its surroundings (Chapter 6) (Rau and Rau 1918, Evans and Matthews 1973c, Spofford et al. 1986, Evans and O'Neill 1988). Upon completing excavation, for example, *Bembix americana* females spend 5–10 minutes leveling the mound before leaving to hunt (Evans 1957b). Such disguises may also be important to female *Methoca*, who, after ovipositing on a tiger beetle larva, carefully close the beetle's burrow and hide the entrance with soil and debris (Burdick and Wasbauer 1959). C. Knisley and colleagues (1989) noted that incompletely closed *Methoca* burrows were later invaded by ants.

Some wasps that keep the entrance closed and concealed also dig short, blind burrows nearby that remain open and attractive to natural enemies (Fig. 7-6). The function of these accessory burrows, which are found only in certain Sphecidae that have outer nest closures, has been the subject of much speculation. In considering hypotheses on their function, Evans (1966a) provided arguments to reject the notions that they evolved as (1) landmarks for orientation, (2) sites for temporary prey storage, or (3) adult resting sites. Although soil for the final closure is sometimes obtained from accessory burrows (Evans 1966a), they are more than just quarries. Some *Bembix*, *Philanthus*, and *Sphex* regularly dig accessory burrows without using the soil for the closure or for concealment of the real entrance. Females may also "reenter and refresh these burrows from time to time" (Evans 1966a) and often reconstruct them if they are destroyed (Tsuneki 1963).

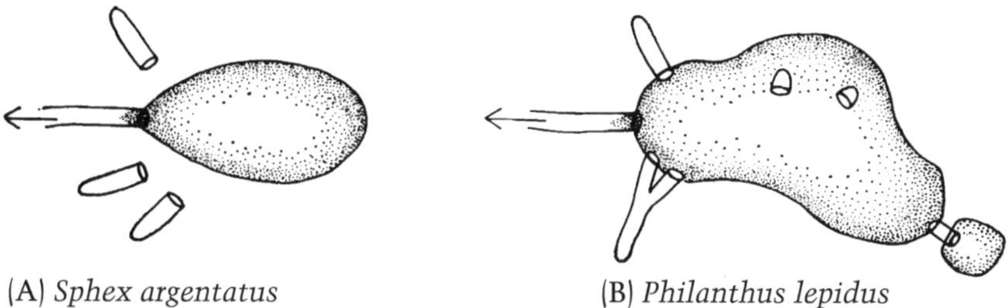

(A) *Sphex argentatus* (B) *Philanthus lepidus*

Figure 7-6. Accessory burrows of sphecid wasps. Arrows indicate true nest burrow, stippling, the soil mounds. (A) *Sphex argentatus.* (B) *Philanthus lepidus.* (A and B from Evans 1966a.)

K. Tsuneki (1965) and H. Evans (1966a) hypothesized that accessory burrows function to distract parasitoids from the real nest entrance. Many bombyliids, chrysidids, mutillids, and miltogrammines are attracted to accessory burrows, apparently treating them as potential oviposition sites (Evans 1966a, Evans and Matthews 1973c). If these parasites are either duped into believing they have oviposited in an appropriate niche or are frustrated by their inability to find a host within the burrow, they may move on and leave the real nest unscathed. The gist of the distraction hypothesis is that accessory burrows further increase the effective number of "nests" among which parasitoids and cleptoparasites must choose. For example, the number of apparent burrow entrances in an aggregation of *Philanthus zebratus* on one day increased from 47 (the number of nests) to 83 owing to the existence of 36 accessory burrows (0–4 per nest) (Evans and O'Neill 1988). The distraction hypothesis can be confirmed only with comparative and experimental studies of accessory burrow function and parasitoid behavior. To date, direct evidence that accessory burrows are antiparasitoid structures is limited to data showing that *Sphex argentatus*, which always builds accessory burrows, suffers lower rates of parasitism than *S. flammatrichus*, which never builds them (Tsuneki 1963).

The hard walls of mud nests and nests within stems or wood cavities provide formidable barriers to entry or oviposition by many potential natural enemies (other than those eulophids, ichneumonids, and torymids that can breach such walls). Similarly, nest cells well below the soil surface are out of the reach of many enemies, particularly after the nest has been completed and closed with a permanent plug. In making their final closures, ground-nesting species fill in much of the main burrow with soil, gravel, and debris (e.g., Evans 1966a, Evans and O'Neill 1988, Weaving 1989), whereas hole-nesters and mud-nesters block the entrance with a sizeable plug of mud, agglutinated sand, resin, wood fibers, tufts of grass, or other plant debris (Krombein 1967) (Chapter 6).

A nest is vulnerable to attack if the entrance is left open while the female is absent. Natural enemies such as ants, miltogrammines, and bombyliids take advantage of open nests to steal prey or to deposit eggs or larvae, and some need only a brief interval to execute an attack. Thus, many ground-nesting species close the entrance temporarily before they leave on foraging trips. However, nests with temporary closures may still be attacked by mutillids and miltogrammines (*Phrosinella*) that can detect and dig through the barriers. Furthermore, at nests with closures, females must drop their prey (briefly exposing them to attack) in order to remove the closure before reentering. This problem may be overcome by the method of prey carriage during nest entry.

PREY HANDLING

While carrying prey to a nest, wasps are especially vulnerable to attack by enemies that oviposit on prey (e.g., *Senotainia* and *Hedychridium*) and to those that find nests by following provisioning females (e.g., *Amobia*). The probability of escaping such attacks during any one provisioning trip may be very high. However, because each cell may be provisioned using multiple foraging trips and because a successful attack during just one of those trips could result in destruction of the cell, the overall probability of a cell's escaping attack may be much lower. For example, even if the probability that a provisioning female will avoid attack on each foraging flight is 0.95, the probability of a cell's escaping parasitism drops to 0.60 if each cell contains 10 prey items and it falls to just 0.28 if each cell is provisioned with 25 prey items. D. McCorquodale (1986) estimated that *Senotainia trilineata* contacted and probably larviposited on prey of *Philanthus gibbosus* during about 1 in 10 foraging trips. Because a typical *P. gibbosus* nest cell is provisioned with 10–14 prey items (Evans and O'Neill 1988), the probability that a cell would escape attack falls to about 0.23–0.35. Thus, there should be a strong selective premium on the ability of females to carry prey to the nest and to enter it quickly while avoiding attack.

When confronted with cleptoparasites that affix larvae or eggs to prey as the wasp approaches the nest, a female has several ways (other than direct counterattack; Spofford et al. 1986) in which she can reduce the probability of a successful attack. These include taking evasive actions while in transit to the nest, concealing prey, and entering the nest quickly.

Evasive Actions

When a wasp returns to its nest with prey, she may have to halt, drop the prey, and remove a temporary closure before entering, turning about, and pulling the prey into the nest. Entry may take just several seconds, but the brief delay caused by the closure may provide an opportunity for a cleptoparasite. During attacks, cleptoparasites such as *Senotainia trilineata* and *Hedychrum intermedium* orient to and follow moving stimuli. Thus, one potential counterstrategy for a prey-carrying female is to evade a parasite either by depriving it of a proper stimulus or by making pursuit difficult. Female *Philanthus* (Evans and O'Neill 1988) and *Tachysphex* (Spofford et al. 1986) use both tactics. If followed by a satellite fly, females may land briefly and remain motionless, a behavior J. Alcock (1975d) referred to as a "freeze stop." During the brief pause in the flight of her original quarry, the fly (which also stops) may respond to and pursue another passing wasp; this is one possible mechanism for lower rates of parasitism in dense nesting aggregations (Wcislo 1984). Females may also throw off their pursuers by making circuitous flights, by rising up and quickly descending directly to the nest, or by temporarily leaving the vicinity of the nest (Spofford et al. 1986, Evans and O'Neill 1988). Similarly, when followed by miltogrammines, *Crabro argusinus* females often increase the number of zig-zags in their flights, reducing the probability of being contacted by a fly to just 10% of that of wasps that do not respond to their pursuer (McCorquodale 1986).

Prey Carriage and Prey Caching

The method of prey carriage may reduce vulnerability to parasitoids and cleptoparasites in one of two ways (Evans 1962). First, prey that are small enough can be carried concealed and protected beneath the body. S. Ristich (1956) suggested that the exposed prey of *Aphilanthops* are more vulnerable to airborne attacks by *Senotainia* than are the prey of *Philanthus*, which are carried closer to the body. Similarly, *Cerceris echo* carries small beetles tucked under the thorax, where they are inaccessible to attack by *Senotainia trilineata* (McCorquodale 1986). Second, species of sphecids that carry prey with their middle legs, hindlegs, or devices on the tips of their abdomens can open and enter their nests with their front legs without dropping their prey, thus speeding their entry (Evans 1962).

Wasps that excavate nests *after* capturing prey face the added problem of protecting the prey while they are digging, because unattended prey may be stolen by ants, tiger beetles, and other wasps (Evans 1962). One potential solution to this problem is for females to cache prey temporarily where they are less likely to be discovered. Various pompilids, as well as sphecids such as *Ammophila*, *Prionyx*, and *Podalonia* that build their nests after hunting, temporarily conceal their prey in the crotches of low plants, within crevices in the soil, or in shallow burrows (Richards and Hamm 1939; Evans 1953, 1966b; Rosenheim 1990a). In a study of *Ammophila marshi*, which cache their caterpillar prey up to 15 cm high on plants, J. Rosenheim (1990a) tested the hypothesis that caching protects prey. He found that, when caterpillars were set out as "baits" both on the ground and on low plants, the majority of those on the ground were taken by ants within 20 minutes, whereas most of those placed on plants were still present several hours later (sufficient time for a female to complete her nest and return for the prey). Thus, aerial caching may provide a wasp with sufficient time to dig a nest. Caching is not a foolproof method of safeguarding prey, however, because some cached prey may still be discovered and carried off or eaten on the spot. I have seen the cached grasshopper prey of *Prionyx* species devoured by other grasshoppers, and H. Evans (1987) reported that cached prey of *Podalonia* are sometimes stolen by ants. Attention must also be given to the hypothesis that aerial caching allows some wasps to store prey away from the soil surface, where high temperatures may desiccate or kill the paralyzed prey (O'Brien and Kurczewski 1982a).

NEST GUARDING

Even prey and eggs within nests may not be safe, because some natural enemies enter nests to forage. The presence of a female in an open burrow probably prevents certain parasitoids and predators from reaching nest cells. For example, *Metopia* females may enter, but then quickly exit nests when the wasp is in the burrow (Olberg 1959, Wcislo et al. 1985). In solitary species, the wasp's presence may often be just a lucky coincidence (rather than nest-guarding behavior per se), but direct counterattacks of varying effectiveness do occur. Female wasps may react aggressively to ants (F. Williams 1919c, Eberhard 1974, Evans and O'Neill 1988) and potential conspecific brood parasites (Field 1992c) near nest entrances. *Pachodynerus nasidens* females may crush the much smaller *Melittobia* with their mandibles (Jayasingh and Freeman 1980), but when a *Philanthus triangulum* attacks a *Hedychrum intermedium* enemy, the chrysidid rolls up in a ball and becomes impervious to the stings of the beewolf (Olberg 1959, Simonthomas and Simonthomas 1972). Mutillids are even more formidable, defending themselves from counterattacks using chemical repellents (Fales et al. 1980) as well as an arsenal of large, pointed mandibles and sharp stings (Brothers 1972, Deyrup 1988).

Nevertheless, even when successful, counterattack cannot be a permanent solution because females must leave their nests to hunt and feed. This constraint has certainly been important in the evolution of nest closures and has perhaps contributed to the evolution of communal nesting in the Sphecidae, Pompilidae, and Vespidae (i.e., those in which multiple females share a nest). Guarding can be an effective deterrent for communal nesters because at all times at least one female is likely to be present in the nest. Communal *Cerceris* nests may contain up to eight females, some of which consistently forage while others guard entrances to repel invading ants and mutillids (Evans and Hook 1986a, Wcislo et al. 1988, McCorquodale 1989a). Females of *Xenorhynchium nitidulum* sit inside of cells in their communal mud nests, with their stings facing outward (West-Eberhard 1987).

The females of a few bethylid species remain with a host after oviposition to guard it from intrusions by other host-seeking females (F. Williams 1919a, Doutt 1973, Hardy and Blackburn 1991). They do so at the expense of delaying finding of another host, but this cost may be offset by the reduction in the likelihood that another wasp will come along and lay its own eggs on the host (often after removing the other female's eggs) (Hardy and Blackburn 1991).

NEST CLEANING AND PREY ABANDONMENT

Apparently, some wasps can determine, to some degree of accuracy, whether a cleptoparasite has attached an egg or a larva to a prey or deposited an egg inside a cell. The possible adaptive responses to the discovery of an attack are either to abandon the cell or prey or to remove the natural enemy. *Tachysphex terminatus* females display both responses (Spofford et al. 1986). When trailed by a satellite fly, a female may halt in a location away from the nest and spend as long as 12 minutes inspecting the surface of the grasshopper prey and removing maggots. Only 3% of the prey inspected by a *Tachysphex* female were later found to contain a maggot, whereas 43% of uninspected prey were infested with at least one maggot. Female *T. terminatus* also abandon prey without cleaning them and desert partially provisioned cells containing as many as six prey items. Similarly, upon detecting a chrysidid waiting at the nest entrance, female *Philanthus triangulum* may abandon their bee prey and enter the nest alone; they also sometimes eject from nests prey that carry maggots (Simonthomas and Simonthomas 1972). Female *Ammophila dysmica* go further, abandoning entire nests during construction when they are attended by cleptoparasitic *Argochrysis armilla* females (Rosenheim 1988). However, although *A. dysmica* frequently cleaned debris from nests and sometimes physically removed chrysidid adults, they were unsuccessful in removing chrysidid eggs, which were firmly attached to prey.

Conclusion

Attacks by natural enemies on wasps and their nests are not only frequent but may also come unpredictably at different times and places from enemies with diverse modes of attack. It is obvious that natural enemies often play an important role in the ecology and evolution of solitary wasps. The importance of natural enemies is especially clear when we consider females of nest-provisioning species. Nest building, nest spacing, nest closure, accessory burrow construction, prey carriage, flight patterns during prey carriage, prey caching, and nest guarding make little sense without at least partial reference to the problems caused by natural enemies. One hopes that hypotheses concerning the functions of specific behaviors will soon be addressed with experimental studies or comparative analyses supported by accurate phylogenies. We must also consider how sets of tactics aimed at different natural enemies are compatible with one another. For example, temporary nest closures during hunting provide protection for the nest in the absence of the wasp, but they also delay entry of the female returning with prey (thus, increasing her susceptibility to attack by species such as *Senotainia trilineata*). Similarly, abdominal prey carriage may reduce the delay in entry, by allowing the female to reopen her nest without dropping the prey, but prey carried in this manner may be more exposed to larvipositing miltogrammines. These issues are important for an understanding of both solitary wasp behavior and the transitional stages in the evolution of social behavior in the Hymenoptera (Evans 1977).

8/ Male Behavior and Sexual Interactions

For most of its history, Hymenoptera ethology has focused on females, while males were ignored, glossed over, and even judged unworthy of serious study (e.g., W. Wheeler 1919, Olberg 1959). To J. Fabre (1921), the male of the bee *Osmia* was an "amorous trifler," and to the Raus (1918), male *Sceliphron* were "lazy, good-for-nothing fellows." E. Reinhard (1929) mirrored the opinions of many of earlier workers in the preface of *The Witchery of Wasps*, where he declares that he will tell "a tale of skilled hunters, . . . adept home-builders, [and] hard working mothers," contrasting these paragons of industrious motherhood with males that he refers to as "shiftless fathers." Thus, it is not surprising that earlier overviews of solitary wasp behavior dealt almost solely with females (e.g., Rau and Rau 1918, Evans and West-Eberhard 1970, Iwata 1976). However, a few detailed descriptions of male wasp behavior began to appear in the 1950s and 1960s (e.g., Kullenberg 1956; Evans 1957b; Lin 1963, 1966, 1967), and J. Alcock and his colleagues were able to offer the first reviews of male wasp behavior by the late 1970s (Alcock et al. 1978, Alcock 1979b). By 1988, 14 papers had appeared on male *Philanthus* alone (Evans and O'Neill 1988).

One reason for the flourishing interest in the behavior of males of all animal species was that, by the 1970s, ethologists had acquired a coherent theoretical context for interpreting sexual differences in behavior. The improved theories grew from a clarification of the role of individual selection in the evolution of behavior (G. Williams 1966) and from the revival and maturation of sexual selection theory (e.g., Trivers 1972, Emlen and Oring 1977).

Conceptual Background: Sexual Selection Theory

The lifetime reproductive output of female wasps is limited by their large investment in each offspring. Such constraints do not pertain to males, who do not manufacture large gametes and do not attend their young (with a few exceptions). In fact, male aculeate Hymenoptera would generally make poor caregivers because they lack a sting, a major tool used by females to provide for young. Because males do not generally contribute to rearing, their prospects for high reproductive success depend upon the number of eggs that they fertilize (Trivers 1972). The males of a population thus compete for a single resource: females (or more strictly, their eggs).

Mate competition can take two general forms (Darwin 1871). First, males compete to find and control receptive females or locations where females are likely to appear. The resulting selection has promoted the evolution of male traits that enhance males' success in competition with other males, including (1) behaviors, such as territoriality and harem defense, that aid in maintaining exclusive contact with potential mates, (2) male structures, such as horns, that enhance their fighting abilities, and (3) male sensory capabilities that promote early detection of distant or concealed females (Andersson 1994). Second, males may also compete for the attention of females by advertising qualities likely to make them better mates. The resulting sexual selection due to female choice has produced male animals with courtship behaviors and ornamentation that increase their attractiveness to females. In insects, sexual selection generally acts more strongly on males than on females, because female success is *usually* linked more closely to the number of eggs they lay than to the number of times they copulate (Trivers 1972). The upshot of all of this is that we can agree with Reinhard (1929) that male wasps are often "shiftless fathers," but that is what we expect them to be in most species. However, philandering does not make for an easy life, particularly when philanderers are common and females are discriminating in mate choice.

Ecological Background

The expected mating success associated with a particular mating tactic is a function of various intrinsic characteristics of a male as well as extrinsic ecological (including social) constraints (Emlen and Oring 1977, Thornhill and Alcock 1983). One especially important ecological factor is the location of receptive females. Because sexually receptive females are usually nonrandomly distributed in the environment, sexual selection should have a strong effect on *where* males search for or wait for females. In nest-provisioning wasps, males often concentrate their activities at places where females emerge (as adults) or nest. However, there are many other rendezvous sites, including flowers, hunting areas, and water-collecting sites (Alcock et al. 1978).

The review of male aculeate mating strategies in this chapter is organized primarily around the theme of the locations where males rendezvous with females. However, the evolution of male mating strategies is obviously driven by more than just the location of females. Even if a male is in the right place, he may not be there at the right time, which varies depending on (1) how many times each female mates, (2) when females become receptive, and (3) which males are selected to father their offspring. Furthermore, the duration of time over which females are available and the density of females and males in the habitat affect the number of competitors that a male faces. Before going on to detail the forms of male mating systems in aculeates, it is worth pausing to consider some of these factors, as they surely affect the evolution of wasp mating strategies (Thornhill and Alcock 1983).

How many times does a female mate? The critical distinction is between wasps with females that mate once and those with females that mate multiple times before laying each egg or clutch of eggs. A single mating may suffice for multiple eggs because females store sperm in their spermathecae. If a female mates just once, the best tactic for a male is to find and mate with virgins as soon as they become receptive, thereby fathering all their female offspring (sons come from unfertilized eggs). Thus, in single-mating species, sexual selection should favor males that quickly find newly receptive females.

If a female mates more than once before each oviposition, then which male fathers her young may depend on the positioning of their sperm in her spermatheca (Eberhard 1991, 1996). Many insects exhibit *last-male sperm precedence*, meaning that the final male that mates with a female before the egg is laid is most likely to be the father of her offspring (Gwynne 1984). Because the spermatheca is a blind sac within which sperm from different males may not mix, the last sperm into it are usually the first ones out when an egg is fertilized. Although last male sperm precedence is typical of many insects (Gwynne 1984), its existence in aculeates has not been substantiated. I will assume that it occurs in wasps, keeping in mind that this is only a working hypothesis. Because most sphecid and eumenine wasp females probably mate just once in their lives (Alcock et al. 1978, Cowan 1986b, Evans and O'Neill 1988), the assumption will not affect most of our discussion. Nevertheless, multiple matings are a common feature of the mating systems of a few genera of nest-provisioning wasps (Cowan and Waldbauer 1984, Cowan 1986b, Brockmann and Grafen 1989, Evans 1989).

At what time after her emergence does a female become sexually receptive? Females of all solitary aculeates are probably receptive before their first oviposition, because mating gives them the option of producing offspring of either sex. This option allows mothers to match the sex of offspring to the amount of food available or to produce clutches with both sons and daughters (Chapter 10). Thus, it is not surprising that females are commonly sexually receptive upon emergence as adults. The extreme form of early receptivity occurs in bethylids whose males enter their sisters' cocoon and mate with them before they emerge. If females mate just once and are sexually receptive at or soon after their emergence as adults, then males should seek out females at this time, particularly when competition among males for mates is intense. Thus, the best tactic for males may be protandry: to emerge as adults before most females so that they are present when potential mates appear. Protandry is a common feature of the life histories of male solitary wasps.

How intense is the competition among males? A male that finds one or more receptive females may also find himself in the company of competitors. Confronted with a sufficient number of competitors, even an extremely tough and speedy male may find it difficult to succeed in mate competition. Therefore, the expected net benefit associated with a particular mating tactic will depend on the level of competition. One measure of the intensity of competition is the operational sex ratio, which in its simplest form is the ratio of the number of mate-seeking males to the number of sexually receptive females (Emlen and Oring 1977). When the ratio exceeds some critical value, a female or territory may be economically undefendable, because a male would spend too much time and energy fighting and risk being distracted by competitors when a receptive female comes along.

The operational sex ratio will often be much higher than the population sex ratio for several reasons. First, when each female mates just once in her life, every mating permanently removes a female (but usually not a male) from the pool of sexually active wasps. Second, even in species with a small degree of protandry, the operational sex ratio is already high when females first appear. For example, by the end of the first day of female emergence, the ratio of males to females was about 5 to 1 for *Bembecinus quinquespinosus* (O'Neill and Evans, unpublished), 8 to 1 for *Philanthus bicinctus* (Gwynne 1980), and over 20 to 1 for *Sphecius grandis* (Hastings 1989b) and *Ancistrocerus adiabatus* (Cowan and Waldbauer 1984). For *B. quinquespinosus*, the number of males fighting for a single female ranged as high as 50, even though the overall population sex ratio was 1:1 (O'Neill et al. 1989).

During mate competition, some males may succeed better than conspecifics because fighting ability is typically correlated with body size, which varies greatly within sexes; body size in wasps is measured in the field by determining maximum head width (Alcock 1979a, O'Neill 1983a), body length (Gwynne 1980), or wing length (Hastings 1989a), all of which are correlated with body mass. Male *Philanthus crabroniformis*, for example, vary in head width from 2.4 to 3.8 mm and in dry mass from 2.2 to 14.7 mg (Evans and O'Neill 1988). Similarly, *Hemipepsis ustulata* males vary in head width from 3.0 to 5.5 mm and in live weights from 6 to 56 mg (Alcock and Bailey 1997).

The case studies of male behavior presented below will provide a framework for a comparative survey in which species are grouped according to similarities in (1) where and when males rendezvous with females and (2) how males compete with one another (e.g., scramble vs. interference competition). The survey reveals a great diversity in mating systems within and between species of aculeate wasps, as well as some interesting convergences in the evolution of the mating systems of distantly related taxa.

Mating at Emergence Sites

Because female solitary wasps are often receptive upon emergence and mate just once in their lives, a premium is often placed on the ability of males to quickly locate freshly emerged females. A variety of mating strategies have evolved in which male solitary wasps mate with virgins at or near their site of emergence, which may be the soil surface, the wall of a mud nest, the entrance of a cavity nest, or a female's cocoon. At these locations, males may defend individuals or groups of emerging females, or they may patrol nonaggressively.

MALE DEFENSE OF GROUPS OF EMERGING FEMALES

Case Study: *Sphecius speciosus* and *Sphecius grandis*

The eastern cicada-killer (*Sphecius speciosus*) was one of the first solitary wasps to be the subject of a study that focused on males. Following up on earlier reports (W. Davis 1920, Reinhard 1929, Dambach and Good 1943), N. Lin (1963, 1966, 1967) observed males on a sandlot baseball field in Brooklyn, New York. Later, male western cicada-killers (*Sphecius grandis*) were studied in Arizona by J. Alcock (1975a) and J. Hastings (1989a,b).

Cicada-killers are truly impressive animals, with robust black, yellow, and rust-colored bodies up to 4 cm long. Females nest in dense aggregations that become sites of intense male competition during the following year's emergence. Male *S. speciosus* (Fig. 8-1) emerge before most females and establish 1.5 to 9 m^2 territories that individual males occupy for up to 12 days. A male's territory perch is usually on the soil surface within several centimeters of an emergence hole. Other wasps will not emerge from the same hole, but the presence of emergence holes is a good predictor of future emergence points since nests are aggregated.

In *The Territorial Imperative*, R. Ardrey (1966) aptly referred to the mating system of male cicada killers as a "rough game." Male *S. speciosus* attempt to maintain exclusive use of areas surrounding their perches, vigorously defending them in confrontations with conspecific males that involve head butting, grappling, and biting (Lin 1963). When a male pursues an intruder beyond his own territory boundary, he may encroach on the domain of an adjacent male. His neighbor may then join the pursuit, so that streams of up to four wasps race through the emergence area. Most battles are between neighbors,

Figure 8-1. Male *Sphecius speciosus* on territory (photo by H.E. Evans).

although such encounters decline with time, apparently as the males habituate to one another. A male's response to other species is much less aggressive, usually involving a brief pursuit that probably represents investigation of a potential mate.

Eastern cicada-killer males do not meet females at their point of emergence from the soil. Rather, they wait for a newly emerged female to fly past. Males apparently distinguish receptive from previously mated, nonreceptive females, which fly through the area in "slow jerky, zig-zag motion" (Lin 1963). A male responds to a nonreceptive female by following only while she is within his territory and, at most, lightly touching her before returning to his perch. In contrast, newly emerged, sexually receptive females fly straight through the area and are pursued and grasped from behind. The pair then flies in tandem, usually to nearby plants, where copulation proceeds. While mating, a pair may be harassed by other males who attempt to disrupt them, but the male's genitalia appear to provide a firm grasp on the female.

According to Lin, copulation lasts from 29 to 51 minutes, with no noticeable courtship before sperm transfer. However, E. Reinhard (1929: 41) describes a more complex courtship: "The paws of his front legs he places on the eyes of his beloved and twists her head from side-to-side. Then he rubs his antennae along the length of hers with a caressing motion; it seems like a friendly gesture, something like the Eskimo's etiquette of rubbing noses. First turning her head like a door-knob, then feelingly stroking her antennae, he at length wins her consent. Together they soar aloft in conjugal flight."

Males of western cicada-killers (*S. grandis*) also establish territories within emergence areas and defend them by butting and grappling intruders (Alcock 1975a, Hastings 1989a,b). In some contests, one male may even grasp another in flight and carry him upward, perhaps as a show of strength (Hastings 1989a). Male *S. grandis* may incur damage to wings and other appendages during fights over territories, perhaps partially explaining why the mean adult life span is just 11–15 days for males (Hastings 1989b).

As it does for *S. speciosus*, the daily peak of male *S. grandis* activity corresponds to the time at which most females emerge (Lin 1963, Hastings 1989a). Most copulations are achieved by large *S. grandis* males that occupy areas of higher female density and hold their territories at the times when the female emergence rate is highest. Although these large "primary residents" are in the best position to intercept virgins, smaller males also have opportunities to mate (Hastings 1989a). When the largest males depart territories in the afternoon, they are often replaced by intermediate-sized, "secondary residents" that mate with the fewer late-emerging females. Furthermore, many small males forgo territoriality altogether, opting to perch in trees adjacent to the emergence area where they can intercept females missed by territory holders.

J. Hastings's (1989b) study of cicada-killers provides the most detailed analysis of protandry in a solitary wasp. In a 3-year study, the seasonal duration of female emergence was 2–5 times longer than the mean male life span. As a result, the date on which a male emerged had a great effect on his potential mating opportunities, determining both the number of receptive females that he was likely to encounter and the number of male competitors he was likely to face. Hastings found that the potential "mating opportunity" varied with emergence date in two of three years. Over the three years, the first cicada-killer males emerged 1–7 days before the first females, and 55% to 95% of the males' emergence period overlapped with that of the females. In 1984, males emerging one to two weeks before the peak of female emergence had the highest mating opportunities. However, in 1981, when females emerged over a longer period and males had shorter life spans, those that emerged late had the highest mating opportunities. Thus, although protandry benefits males in some years, variation in extrinsic factors outside of male control favors later emergence in other years. Such yearly variation in selection pressures may account for incomplete protandry in *S. grandis* and other wasps. H. Evans and I (Evans and O'Neill 1988) reported that protandry occurred in 17 of 18 species of Sphecidae for which emergence data were available but that the seasonal emergence periods of the two sexes overlapped in all of the species. Protandry has also been observed in other families of nest-provisioning wasps (King and Holloway 1930, Jayakar 1963, Brothers 1972, F. Gess and Gess 1980, Cowan 1981) as well as in the cleptoparasite *Hedychrum intermedium* (Simonthomas and Simonthomas 1972).

Defense of Groups of Emerging Eumenine Females

As do male cicada-killers, males of some hole- and mud-nesting eumenines practice emergence area defense, centering their territories around clusters of nest cells. Using trap nests, D. Cowan (1979, 1981) found that males of *Euodynerus foraminatus* emerge from the outer set of cells in a nest several days before their sisters. When a nest is isolated from others, one male (usually the largest) permanently expels his brothers and mates with his emerging sisters. Where nests are clumped, one male also drives off rivals from nearby nests and mates with females from other broods. When a male senses an impending female emergence, he straddles the nest hole and mates with her as she emerges, ultimately mating with up to five sisters (and sometimes with nonsisters from nearby nests). The territory of a successful male of one species of *Epsilon* can harbor brood cells of several dozen potential mates (A. Smith and Alcock 1980). The West African mud-nesting species *Synagris cornuta* has a similar mating system, in which larger males fight using large horns on their mandibles (Roubaud 1911). The common features of *Sphecius*, *Euodynerus foraminatus*, *Epsilon*, and *Synagris cornuta* are (1) emergence sites that can be easily found by males and (2) size variation that allows large males to dominate an area containing more than one preemergent female.

SIB MATING WITHOUT DEFENSE AT EMERGENCE SITES

Whereas male siblings of *Euodynerus foraminatus* fight for access to emerging sisters, males of other Eumeninae and many Bethylidae that mate with their sisters do not engage in combat with brothers even when more than one male is present.

Case Study: *Goniozus* spp.

Sib mating has been best-studied in the bethylid genus *Goniozus* (Table 8-1). *Goniozus* females deposit clutches of eggs on hosts, where their larvae feed gregariously and then pupate in groups near the host. Male *Goniozus* emerge 10 hours to several days before their sisters and remain with the brood. A male chews his way into the cocoon of one of his sisters. After copulating, he moves on to another cocoon or mates with sisters that have emerged unassisted.

The high mating success experienced by many male *Goniozus* results from their mothers' producing isolated clutches in which sisters predominate. As will be discussed in Chapter 10, female Bethylidae often regulate sex ratios precisely (Green et al. 1982, Morgan and Cook 1994). They use this ability to produce strongly female-biased broods that often contain only a single male (Table 8-1). Although a lone male can often mate with all his sisters in a small brood, mothers ensure an adequate sperm supply for their daughters in larger broods by increasing the number of males (unfertilized eggs) in a brood as the clutch size increases (Gordh 1976, Gordh et al. 1983, Gordh and Medved 1986). When more than one brother is present in large clutches, however, they do not fight for females, perhaps because of the cost of harming relatives in terms of indirect fitness losses. Furthermore, there may be a physiological limit to the number of sisters a male can inseminate. Nevertheless, males are no slouches: J. Gifford (1965) observed one male *G. indicus* mate 23 times with 14 females.

Sib Mating in Other Species

Sib mating without aggression between brothers has been observed in at least three other genera of Bethylidae and in several species of Eumeninae (Table 8-1). Common features of these species include protandry, a lack of direct competition among brothers (apparently), and female-biased sex ratios (usually). However, female *Cephalonomia tarsalis* and *Paraleptomenes miniatus* lay two-egg clutches consisting of one daughter and one son, who mate with one another. Sib mating at nests has not been documented in ground-nesting wasps, whose males probably lack information on the number, location, and emergence schedules of sisters lying deep within their subterranean cells.

Table 8-1 Some species of solitary wasp in which mating between siblings occurs

Species	Location of matings	Protandry	No. of females/male in brood	Competition between brothers	References
Bethylidae					
Cephalonomia gallicola	Inside cocoon	Yes	—	—	Kearns 1934
C. tarsalis	Inside cocoon	Yes (2 d)	1	—	D. Powell 1938
Goniozus aethiops	Inside cocoon	Yes (1–2 d)	1–7[1]	No	Gordh & Evans 1976
G. claripennis	Inside cocoon	—	—	—	Voukassovitch 1924
G. emigratus	Inside or outside cocoon	Yes	5.5[2]	—	Gordh & Hawkins 1981
G. gordhi	Inside or outside cocoon	Yes (10 h)	0–9	—	Gordh 1976
G. legneri	Inside or outside cocoon	Yes (10–12 h)	3–8[2]	No	Gordh et al. 1983
G. natalensis	Inside cocoon	Yes	8	—	Conlong & Graham 1988b
Laelius pedatus	Outside cocoon	Yes	2	—	Mertins 1980
Sclerodermus immigrans	Inside cocoon	Yes	≥5	—	Bridwell 1919
Vespidae (Eumeninae)					
Ancistrocerus adiabatus	Outside nest	Yes	3.4–6.3[3]	No	Cowan 1981
Paraleptomenes miniatus	Outside nest	Yes	1	—	Jayakar & Spurway 1966

[1]Also occasional all-female broods. [2]Data for females' early broods (percentage of female offspring declines as the mother's sperm supply is depleted). [3]Varies with nest hole diameter.

Whether or not it involves a male's defense of his natal nest or clutch, sib mating may be frequent enough to result in high levels of inbreeding in bethylids and eumenines. Cowan (1979), for example, observed that 40% of emerging *Euodynerus foraminatus* females mated with their brothers. More recently, T. Chapman and S. Stewart (1996), who used electrophoresis to examine seven marker loci in *Ancistrocerus antilope*, concluded that about 90% of matings occurred between brothers and sisters. This level of inbreeding is extreme even within the frequently incestuous Hymenoptera.

MALE DEFENSE OF INDIVIDUAL EMERGING FEMALES

Male wasps cannot always control access to more than one female at a time. Rather, their best chance of being polygynous may lay in sequentially defending a series of mates.

Case Study: *Bembecinus quinquespinosus*

Bembecinus quinquespinosus has been studied in detail at the Pawnee National Grasslands in Colorado (O'Neill and Evans 1983a, O'Neill 1985, Evans and O'Neill 1986, Evans et al. 1986, O'Neill et al. 1989). Female *B. quinquespinosus* nest in dense aggregations within patches of coarse, sandy gravel surrounded by expanses of short, sparse grasses and forbs. Because their nest entrances are often within 5–10 cm of one another, adults of both sexes emerge in great numbers during the following July, with the first males preceding the first females by several days. Males search for virgin females about to emerge from the soil, apparently attracted by the odor and perhaps the sound of the preemergent female. Because of protandry, mate-seeking males far outnumber emerging females, and the ensuing competition is very intense. In fact, interactions among male *B. quinquespinosus* can be described as melees without risking hyperbole.

Between 7:30 and 8:30 A.M. on clear days, males congregate in the emergence area, remaining there until late morning or early afternoon. While in the emergence area, they fly about in low sinuous patterns 5–10 cm above the soil surface, stopping to examine the soil surface and conspecific males with their antennae. On occasion, males are so intent on finding females that they are ambushed by the foraging horned lizards *(Phrynosoma douglassi)* that frequent the emergence area. At the peak of seasonal activity, the thousands of males swarming in a 20–100 m^2 emergence area present a bewilder-

Figure 8-2. Males clustered around a female in a "mating ball" (the dorsum of the black-and-white–striped abdomen of the female is in the middle of the cluster). (Photo by H.E. Evans.)

ing pattern of movement in which individual males are difficult to follow. At these times, one can also see groups of up to 15 males giving their full attention for up to 10 minutes to outwardly nondescript points on the soil surface. These are the points from which virgin females are about to emerge. Close inspection of these males reveals that they are digging synchronously with their front legs. When more than one male is present, the diggers are not cooperative and their mutual interference results in the excavation of no more than a shallow depression in the sand. At this time, the closest that males come to actually fighting is mutual jostling as each attempts to get closest to the site of the female's emergence.

When a female finally appears at the surface, there is an abrupt transition from a loose group of diggers to a frenzied cluster of 2 to 50 males balled around her (Fig. 8-2). The largest and most active of these "mating balls" may drift several centimeters and seem to boil as the males on the outside of the cluster try to work their way inward to the female. The mating ball grows as nearby males dive into the cluster; the tumult continues for up to 25 minutes as the group breaks up only to reform again. At some point during a typical contest, two males become tightly wrapped around the thorax of the female and continue to struggle, until one gains a dorsal position and attempts to fly away carrying the female beneath him. If he is capable of carrying her and is not overtaken by pursuing males, the male lands in the surrounding vegetation and only then does the pair copulate for several seconds. If he has difficulty carrying her because he flies too slowly or lands again while still within the emergence area, the mating ball reforms and the battle resumes. Although few males were given individual marks in the studies cited, one large marked male was seen mating twice, so polygyny is possible (and perhaps common).

Although females seem to have little control over which males win battles, they are not passive participants in these free-for-alls. A female sometimes eludes her suitors when a short-lived mating ball forms prematurely before she emerges. At this point, she does not wait for a mate but immediately flies away. Other females escape from mating balls before they are mated, probably because males interfere with each other. However, one male is usually successful: only 5% of females seen emerging in the presence of males left the emergence area not in the company of a male (undoubtedly, some females also emerge without being detected by either wasp males or human observers) (O'Neill et al. 1989). Avoiding males in the emergence area may benefit females because they can be injured in contests, particularly when many males are involved. One female in a mating ball lost about 75% of her wings, and another died when her head and forelegs were partially torn from her thorax. The relatively few females that leave the emergence area alone and unmated are available to males active in lower densities in the vegetation and open sand surrounding the emergence area. As do males in the emergence area, these males patrol in sinuous flights, landing intermittently and pouncing on other insects. However, they do not stop to dig, and they never fight. When they encounter receptive females, a brief mating occurs, without the preliminary tandem flight, which is unnecessary because low male densities outside of the emergence area reduce the chance of interference.

Males adopting this nonaggressive patrolling tactic do not differ in body size from those in the population as a whole (mean body mass = 8.7 mg). However, those found in the emergence area (11.6 mg) are much larger than the average for the population, and those found in mating balls are larger still (12.7 mg) (O'Neill et al. 1989). Even in this group of combatants, larger males have an advantage, as judged from observations of males attempting to carry females away. Males that transport females without apparent difficulty, leaving the emergence area without landing again and without further interference, are always larger (30% heavier on average) than their potential mates and thus are able to lift and carry them fairly easily. Males that have difficulty are usually smaller than the female (6% on average) and are often caught by competitors who renew the fighting and sometimes usurp the female.

Among the hundreds of males captured while carrying females away from mating balls, none were smaller than the average for the population. Thus, small males stand little chance of succeeding within the emergence area, simply because they lack the strength to carry most females to a safe haven where mating can occur unimpeded. Because mated female *B. quinquespinosus* are subsequently unreceptive and most females apparently mate with a male in the emergence area, few virgins are available to the smaller males that patrol the periphery. However, a small chance of achieving a mating via the nonaggressive patrolling tactic is probably greater than the essentially zero chance of competing successfully against the largest males in the emergence area.

Defense of Individual Emerging Females in Other Species

The mating behavior of *Bembicinus quinquespinosus* closely resembles that of *Bembix rostrata* (Schöne and Tengö 1981). Male *B. rostrata* emerge 1–5 days before females and patrol an emergence area, where females of the previous generation nested in densities as high as 35 nests/m^2. Groups of up to 50 males congregate and "dig in the sand, often in a furious way, trying to supplant each other in the centre of the crowd." A cluster of males forms around the emerging female, and one male attempts to carry her away. Males of another bembicine, *Glenostictia satan*, also form clusters around emerging females (Longair et al. 1987), but, unlike *B. quinquespinosus* and *B. rostrata* males, *G. satan* males do not carry females away. Rather, once the successful male grasps a female by the neck and the petiole, he is left unharassed to complete mating. Perhaps, by locking on to his mate, he is impossible to dislodge and is therefore left alone. Male bees, which possess devices for tightly grasping receptive females, exhibit similar forms of mating (O'Neill and Bjostad 1987, and references therein). Males of the sphecid *Trigonopsis cameronii* (Eberhard 1974) exhibit a variation on the theme of defense of individual emerging females. Males patrol clusters of multicelled mud nests, using wet cell caps on individual cells as cues to detect virgin females about to emerge. They vigorously defend these cells and mate with the female immediately upon her appearance.

The biologies of *Bembecinus quinquespinosus*, *Bembix rostrata*, *Glenostictia satan*, and *Trigonopsis cameronii* (as well as the bee *Centris pallida*; Alcock et al. 1977) show variation in details, but they share certain features that make defense of emerging females a profitable tactic for some males. First, females of these species are receptive upon emergence and appear to mate just once. Second, preemergent females are easy to find because of high densities, short emergence periods, or cues that betray their precise location.

The studies of *B. quinquespinosus* and *B. rostrata*, in particular, reveal how sexual selection affects the ability of males to quickly locate receptive virgins. Males are so attuned to finding potential mates that they display a low threshold of response to anything that remotely resembles females. Males of both species attend to nonvisual cues when they search the soil surface for evidence of an impending emergence. *Bembix rostrata* males locate preemergent females using chemical and acoustical cues (Schöne and Tengö 1981, Larsen et al. 1986). Buzzing noises made by the subterranean females are at an intensity and frequency audible to human observers from at least several meters, but they also extend into the ultrasonic range. The sounds may actually result from the females' use of vibration to loosen soil while digging upward (Spangler 1973). Acoustical cues are apparently unnecessary for *B. quinquespinosus* males, because they will dig out freshly killed females buried by researchers (O'Neill and Evans 1983a). Male *B. quinquespinosus* are discriminating enough that they do not respond to insects of other species that are buried and only briefly investigate sites of natural emergence of other species (O'Neill and Evans 1983a). However, males of both *B. rostrata* and *B. quinquespinosus* often make the mistake of digging out conspecific males.

Misdirected mate searching and even misguided copulation attempts probably result from (1) the sensory limitations of males that could cause them to confuse various objects and organisms with conspecific females and (2) the eagerness of males to take advantage of all potential mates. Because conspecific males are usually similar in appearance to females and are active in the same areas, male wasps often pounce on one another. In many cases, it is easy to distinguish these homosexual interactions from aggressive contests, especially when the perpetrator attempts copulation. In the wild contests surrounding recently emerged females, male *B. quinquespinosus* sometimes mount other males and attempt to carry them away (O'Neill and Evans 1983a). Such misdirected mating attempts have also been observed in *B. rostrata* (Schöne and Tengö 1981), *G. satan* (Longair et al. 1987), *Crabro cribrellifer* (Low and Wcislo 1992), and *Pseudomasaris zonalis* (Longair 1987). Male *Dasymutilla foxi* court their fellow males (Spangler and Manley 1978) even though winged males differ greatly in appearance from the wingless females. Homosexual mountings are maladaptive for male wasps, but their brevity is such that they probably cost little. Furthermore, the value of a higher response threshold that allowed males to avoid such reactions could be offset by a loss of genuine mating opportunities (Thornhill and Alcock 1983).

A male wasp's threshold for mating attempts is often so low that he may try to mate with dead insects or noninsects. Male *Glenostictia satan* attempt to mate with "dead conspecifics, fly pupae, *Eriogonum* leaves, *Solanum* fruits, and the reddish basal stems of *Talinum* plants" (Longair et al. 1987). Male *Bembecinus quinquespinosus* pounce on rabbit pellets and the droppings of the horned lizard *Phrynosoma douglassi*; the latter contained the remains of conspecific wasp prey of the lizard in the form of exoskeletal fragments so perhaps emanated odors that males associated with potential mates (O'Neill, personal observation). Male *Bembix rostrata* attempt copulation with severed abdomens of females, although they are discriminating enough to prefer those of virgins over older females (Schöne and Tengö 1981). All three of these examples come from species in which competition for mates occurs in areas of high male density. Because males facing intense competition in high-density areas must have split-second reactions to potential mates, they are prone to make recognition errors.

MALE PATROLLING OF EMERGENCE SITES

Interactions like those observed in *Sphecius speciosus* and *Bembecinus quinquespinosus* are not a feature of all species with males that seek females in emergence areas. Males of many species patrol

Table 8-2 Examples of locations in which male wasps nonaggressively patrol for receptive females

Location of patrolling	Family	Species	References
Emergence/ nesting area	Pompilidae	*Evagetes subangulatus*	Barrows 1978
	Sphecidae	*Bembecinus nanus*	Evans & O'Neill 1986
		Bembix spp.	Evans 1957b, 1966c
		Cerceris watlingensis	Elliott 1984
		Microbembix monodonta	J. Parker 1917
		Oxybelus bipunctatus	Peckham et al. 1973
		Philanthus zebratus	Evans & O'Neill 1978
		Plenoculus boregensis	Rubink & O'Neill 1980
		Stictia carolina, S. heros	Evans 1966c, Larsson 1989b
		Tachytes intermedius	Kurczewski & Kurczewski 1984
	Vespidae: Masarinae	*Jugurtia confusa*	F. Gess & Gess 1980
Emergence/ hunting sites	Bethylidae	*Epyris subangulatus*	F. Williams 1919a
	Mutillidae	*Dasymutilla bioculata, D. foxi*	Brothers 1972, Manley & Taber 1978
		Pseudomethoca propinqua	Jellison 1982
		Sphaerophthalma orestes	Mickel 1938
	Scoliidae	*Scolia manilae*	F. Williams 1919c
	Tiphiidae	*Tiphia popilliavora*	King & Holloway 1930
	Sphecidae	*Larra bicolor*	Castner 1988
Flowers	Pompilidae	*Pepsis thisbe*	Alcock & Johnson 1990
	Sphecidae	*Ammophila dysmica*	Rosenheim 1987b
		Cerceris graphica	Alcock & Gamboa 1975
	Vespidae: Eumeninae	*Ancistrocerus antilope*	Cowan & Waldbauer 1984
	Vespidae: Masarinae	*Ceramius tuberculifer*	Mauss 1996

Table 8-2 *Continued*

Location of patrolling	Family	Species	References
		Quartinia vagepunctata	F. Gess & Gess 1992
		Pseudomasaris vespoides, P. zonalis	Longair 1987
Water sources	Vespidae: Eumeninae	*Abispa ephippium, Paralastor tricarinulatus*	A. Smith & Alcock 1980
	Vespidae: Euparagiinae	*Euparagia richardsii*	Longair 1985a
	Vespidae: Masarinae	*Ceramius capicola*	F. Gess & Gess 1980
Sleeping clusters	Sphecidae	*Steniolia obliqua*	Evans 1966c

Note: Many of the publications cited provide only brief and sometimes anecdotal accounts of male behavior.

emergence sites nonaggressively (Table 8-2). Two well-studied examples are the high-flying males of *Philanthus zebratus* and the sun-dancing males of many *Bembix* species.

Case Study: *Philanthus zebratus*

Male *Philanthus zebratus* were studied at a site called Deadman's Bar (not a drinking establishment) near the banks of the Snake River in Grand Teton National Park (Evans and O'Neill 1978, O'Neill and Evans 1983a). Usually between 10:30 and 11:00 A.M., they gather in the nesting area, perching on the ground or on low plants, but not interacting with other males or digging for females. However, they soon rise from their perches to heights of 3–5 m above the nesting area (see Fig. 8-8), where they hover and make slow horizontal flights over distances up to 10 m. The flights last less than 20 seconds but are repeated many times during the day, interspersed with resting periods of similar duration. While patrolling above the nesting area, males chase passing insects, including females that rise to the same heights while making orientation and foraging flights. Males intercept females during these flights, and mating pairs descend to the ground.

Case Study: *Bembix* spp.

Similar flights, much closer to the ground, are made by mate-searching male *Bembix* of many North American species. Females of these *Bembix* nest in dense aggregations, often in the open sandy expanses of beaches and inland sand dunes (Evans 1957b, 1966a). Soon after emergence each year, males patrol the emergence/nesting areas using flights often referred to as "sun dances" (Rau and Rau 1918). The sun dance "consists of a more or less continuous flight in circles, figure eights, and irregular patterns" (Evans 1957b) that resembles the "rhythmical, gliding motions" of skaters (Rau and Rau 1918). Males apparently range widely within the bounds of the nesting area, generally remaining within 10 cm of the soil surface, where they mingle with other males. Sun dances of different species vary in height, in speed, and in the duration of pauses between flights. The most distinctive flights are those of male *Bembix occidentalis* and *B. pruinosa,* which combine their sun dances with "hopping dances," series of short hopping flights that make groups of males look like "aggregations of very small toads" (Evans 1957b). Sun dances have also been reported for *Stictia,* who patrol at slightly greater heights (Evans 1966c, Larsson 1989b), and *Dasymutilla vesta,* who patrol mixed with males of their host species, *Bembix cinerea* (Evans 1957b).

Patrolling male *Bembix* pounce on other insects, including conspecific males, but they display no overt aggression during their patrols. When a receptive female is intercepted, they mate briefly while the male sits atop his mate and holds her firmly with the aid (in some species) of stout spines or teeth on the femur of his midlegs. For most species, it is not known whether mating females have begun nesting, but because nesting females of *B. cinerea* and *Stictia carolina* always reject males (Evans 1957b, 1966c), mating females are probably recently emerged virgins.

Patrolling of Emergence Areas by Other Species

The high flights of *Philanthus zebratus* and the sun dances of *Bembix* are classic examples of scramble competition. Patrolling tactics in emergence/nesting and emergence/hunting areas have been reported in a variety of families of solitary wasps (Table 8-2), although detailed studies are rare. In addition, it may not always be possible to reliably classify mating sites as either emergence or nesting areas,

Table 8-3 Variation in the occurrence of overt aggression among males of species that mate in emergence areas

Species	Stage at which fighting for female occurs		
	Before emergence of female	At emergence (appearance) of female	After male pairs with female
Ancistrocerus adiabatus	No	No	No
Bembecinus nanus	No	No	No
Goniozus legneri	No	No	No
Philanthus zebratus	No	No	No
Bembix nubilipennis	No	No	Sometimes
Euodynerus foraminatus	Yes	No	No
Trigonopsis cameroni	Yes	No	No
Glenostictia satan	Yes	Yes	No
Bembecinus quinquespinosus	Yes	Yes	Yes
Bembix rostrata	Yes	Yes	Yes
Sphecius speciosus	Yes	Yes	Yes

Sources: See text and Table 8-2 for references.

because females often nest in the same locations in successive years and some females nest before others have emerged. Similarly, female parasitoids may hunt in areas where other females are still emerging (e.g., Mutillidae in Table 8-2). The time in the life cycle that females mate can be determined either by observing when marked females of known age copulate or by assessing the physiological condition of the females found mating. Recently emerged females usually lack mature oocytes and have an abdomen replete with fat bodies.

MATING WITH NEWLY EMERGED FEMALES: A SUMMARY

Tactics involving mating with newly emerged females span the spectrum from completely nonaggressive patrolling to intensely fierce fighting that may occur before, during, or after a female's appearance (Table 8-3). Not unexpectedly, no one characteristic of a species turns out to be a good predictor of its mating tactic. For example, although the density of competing males varies among species, males of some species with high density never fight whereas others battle with one another over receptive females. Further field studies and comparative analyses will be necessary before we understand the relative roles of and interactions among factors such as (1) female density, locatability, and emergence schedules; (2) female choice; (3) male density; and (4) variation in competitive abilities among conspecific males.

Protandry is a common feature of species whose males seek newly emerged females and probably represents an adaptive aspect of male life-history strategies. We should also consider whether protandry is a nonadaptive effect of differential developmental rates of offspring of different size: because males are nearly always smaller than females, perhaps they emerge sooner because they have shorter developmental periods. However, it appears that protandry is not just a by-product of differential developmental rates. Both sexes of most species of temperate-zone wasps complete most of their growth before overwintering, so both sexes have finished growing before breaking diapause in the spring (Evans and West-Eberhard 1970). In addition, adults of cavity-nesting eumenines often do not emerge from their natal nests until several days after they have emerged from their cocoons. Another point to note is that protandry occurs in *Bembecinus quinquespinosus* and *Bembix rostrata*, both of which lack sexual size dimorphism (O'Neill and Evans 1983a, Ghazoul and Willmer 1994). Furthermore, females of those cavity-nesting species whose outermost nest cells contain females emerge before their smaller brothers (Krombein 1967).

Male Defense of Individual Nests

Although it is often difficult to determine whether males rendezvous with emerging or nesting females, the distinction is clear in certain genera of sphecid wasps whose males defend active nests and mate with provisioning females.

Case Study: *Trypoxylon politum*

The males of *Trypoxylon politum*, the pipe-organ mud-dauber, have been studied by H. Brockmann and A. Grafen in Florida (Brockmann 1988; Brockmann and Grafen 1989, 1992). The common name of these slender, glossy black wasps derives from their shape and the makeup of the tubular mud nests that females construct on vertical surfaces, such as cliff faces, tree branches, palm fronds, and bridges (Cross et al. 1975, Brockmann 1988). Within these mud tubes, females provision a linear sequence of cells stocked with paralyzed spiders. After fully stocking a cell, they lay an egg on the mass of spiders and seal the cell with a mud partition.

When a male *T. politum* finds a provisioning female's nest, one unattended by another male, he perches inside the nest hole with his head oriented outward and downward so that he can scan for potential intruders. Males are especially aroused by the arrival of a conspecific male, which may climb onto the nest, grasp the resident male with his mandibles, and attempt to pull him out. As the rivals struggle, they may not only continue to bite, but also slap one another with their abdomens. Fights are usually short, though 17% last eight or more minutes. Neither larger males nor resident males have a consistent advantage in these potentially deadly encounters, during which males may fall to their death in the water below territories.

A male that controls an active nest can mate with the provisioner and does so repeatedly (the 2,694 *T. politum* matings recorded by Brockmann and Grafen must surely be a world record for solitary wasp watchers). Matings occur at almost any time while the female builds or provisions the nest, although most occur during the crucial interval between completion of cell provisioning and oviposition. When the female maneuvers the last prey spider into place and turns to lay an egg, the male grasps her by the antennae with his forelegs and pulls her toward him (Fig. 8-3) until she turns around and copulates. This sequence is repeated up to eight times before the male allows the female to lay an egg while still holding onto her. While mating, the male holds the female at an oblique angle inside the nest tube, thus preventing other males from entering and usurping his position.

Mate guarding of the type observed in *T. politum* is often associated with male-biased operational sex ratios (Thornhill and Alcock 1983), which may be enhanced by male-biased population sex ratios. In the southern United States, *T. politum* populations have two generations each year. In the first, males constitute 50% to 72% of the population (mean = 65%). In the second generation, sex ratios are often 1 : 1 or even female-biased (Brockmann and Grafen 1992). However, some of the late-emerging males of the first generation live

Figure 8-3. Male *Trypoxylon politum* (A) holding onto female and (B) pulling her from the nest prior to copulation (from Brockmann and Grafen 1989).

long enough to interact with males and females of the second generation. Thus, although the primary sex ratio of the second generation may not be male-biased, mate-seeking males may outnumber nesting females in both generations, thus leading to intense competition for the relatively few mates.

Female *T. politum* mate multiply (and presumably use the sperm of the last mate to fertilize an egg), with the consequence that the most critical time for a male to be with a female on a nest is just before oviposition. Male *T. politum* prefer to guard active nests that have spider prey present and that are within 24 hours of having an egg deposited. Males exhibit particularly high nest fidelity when the female is about to complete provisioning of a brood cell, and males invariably terminate guarding immediately after oviposition. They may return to the same nest when a later cell is being provisioned, or they may move on to another active nest. Furthermore, when a male usurps a nest before oviposition, he repeatedly copulates with the female, an action that may displace the previous resident's sperm. Although large *T. politum* males are not better fighters, they probably have higher mating success because they have more preoviposition copulations than their smaller rivals, and it is these matings that probably provide the sperm that fertilizes the egg.

The form and timing of guarding help ensure that a female's egg will be fertilized by the resident male (again assuming last-male sperm precedence). However, because Hymenoptera have haplodiploid sex determination, males are related only to the daughters produced by their mate (because her sons are produced from unfertilized eggs). Thus, a guarding male benefits if he can influence the probability that his mate will produce daughters. Brockmann and Grafen found that the probability that an offspring would be a female was nearly six times higher on guarded than on unguarded nests. However, a guarding male does not gain this benefit by directly forcing the female to lay a fertilized egg. Rather, a female on a guarded nest forages more efficiently, and more heavily provisioned cells tend to produce daughters (which are larger on average than sons).

Females at guarded nests have higher provisioning rates because males not only fend off conspecific male rivals but also defend the nest against cuckoo wasps, miltogrammine flies, and ants. Upon the appearance of one of these nest enemies, the male buzzes loudly, lunges outward, and attempts to bite the intruder. In the absence of male guards, females spend more of their own time guarding and consequently less time provisioning. Occasionally, females also receive other forms of aid. Some males consolidate the wet mud on the inner walls of the nest, push spiders to the back of the cell, where they are less likely to fall out of the vertical cell, and malaxate prey (malaxation may soften prey to allow more efficient feeding by the wasp larva).

Although males remain with females and aid in nesting, their "devotion" to their partners has its limits. Although males drive off potential nest enemies, they will copulate with nonresident females that arrive at the nest to steal prey. A male presumably "gains more by placing his sperm in another female's storage sac than he would from exerting himself to defend a spider or two" (Brockmann 1988). Where nest density is high, guarding males sometimes mate with females in nearby nests (Hartmann 1944b). Matings (or attempted matings) between nest guarders and nonresident females has also been seen in other *Trypoxylon* (Rau 1928, Kurczewski 1963b, Krombein 1967, Coville and Coville 1980, Brockmann and Grafen 1989).

DEFENSE OF INDIVIDUAL NESTING FEMALES BY OTHER SPHECID WASPS

Nest guarding and repeated mating with a single provisioning female have evolved independently in at least four sphecid subfamilies, and they are not restricted to species of any one nest type (Brockmann and Grafen 1989) (Table 8-4). Other than *Trypoxylon,* nest guarding has been best studied in several species of *Oxybelus* (Fig. 8-4). Nest-guarding males of different species vary in how they contribute to nesting and in whether they guard from positions inside or outside the nest (Brockmann and Grafen 1989). Some apparent differences among species may result from the varying duration of different studies. Although brief observations suffice to confirm that males guard nests, only prolonged studies (Coville and Coville 1980, Brockmann and Grafen 1989, Brockmann 1992) are likely to chronicle infrequent male behaviors. Some "helping" behaviors of *T. politum* males were seen only a handful of times in several thousands of hours of observation (Brockmann and Grafen 1989).

Nest guarding is one potential tactic for males of species with multiple mating by females, male-biased operational sex ratios, and last-male sperm precedence (though the last may not be critical for the evolution of nest guarding). Consider that a male might have two options: to guard a nest (and mate with the provisioner) or to patrol widely without nest defense (and mate with receptive females he encounters). Females of both *Oxybelus sericeus* (Hook and Matthews 1980) and *T. politum* (Brockmann and Grafen 1989) are known to copulate while away from the nest. On average, patrollers may contact and mate with more females, but under some circumstances sedentary guarding may be more profitable. Patroller matings may not often result in offspring, if most nests are guarded and if nest guarders have the last chance at mating before fertilization of the egg. Even if last-male sperm precedence is weak or lacking, a nest guarder may help ensure paternity simply by mating more often than his potential rivals, with the result that rival sperm are outnumbered. Thus,

Table 8-4 Selected species of Sphecidae in which males guard (or are suspected to guard) nests and mate with provisioning females

Species	Location of guard	Male defends nest from conspecific males	Other insects repelled	References
Sphecinae				
Dynatus nigripes	Outside nest	Yes	Not known	Kimsey 1978
Larrinae				
Pison strandi	Inside nest	Not known	Not known	Tsuneki 1970
Trypoxylon monteverdae	Inside or outside nest	Yes	Chrysidids, ichneumonids	Brockmann 1992
T. nitidum	Inside nest	Yes	Other *Trypoxylon*	Coville 1981
T. politum	Inside nest	Yes	Ants, chrysidids, miltogrammines, other *Trypoxylon*	Brockmann & Grafen 1989, Cross et al. 1975
T. spinosum	Inside nest	Not known	Ants	Hook 1984
T. superbum	Inside nest	Not known	Ants	Coville & Griswold 1984

T. tenoctitlan	Inside nest	Yes	Ants, chrysidids, conspecific females	Coville & Coville 1980
T. tridentatum	Inside nest	Not known	Chrysidids	Paetzel 1973
Crabroninae				
Oxybelus sericeus	Inside or outside nest	Yes	Miltogrammines	Bohart & Marsh 1960, Hook & Matthews 1980
O. subulatus	Inside nest	Yes	Miltogrammines	Peckham 1977
Philanthinae				
Trachypus denticolli	Inside nest	Not known	Not known	Janvier 1928

Source: Modified from Brockmann and Grafen 1989, where more complete list of references can be found.

Figure 8-4. Copulating pair of *Oxybelus sericeus* at nest entrance; the female has dropped her prey on the ground in order to mate (from Hook and Matthews 1980).

patrollers are less likely to succeed if most nests are guarded, which is more likely if the number of mate-seeking males exceeds that of nesting females. Population sex ratios are male-biased, not only in *T. politum* (Cross et al. 1975, Brockmann and Grafen 1992) but also in *T. superbum* (Coville and Griswold 1983) and in one population of *T. tenoctitlan* studied by R. Coville and P. Coville (1980). However, Coville and Colville found no evidence of male-biased ratios in other *T. tenoctitlan* populations, and R. Coville and C. Griswold (1983) reported female-biased ratios in *T. xanthandrum*. Note, however, that the historically typical sex ratios of these species may differ from those in the trapnests of the cited studies, because trap-nest sex ratios are sensitive to the particular hole diameter chosen for the study (Chapter 10). Furthermore, population sex ratio is only one determinant of operational sex ratio.

Nest guarding may enhance the reproductive success of females, as well as that of males. For example, male *Oxybelus subulatus* chase away all intruding insects, including conspecific males and parasitoid flies, from the nest. Consequently, nests guarded by males have about half the rate of miltogrammine fly parasitism as unguarded nests (Peckham 1977). *Trypoxylon* males sometimes behave in ways that suggest that benefits to females are not simply side effects of the male mating tactic. Male *T. superbum* not only repel parasitoids during provisioning but also stay until the larvae have spun cocoons and are resistant to attack by ants (Coville and Griswold 1984). Furthermore, male *Trypoxylon* sometimes exhibit decidedly unmale-like behaviors by participating directly in brood rearing. Male *T. rubricinctum* help females by cleaning out nest holes and moving prey into position (Paetzel 1973), and *T. politum* and *T. monteverdae* assist females by smoothing fresh mud on the nest walls (Brockmann and Grafen 1989, Brockmann 1992). Female *T. tenoctitlan* do not even begin to provision nests until they have attracted a male (Coville and Coville 1980). It seems likely, however, that male nest guarding in sphecids evolved as a mating strategy and that any behaviors that improve nest defense evolved later. Once guarding of nests from conspecific rivals evolved because of its effect on the number of offspring fathered, males that lowered their threshold of aggression and attacked nest parasites and marauders would have more surviving offspring (Brockmann and Grafen 1989). Under this scenario, helping would become more extensive until its benefits were outweighed by its costs in terms of energy expenditure and missed matings elsewhere. It is instructive that copulations of *T. politum* are 23 times more frequent than their (nonguarding) helping behaviors: males may be helpful, but they have their sexually selected priorities straight.

Given that nest guarding has evolved repeatedly in sphecid wasps, why has it not appeared in eumenines with similar nesting strategies? The answer may lie in the provisioning patterns of the two groups. Females of the nest-guarding sphecids lay an egg after the cell has been completely provisioned, whereas female eumenines lay it in an empty cell before provisioning. Thus, a male sphecid nest guarder benefits by remaining at the nest during provisioning, awaiting the time when the female will lay an egg. When she finally does lay a fertilized egg, she will probably use sperm from the nest guarder, because he has packed her spermatheca with his gametes. On the other hand, because eumenines provision after fertilization and egg laying, nest guarding during provisioning does not enhance the male's probability of fathering the offspring in the cell and it decreases his chance of finding other females.

Mating at Resources

The food or nest materials gathered by females are often either very abundant or widely dispersed, so males usually cannot gain much by seeking females at particular resource sites. However, there are a few exceptions.

TERRITORIALITY AT RESOURCES

Many male bees defend pollen and nectar sources, and females gain access to food only by mating with the resident territorial male (Alcock et al. 1978). Thus, males defending particularly rich flower resources have the chance of mating with multiple females. Defense of territories at resource sites is uncommon in wasps, however. In the few known examples, the resources that attract females are patchily distributed and rare enough to warrant a male's restricting his mate-seeking activities to these sites. Males of these wasps do not control the females' access to resources but simply prevent other males from gaining access to resource-based hotspots of female activity.

One example among wasps is *Stenodynerus taos*, whose males defend territories at sites where females collect water for nests. Such defense is feasible because water is rare in the desert habitat of this species (Longair 1984). Similarly, males of *Aphilanthops subfrigidus* establish territories within mating swarms of the ant *Formica subpolita,* where the females hunt ant queens (O'Neill 1990, 1994). Perhaps the most unusual resource-based mating sites are those of *Mellinus rufinodus* (Evans 1989). Females of this species hunt flies at piles of mammalian feces. Males establish territories at these patchily distributed sites, driving off conspecific male intruders and mating with visiting females. One male copulated 11 times in 30 minutes, impressive sexual stamina on the part of the wasp, as well as an admirably long interval for Evans to have stationed himself next to dog excrement.

COURTSHIP FEEDING BY THYNNINE WASPS

Some males of the tiphiid subfamily Thynninae do control access of females to resources. However, male thynnines control female access to nectar in a way very different from that of territorial bees. Rather than wait for the females to come to them, they bring the wingless females to the nectar (or the nectar to the females) as a means of inducing the females to mate.

A B

Figure 8-5. (A) Calling female *Megalothynnus klugii.* (B) Male *M. klugii* grasping a calling female. (A–B from Alcock and Gwynne 1987.)

Case Study: *Megalothynnus klugii*

The mating system of *Megalothynnus klugii*, which was studied in western Australia by J. Alcock and D. Gwynne (1987), follows a pattern typical for many Australian thynnines: scramble competition among males followed by courtship feeding of the female by the male. When female *M. klugii* are ready to mate, they emerge from the soil, ascend to the top of vegetation, and release a pheromone that quickly attracts one or more males, who are usually larger than the female (Fig. 8-5A). A male detecting a female's odor approaches her from downwind in a zig-zag flight pattern. Upon reaching the female, he immediately grasps her (Fig. 8-5B), pulls her from the perch, and carries her to nearby flowering shrubs.

Upon coupling with the male, the female curls up beneath his abdomen and allows herself to be carried from flower to flower as the male forages. At various intervals, the male regurgitates nectar to the female, who briefly uncoils in order to imbibe it. After a series of feedings, the pair uncouple and the female drops to the ground, where she burrows in to hunt for scarab larvae. She presumably remains underground until she finds one or more prey, laying an egg on each victim, so the male that has just fed her has a good chance of fathering her next few offspring. Her subsequent young may be fathered by other males because female thynnines may mate several times, perhaps to obtain multiple feedings. Because large males were better able to carry large females, and because large females are more fecund, large males may ultimately enjoy greater reproductive success (Alcock and Gwynne 1987).

Courtship Feeding by Other Australian Thynninae

Courtship feeding has evolved independently in insects as diverse as scorpionflies (Mecoptera), dance flies (Diptera: Empididae), and lygaeid bugs (Hemiptera: Lygaeidae) (Thornhill and Alcock 1983). Within the Hymenoptera, however, only male thynnines provide females with food as an inducement to mating. Furthermore, courtship feeding has been observed only in Australian species of Thynninae, although the subfamily is also represented in South America. *Megalothynnus klugii* represents one of the relatively few species of the Thynninae whose mating system has been well studied, but information on other Australian species reveals a great diversity of nuptial feeding styles (Table 8-5). Males feed females with liquid from flowers, extrafloral nectaries, or homopteran honeydew secretions (Given 1954), and males of one species feed females with droplets exuded from their anuses (Alcock 1981b). Female behavior also varies (Alcock 1981b); they use different postures to call from exposed perches, from burrows in the soil, or from beneath the soil.

Morphological modifications associated with the mating system of thynnines include various modifications of the males' heads that facilitate feeding, reversed sexual size dimorphism, and a coupling mechanism allowing males to carry females while *in copula* (Given 1954, Evans 1969b). After reaching flowers and coupling, male *M. klugii* continue to carry the wingless female. Such "phoretic copulation" is common in the Thynninae (Burrell 1935, Evans 1969b, Alcock 1981b) as well as in *Timulla vagans* (Sheldon 1970) and at least three genera of Bethylidae in which males apparently do not carry females to resources (*Apenesia*, *Dissomphalus*, and *Pristocera*)

Table 8-5 The diversity of forms of nuptial feeding in Australian Tiphiidae, subfamily Thynninae

Species	Form of nuptial feeding	Reference
Megalothynnus klugii	Female feeds from droplet regurgitated directly to her by the male	Alcock & Gwynne 1987
Neozeleboria proximus	Female feeds herself after being carried by male to honeydew source	Burrell 1935
Rhagigaster aculeatus	Female feeds herself after being carried by male to flowers	Burrell 1935
"Species I"	After being carried to plant, female feeds on fluid regurgitated onto leaf surface by the male	Alcock 1981b
"Species 24"	Female feeds on droplet male regurgitates onto ventral surface of distal segments of abdomen	Given 1954
Tachynomyia pilosula	Female feeds from bolus of food in concavity beneath the male's head; male carries food to female	Given 1954
Thynnoides spp.	Female feeds on droplet exuded from male's anus	Alcock 1981b

(Evans 1969b). Both sexes of these thynnines and bethylids have genitalia uniquely modified to hold the female securely (Evans 1969b). The clypeus and mandibles of male *Timulla vagans* appear to be structured to aid in holding females during nuptial flights (Sheldon 1970). I also noted earlier that male *Bembecinus quinquespinosus* (O'Neill et al. 1989) and *Bembix rostrata* (Schöne and Tengö 1981, Ghazoul and Willmer 1994) carry females before mating. These diverse species of Tiphiidae (Thynninae), Mutillidae, Bethylidae, and Bembicinae (Sphecidae) provide one of the clearest comparative trends in interspecific body size in the solitary wasps. Females of virtually all solitary aculeate species are larger than conspecific males. Most of the exceptions occur in those groups just discussed. There is no difference in mean size between the sexes of *B. quinquespinosus* (O'Neill et al. 1989) or *B. rostrata* (Ghazoul and Willmer 1994), whereas males average larger than females in the thynnines, mutillids, and bethylids in which males carry females during courtship or mating. It appears that, along with the evolution of female winglessness, the major phenomenon associated with an increase in the size of males relative to females is higher reproductive success of males that can carry females (O'Neill 1985). In *Microbembex cubana*, males are also larger than females, but mating has apparently not been observed (Toft 1987).

Although a formal comparative analysis has not been done (and will have to await a more solid systematic foundation in this relatively little-known group), several tentative hypotheses concerning the evolution of nuptial feeding can be proffered. The reduction of the female's mouthparts in the Thynninae is thought to be a derived character within the Tiphiidae. Within the Thynninae, females of the genus *Rhagigaster* have relatively well-developed mouthparts and can feed on honeydew without male assistance (Given 1954). Thus, male *Rhagigaster* carry the female to food on which she can feed on her own. This form of courtship probably represents the ancestral form of nuptial feeding for thynnines. We can compare the behavior of this species with what appears to be one of the most evolutionarily derived forms of nuptial feeding. In wasps of the genus *Tachynomyia*, the females' mouthparts are extremely reduced, whereas the males' heads are highly modified for transport and transfer of nectar or honeydew (Given 1954). The ventral portion of the male's head bears a deep concavity bordered by a dense fringe of recurved hairs that allow the male to carry a droplet of nectar "considerably larger than the head" (Fig. 8-6) to a calling female.

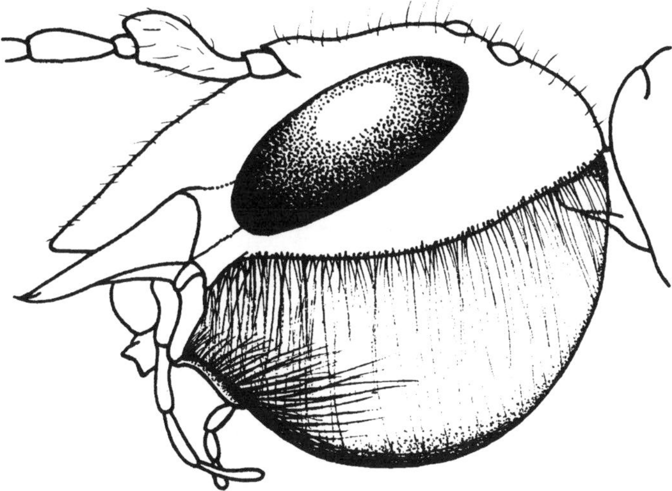

Figure 8-6. Head of male *Tachynomyia*, showing position of nectar droplet carried by male (redrawn by Catherine Seibert from Given 1954).

Female thynnines benefit from being transported and fed, because all are wingless and many have vestigial mouthparts. However, males of a thynnine species probably vary in their quality as mates because larger males probably can carry females farther and faster. Several aspects of female thynnine behavior suggest that they exhibit mate choice at some level. First, females may resist being carried off by males by clinging to their calling perches (Alcock 1981b, Alcock and Gwynne 1987). Second, *Megalothynnus klugii* females can abruptly terminate copulation by withdrawing their genitalia (Alcock and Gwynne 1987). Third, females that are not fully fed may not lay a batch of eggs immediately, but resume calling to attract another male. If the last male to mate with her fertilizes her eggs, she may have the option of using the sperm of only those males capable of satiating her. These forms of control by the females, combined with the facts that a successful male must win the competition to find females (Alcock 1981b) and must be strong enough to carry a female in flight (Alcock and Gwynne 1987), mean that males that father young are likely to be a nonrandom subset of the population. Theoretically, such mate choice can benefit the female by providing her with a superior genetic endowment for her young, as well as providing the obvious direct material benefits of more food (Alcock 1988).

PATROLLING RESOURCES

Case Study: *Pseudomasaris vespoides* and *Pseudomasaris zonalis*

Because female masarines forage for pollen and often specialize on particular plant species (Chapter 5), one might expect males to seek females at flowers. This general prediction was confirmed by R. Longair (1987) for two species of the genus *Pseudomasaris* that he studied in a Colorado mountain meadow containing patches of flowering *Penstemon unilateralis* (a pollen source for *P. vespoides*) and *Phacelia* sp. (the host plant of *P. zonalis*).

Male *P. zonalis* restrict their searches to patches of flowering *Phacelia*, flying circuits both within and between patches and never engaging in aggressive encounters with rivals. Male *P. vespoides* patrol clumps of flowering *Penstemon*, sometimes making only brief visits to each patch, where they hover and investigate other insects. The duration of these stops is always less than 5 seconds and is correlated with the number of flowers in the clump. At other times, males make longer sojourns in a patch, perching on one plant at the edge of the clump. Between occasional investigative flights, they also briefly defend the patches from other males in encounters that may escalate to grappling contests. However, within 5–8 minutes, they switch back to between-patch patrols. Male *P. vespoides* mate with females they encounter on flowers; Longair observed one marked male mating on six occasions.

Patrolling of Resources by Other Species

Examples of male wasps' patrolling resources for females are much more common than known cases of territoriality at resources.Certain male wasps nonaggressively patrol flowers or water sources, with most examples occurring in the three subfamilies of solitary Vespidae: Eumeninae, Masarinae, and Euparagiinae (Table 8-2).

Mating at Scent-Marked Territories

Males of many Philanthinae display a form of territoriality not found in other solitary wasps. Males of all territorial Philanthinae exclude rivals from the vicinity of their perches and scent-mark nearby plants using a behavior referred to as "abdomen dragging" (Alcock 1975d).

Figure 8-7. Territorial defense and scent-marking in male *Philanthus.* A scent-marking male is shown in the upper left, while two males about to butt heads are shown on the right (drawing by Byron Alexander from Evans and O'Neill 1988).

As often as two or three times per minute (Evans and O'Neill 1988), a male lands on a stem or leaf within his territory and walks up a short distance (and often back down again), pressing his head and abdomen against the plant surface while adopting a shallow V-shaped posture (Fig. 8-7). Abdomen dragging is exhibited only by territorial males of *Aphilanthops, Cerceris, Clypeadon, Eucerceris,* and *Philanthus* (Evans and O'Neill 1988, O'Neill 1990). By far, the best-studied genus is *Philanthus* (the beewolves) (Evans and O'Neill 1988, 1991).

Biochemical analyses confirm Alcock's (1975d) supposition that abdomen dragging serves to deposit chemicals on plants. If plant parts on which males have just dragged their abdomens are analyzed for the presence of volatile chemicals, a rich array of compounds is revealed (Schmidt et al. 1985, McDaniel et al. 1987, Schmidt et al. 1990, McDaniel et al. 1992). The same chemicals found on the plant also occur in extracts of the heads of males but not in extracts of male thoraces and abdomens, of females, or of plants not visited by males. Each beewolf species deposits a specific blend of three to eight hydrocarbons, each containing from 13 to 20 carbons. Males produce relatively large quantities of these chemicals. For example, the 220 μg of chemicals in the heads of male *P. basilaris* (Schmidt et al. 1985) represents about 2% of the dry mass of an average-sized male (mean dry mass = 0.01 g). The chemicals deposited on territories by *P. crabroniformis* (Evans and O'Neill 1988), *P. triangulum* (Schmidt et al. 1990), and *Eucerceris arenaria* (Alcock 1975b) are detectable by human noses.

The anatomical source of the volatile chemicals proves to be large mandibular glands (McDaniel et al. 1987). The glands fill major portions of the heads of males of most species of beewolves, but they are small in all females and in males of the non-scent-marking *P. albopilosus* (Evans and O'Neill 1988). D. Gwynne (1978) found that the mandibular glands of *P. bicinctus* communicate by way of a duct to a pair of clypeal hair brushes near the bases of the mandibles. In *Philanthus*, the brushes consist of tufts of hairs whose longitudinal grooves presumably serve to channel the chemicals along the hairs as the male rubs them on plants (Evans and O'Neill 1988). As a scent-marking male walks forward, he also drags the larger and more diffuse hair brushes on the venter of his abdomen over the area on which the clypeal brushes were applied. This action probably spreads the secretion over a broader surface, promoting rapid evaporation and diffusion downwind from the territory.

Because of the strict association of scent-marking with territoriality in philanthines, two alternative functions of scent-marking have been considered: (1) that they are sex pheromones attracting receptive females or (2) that they are warning signals alerting competitors to the presence of another male. Bioassays have not been done, but observations lend support to the first hypothesis. Just before mating with a territory resident, females of *P. basilaris*, *P. psyche*, and *P. pulcher* approach the territory from downwind, often flying in a zigzag pattern (Evans and O'Neill 1988), a behavior typical for insects orienting to an airborne pheromone. Moreover, if the chemicals were warning pheromones, they would be expected to reduce the frequency of fights over territories. In fact, fights are very common and, in *P. basilaris*, most frequent soon after a resident starts scent-marking (O'Neill 1983b). Rather than being repelled, intruders may actually be attracted to a territory because of the chemicals deposited on it. By usurping a scent-marked territory, males could forgo some of the cost of producing pheromone (Alcock 1975b).

A large pool of nonterritorial males await the chance to take over a territory, as has been demonstrated using removal experiments with *Philanthus* and *Eucerceris*. In these experiments, a male is removed from a territory and sequestered, after which the territory is monitored for the arrival of new males. The prevalence of such "floaters" can be confirmed by repeating the removals until no males occupy the territory. Using sequential removals, Evans and I (Evans and O'Neill 1985) induced up to 14 males of *Eucerceris flavocincta* to occupy a territory in one day, with a mean of about four males per territory. Thus, there were at least three nonterritorial males for every male holding a territory. The number of floaters also exceeded the number of territorial males in studies of *P. pulcher* and *P. psyche* (O'Neill 1983a). The numerous potential challengers to a resident's hold on a territory account for the high frequency of fights on territories.

The forms of fighting among male philanthines vary among species, but they fall within several definable categories: swirling flights, butting, and grappling. In swirling flights, two or more males fly about one another in tight circles at speeds that make it difficult to keep track of their identity (like trying to follow the pea in a very fast shell game). Swirling flights were first described for *P. multimaculatus* (Alcock 1975d) and are the only form of fighting in one population of *P. psyche* (O'Neill 1979). However, they undoubtedly represent contests (rather than mutual investigation), because they are sometimes followed by territory usurpation (Evans and O'Neill 1988).

In butting, as if they were airborne bighorn sheep, two males fly at one another (Fig. 8-7) and butt heads, sometimes producing a clicking sound audible several meters away (O'Neill 1983a,b). Flying male *P. basilaris* may also butt perched opponents, and a pair of males may clash up to four times in quick succession (O'Neill 1983b).

In the most intense form of interaction, grappling, males clasp one another and may try to bite their opponent. Grappling is usually initiated as the males grasp one another in midair, but it may continue after a pair fall to the ground. Bouts of grappling and butting may be intermingled in prolonged battles (O'Neill 1983a,b).

Butting and grappling are common among territorial male solitary wasps, and larger contestants usually have a clear advantage over their smaller rivals (Table 8-6). The best examples come from observational studies of *Philanthus* and *Eucerceris* (O'Neill 1983a,b, Evans

Table 8-6 Some species of solitary wasp in which conspecific males on territories engage in aggressive encounters that escalate beyond simple pursuit

Species	Form of interaction[1]	Larger males more successful? (research method or evidence[2])	Frequency of usurpations	References
Pompilidae: Pepsinae				
Hemipepsis ustulata	Sp	Yes (RE), but there is also an advantage to being a resident on a territory	Common	Alcock 1979a, Alcock & O'Neill 1987, Alcock & Bailey 1997
Sphecidae: (Sphecinae)				
Sphex pennsylvanicus	G, Bi	—	—	Kurczewski 1998
Trigonopsis cameronii	G, Bi	—	—	Eberhard 1974
Sphecidae: Crabroninae				
Oxybelus sericeus	G	—	Common	Hook & Matthews 1980
Sphecidae: Larrinae				
Tachytes tricinctus	Bu, G	Yes (PO)	Common	Elliott & Elliott 1992
Trypoxylon politum	As, G, Bi	No, larger males won 50% of fights (O)	45% of contests	Brockmann & Grafen 1989
T. monteverdae	As, G, Bi	—	—	Brockmann 1992
T. tenoctitlan	Bi	Yes, larger males often displaced smaller males even without fights, but fights between males of the same size often escalated (O)	Common	Coville & Coville 1980

Sphecidae: Bembicinae				
Sphecius grandis	Bu, G	Yes, outcome highly correlated with size difference (O)	—	Alcock 1975a, Hastings 1989a
S. speciosus	Bu, G, Bi	—	Common	Lin 1963
Sphecidae: Philanthinae				
Aphilanthops subfrigidus	Sw, Bu, G	—	—	O'Neill 1990
Eucerceris arenaria	Sw, Bu, G	—	—	Alcock 1975b
E. flavocincta	Sw, Bu, G, Bi	Yes, larger males won 100% of fights (O, RE)	35% of contests	Evans & O'Neill 1985
Philanthus basilaris	Bu, G, Bi	Yes, larger males won 100% of fights (O, RE, PO)	37% of contests	O'Neill 1983a
P. bicinctus	Bu, G, Bi	—	Common	Gwynne 1978
P. crabroniformis	Bu, G	Yes, larger males won 100% of fights (O, RE)	54% of contests	O'Neill 1983a
P. psyche	Sw, G (rare)	Yes (RE)	Common	O'Neill 1979
P. pulcher	Sw, Bu, G	Yes, larger males won 97% of fights (O, RE, PO)	16% of contests	O'Neill 1983a
Vespidae: Eumeninae				
Epsilon sp.	G (rare)	—	—	A. Smith & Alcock 1980
Euodynerus foraminatus	"Interact aggressively"	Large size advantage inferred (PO)	—	Cowan 1981

[1]As, slapping with abdomen; Bi, biting; Bu, butting with heads; G, grappling with legs; Sp, spiral flights; Sw, swirling flights (see text for explanation of forms of encounter). [2]O, direct observation of fights (interactions between males of equal size excluded); PO, patterns of territory occupancy; RE, removal experiments.

and O'Neill 1985). In observations of over 150 aggressive encounters in four species, the larger male of a pair won 99% of the fights, even though differences in size between combatants were often very small. Such a size bias is not surprising because the largest males of some *Philanthus* outweigh the smallest by 3 to 6 times (on a human scale, some beewolf fights are analogous to contests between a heavyweight boxer and a 5-year-old boy).

Because of this size advantage, males that occupy territories should be larger on average than conspecifics excluded from territories, a prediction confirmed in several studies. Males on territories tend to be larger than the average in *P. pulcher* (O'Neill 1983a) and *P. basilaris* (O'Neill 1983b). In *P. basilaris*, males range in head width from 2.2 to 3.7 mm. However, although males <3.2 mm constituted 34% of the population, they occupied only 11% of the territories. Removal experiments also confirm that territory residents tend to be larger than the nonterritorial males. Only one *P. pulcher* male among the 38 replacements was larger than the original resident (O'Neill 1983a), and none of the 30 *P. basilaris* replacements were larger than the original male (O'Neill 1983b).

Although territoriality occurs in many wasps, an obligate association of territoriality and scent-marking is unique to the Philanthinae and is probably an ancestral trait within the subfamily, as are the clypeal brushes used to apply chemicals (Bohart and Menke 1976). Territoriality and scent-marking have been observed in over two dozen species in five genera of Philanthinae: *Aphilanthops* (O'Neill 1990), *Cerceris* (Evans and O'Neill 1985), *Clypeadon* and *Eucerceris* (Alcock 1975b, Evans and O'Neill 1985), and *Philanthus* (reviewed in Evans and O'Neill 1988). D. Banks (1995) argues that male *Cerceris binodis* are territorial without scent-marking, but it is not clear whether males are actually territorial since no description of fights is given. Within the Hymenoptera, scent-marked territories are also found in certain bee species (Raw 1975, Vinson et al. 1982, Vinson and Frankie 1990, O'Neill et al. 1991).

DENSITY AND LOCATION OF PHILANTHINE TERRITORIES

Two major ways in which territoriality differs among philanthines are in the density of territories and in their location relative to nests. The description of the behavior of male *Philanthus bicinctus* and the following comparative survey illustrate the differences.

Case Study: *Philanthus bicinctus*

The mating system of *Philanthus bicinctus*, the bumblebeewolf (the largest North American beewolf), was studied from 1975 to 1978 at Great Sand Dunes National Monument in southern Colorado (Gwynne 1978, 1980) and in northern Colorado and Yellowstone National Park (Schmidt et al. 1985, Evans and O'Neill 1988). At all three places, females nest in large aggregations and probably mate just once, sometime after they start their nests (Gwynne 1981, O'Neill and Evans 1982).

The activities of male *P. bicinctus* at Great Sand Dunes are synchronized with the nesting rather than with the emergence activities of females. Males start to emerge one week before the first females, but they do not establish territories until the females begin digging. The distribution of territories, which are centered on small bare spots amid grasses, snakeweed, yucca, and prickly pear cacti, closely correspond with nest distribution. Each year, the first territorial males associate themselves with the first cluster of nests, and the territorial area expands outward with the growth of the female nesting aggregation. Males in Yellowstone (Fig. 8-8A) and in northern Colorado also establish territories in nesting areas (Evans and O'Neill 1988).

After spending each night in sleeping burrows, males at Great Sand Dunes arrive on territories in midmorning, immediately scent-mark the surrounding plants, and await the (possible) arrival of receptive females. They react quickly to passing insects, investigating any intruders that enter their territories and responding most strongly to conspecific females not carrying prey. Although this behavior increases their chances of intercepting receptive females and repelling intruding male challengers, it also attracts the attention of predatory robber flies (Gwynne and O'Neill 1980) and foraging *P. basilaris* females, which feed male *P. bicinctus* (and *P. basilaris*) to their young (O'Neill and Evans 1981).

The rate of scent-marking by male *P. bicinctus* declines during their daily tenure on territories, but the males aggressively defend their small domains throughout the day, pursuing, butting, grappling, and biting intruders. Mating, a relatively brief affair lasting 0.5 to 6.5 minutes, is initiated when a female interrupts her digging and flies to a nearby territory. Most *Philanthus* matings occur early in the season at about the time females start their first nests; those females captured while mating have ovaries that are not fully developed

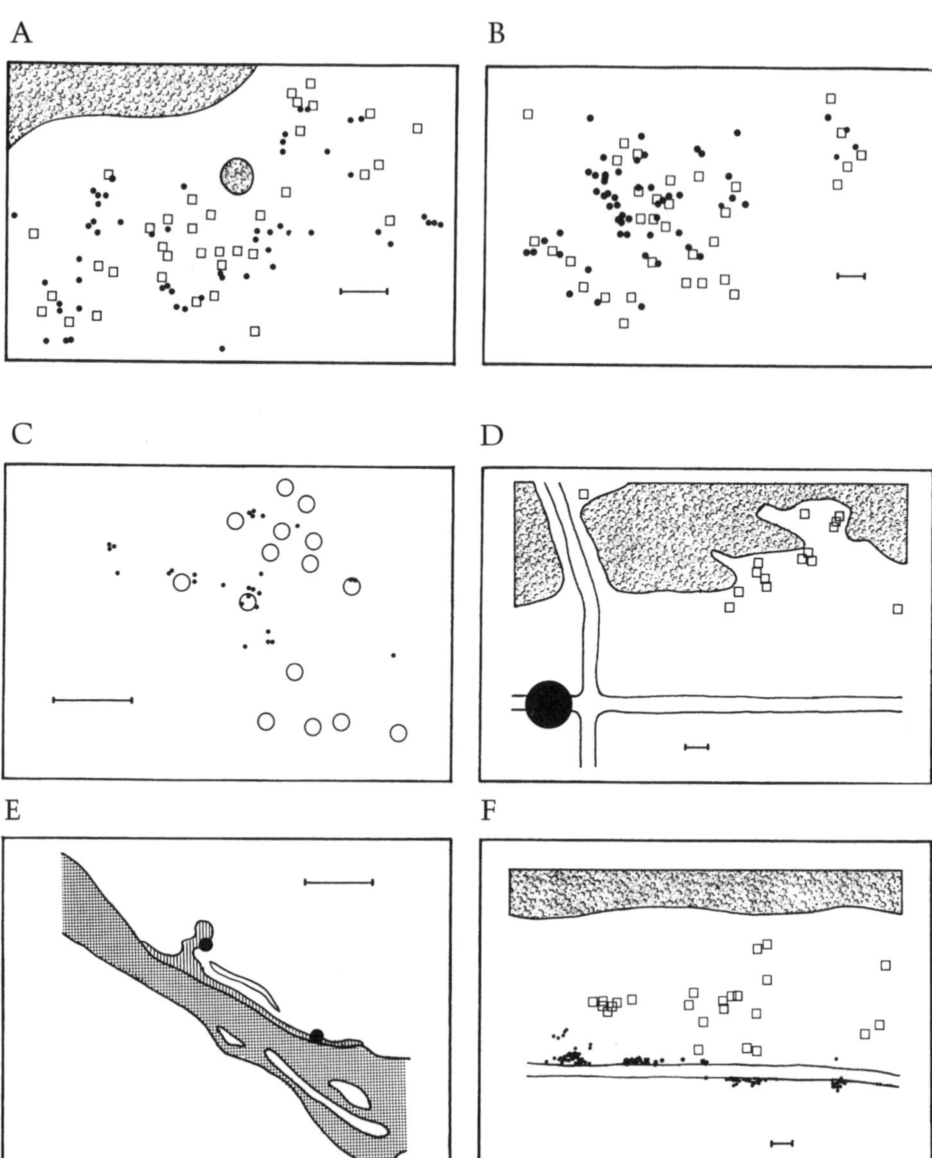

Figure 8-8. Maps of nest and territory distribution for six species of *Philanthus*. Except where indicated, solid circles are individual nest entrances, and hollow squares are individual territories. Species (and scale of map) are as follows: (A) *P. bicinctus* (5 m). (B) *P. psyche* (1 m). (C) *P. basilaris* (250 m; numbered circles are groups of territories [leks]. (D) *P. crabroniformis* (10 m; large circle is a nesting area). (E) *P. pulcher* (150 m; large circles are nesting areas; vertical hatching represents area containing dispersed territories, and stippling indicates the Snake River). (F) *P. zebratus* (5 m). (A–F redrawn from Evans and O'Neill 1988, O'Neill 1979, O'Neill 1983b, O'Neill and Evans 1983a.) Approximate distribution of trees is shown on three of the maps.

(Evans and O'Neill 1988). Thus, the best place to establish a territory is probably near active, newly established nests of recently emerged females. Gwynne (1980) confirmed this hypothesis, demonstrating that there was a significant correlation between the number of nests within 3 m of a territory and the proportion of a season in which the territory was occupied. Furthermore, those territories closest to many active nests tended to be occupied by the largest males. However, other territories are also established within areas of flowering snakeweed (where females forage) as far as 300 m from the main nest aggregation.

TERRITORY LOCATION AND DENSITY IN OTHER PHILANTHINAE

To illustrate the extremes of variation in the location of territories relative to nests, we can compare the mating system of *Philanthus bicinctus* with that of species with both higher and lower nest densities. The nest densities of *P. bicinctus*, as well as those of *P. psyche* (O'Neill 1979), are such that sufficient space is available within a nest aggregation for territories of a substantial proportion of males of the population (Fig. 8-8). However, nests of *P. basilaris* are widely dispersed and, thus, more difficult for males to find (O'Neill 1983b). Although female *P. basilaris* do not congregate, males cluster their territories in groups functionally equivalent to the leks of such birds as sage grouse, with the exception that the male wasps use scent rather than visual and acoustic signals to attract mates. As do males of lekking birds, male *P. basilaris* gather in "arenas," defend individual territories, and attract females who come to the territories for the sole purpose of mating. The leks of *P. basilaris* tend to be elliptical and up to 30 m long, with the average distance between the centers of neighboring territories being about 2 m. In 1979, one lek had 5–47 territories per day, while another had 4–16 territories. Many smaller leks also appeared, but for only a few days during a season. Curiously, different generations of males form leks in the same locations every year, even though no obvious landmarks or concentrations of nests and resources distinguish the sites occupied from those not used by males. One *P. basilaris* lek, for example, was occupied every year from 1978 to 1981.

The leks contain no significant resources that would cause females to converge on the sites occupied by males. Because territories are not associated with nesting areas, any one territory would be unlikely to include more than a single nesting female (and almost all contain no nests). Furthermore, territories contain no unusual concentrations of flowers on which females hunt or take nectar. Thus, we are left with the hypothesis that males defend scent-marked plants. As do males of *P. bicinctus*, those of *P. basilaris* vigorously defend their territories (Table 8-6), and some are very successful: one male *P. basilaris* maintained control of his territory for at least 20 days (O'Neill 1983b). It appears that female *P. basilaris* are attracted to leks only to find mates.

Nests of *P. basilaris* are so widely dispersed that males probably cannot monopolize prime nesting sites with abundant virgins. Nests of other *Philanthus* are easily found by males, but densities are so high that it is likely that only a small fraction of the males could fit territories within a nesting area. The result is that competition for territories would be extremely high and the cost of defense probably prohibitive. Perhaps it is for this reason that males in species with very high nest densities avoid the nesting areas and instead establish territories nearby in groups of intermediate (e.g., *P. pulcher*, *P. triangulum*) to high (e.g., *P. crabroniformis*) density that could also be referred to as leks or dispersed leks (Fig. 8-8).

Comparative patterns reveal that the relationship between female nest density and territory location in philanthines is complex (Table 8-7). Males apparently stay near nest aggregations if possible, but they avoid dense aggregations if competition is too intense. In some species, a territory is valuable because of its strategic position near a nesting/emergence area (e.g., *P. psyche*) or female hunting sites (e.g., *Aphilanthops subfrigidus*). In others, however, the plot of ground defended by a male probably has no intrinsic value independent of the chemical investment, because territories do not appear to be particularly good nesting or foraging areas (e.g., *P. basilaris* and *P. crabroniformis*). Thus, the main value of each occupied territory in all species may derive almost entirely from the chemical investment made by the resident (or by the previous resident if the territory was recently usurped). Several other facts also support this conclusion. First, male beewolves often attempt to usurp the scent-marked territory of a conspecific male even when nearby suitable, unoccupied plots are available (Evans and O'Neill 1988). Second, if it was actual real estate that was important, one would expect interspecific fights when two species' territories mix. Although *P. basilaris* and *P. bicinc-*

Table 8-7 Location of male territories relative to female nests in selected species of Sphecidae, subfamily Philanthinae

Species	Relative nest density	Relative territory density	Territory location	Reference
Philanthus crabroniformis	Very high	High	Near nests	Evans & O'Neill 1988
P. pulcher	Very high	Intermediate	Near nests	Evans & O'Neill 1988
P. triangulum	Very high	Intermediate	Near nests	Simonthomas & Poorter 1972
P. zebratus	Very high	Intermediate	Near nests	O'Neill & Evans 1983a
P. bicinctus	High	High	Within nesting area	Gwynne 1978, 1980; Evans & O'Neill 1988
P. psyche	High	High	Within nesting area	O'Neill 1979
Aphilanthops subfrigidus	Intermediate	High	Hunting sites	O'Neill 1990; O'Neill, unpublished
Philanthus barbiger	Intermediate	Intermediate	Near nests	Evans & O'Neill 1988
P. serrulatae	Intermediate	Low	Near nests	Evans & O'Neill 1988
P. basilaris	Very low	High	Leks without nests	O'Neill 1983b
Eucerceris flavocinta	Very low (?)	High	Leks without nests	Evans & O'Neill 1985

tus territories intermingle at the several sites, males of the two species do not battle for possession of territories even though they fight vigorously with conspecifics (O'Neill 1983b, Evans and O'Neill 1988).

Mating at Landmarks: Hilltopping

Male philanthines apparently use pheromones to attract females. However, males of some solitary wasp species, as well as males of other insect orders, seem to use existing landmarks as "signals," stationing themselves on prominent rises in the habitat and mating with females who come to these landmarks. The general term for this phenomenon is *hilltopping* (Shields 1967).

Case Study: *Hemipepsis ustulata*

One of the best-studied hilltopping insects is the tarantula hawk of the species *Hemipepsis ustulata*, which is active during the spring in southwestern U.S. deserts (Alcock 1979a, 1981a, 1983a,b, 1984a,b; Matthes-Sears and Alcock 1985; Alcock and O'Neill 1987; Alcock and Carey 1988; Alcock and Bailey 1997). Female *H. ustulata* are among the largest and most imposing Hymenoptera, being up to 4 cm long and bearing an imposing sting used to paralyze adult tarantulas. Both sexes have a shiny black body and reddish orange wings (Fig. 8-9A).

As befits such large animals, male tarantula hawks practice territoriality on a large spatial scale, defending plants on prominent landmarks in contests of unusual structure and duration. The territories center around jojoba bushes, creosote bushes, saguaro and cholla cacti, or palo verde trees on the spines of ridges (Fig. 8-9B). In March or early April, males arrive on trees as early as 5:30 A.M. and occupy the territories well into the afternoon or early evening (maximum of 5 hours residency/day). Territoriality may continue through early June, but as the season progresses and temperatures rise, males arrive

Facing page
Figure 8-9. (A) Male *Hemipepsis ustulata* on territory perch. (B) A ridge in Sonoran Desert (Arizona), with palo verde trees used by male *H. ustulata* standing out against sky. (Photos by J. Alcock.)

A

B

and depart earlier each day, sometimes returning as it cools late in the afternoon.

The most successful male tarantula hawks occupy territories for long periods during a season, often returning to the same perch day after day. The record is held by a male that was present on the same territory for 40 days (Alcock 1984b). Some short-term residents hold territories only once, often because they are ejected by larger males after occupying territories early in the day. Thus, as many as 12 residents may occupy a territory over the course of a season, the average duration of residency being 8 days. A male on a territory spends most of his time perched where he can survey his domain and pursue intruders. The males are very attentive to possible interlopers and potential mates, chasing other males coming within 10–20 m of their perch, but they also react to passing butterflies and distant turkey vultures. As it does for male beewolves, the length of the pursuit flight by tarantula hawks increases with the similarity of intruders to conspecifics. Next to other tarantula hawks, the insect that elicits the greatest response is the fly *Mydas xanthopterus*, whose males resemble *H. ustulata* in color, size, and shape (Nelson 1986).

Most conspecific male intruders immediately flee upon the approach of a resident, but some stay to engage in combat consisting of repeated "spiral flights." Dueling males ascend in flight while spiraling about one another, sometimes so close as to clash wings. They continue rising to a height of 10 or more meters above the territory, often moving out of an observer's binocular range. At the apices of their flights, they dive back down only to engage in further spiral flights on occasion. The duration of interactions between pairs of males is usually less than 1 minute, but some encounters last over 30 minutes, as males continue their bloodless quarrel with up to several hundred sequential spiral flights. One intruder usurped a territory after a battle consisting of 138 spiral flights over a 43-minute period. Nevertheless, prolonged battles constitute fewer than 5% of all contests. Even though there is little physical contact between contestants, fights are generally won by larger males, who may be not only stronger but also faster. Alcock (1979a) found that males that replaced 30 residents he had removed from territories were smaller than the original resident 20 times, equal in size 2 times, and larger 8 times. However, there appears to be a strong residency effect in contests because many smaller residents win contests (Alcock and Bailey 1997).

Whereas contests among male beewolves are equivalent to wrestling matches, those of male *H. ustulata* may be more akin to track meets. Alcock suggests that the contests function as mutual assessments of an opponent's flight speed; a male could not benefit from remaining on a territory with a faster rival who would usually win a race to a passing receptive female. Mating occurs when a flying female enters a territory, where she sometimes circles the resident's palo verde before being captured in midair by a male as he quickly darts out from his perch. The pair falls to the ground, sometimes far from the territory, and then mate for about 1 minute. Males generally return to their territory immediately after mating.

Certain territories are apparently more attractive to males than others: they are occupied (1) earlier in the season, (2) more often during the season, and (3) over longer periods each day. Correlations between these three measures of attractiveness were highly significant. Data on territory occupancy allowed Alcock to calculate "preference scores" for each territory and determine that preferences were consistent across generations. Further, when males switch territories by moving on to one left vacant or by usurping one from a smaller male, they almost always shift to one of the territories with a higher score. Territories with higher scores tend to be (1) occupied by larger males, (2) more quickly reoccupied when the resident is removed by an observer, and (3) fought over more intensely as judged from the frequency and duration of spiral flights and frequency of usurpations at different territories.

But why these particular territories, and is occupancy of the preferred sites correlated with higher mating success? Sites most attractive to males are *not* frequented by females for hunting or feeding: the tarantula prey of females are widely scattered through the desert, and blossoms on defended trees are not sources of nectar. However, the trees and shrubs defended by males are not randomly distributed along a ridge. Males occupy only those "trees and shrubs silhouetted against the sky when approached from below" (Alcock 1979a), avoiding many trees below the spines of the ridges. Furthermore, among those palo verde trees along the backbone of a ridge, males favor bulkier trees and those of higher altitude, either at the top of the ridge or on "prominent rises" along the ridge (Fig. 8-9B) (Matthes-Sears and Alcock 1985). In considering why males and females would prefer conspicuous high points, Alcock proposed that the home ranges of many foraging females overlap at ridgetops, where territory residents would enjoy high mating success. Unfortunately, testing this hypothesis has proved difficult, given the low frequency of observed matings: just 4 in 270 hours of observation in 1981 and 1985 (Alcock 1983b, Alcock and O'Neill 1987). The data hint that females more often visit territories most attractive to males, but procuring sufficient data to test the hypothesis would require a small army of field assistants.

OTHER HILLTOPPING SPECIES

Hilltopping has been observed in two other species of solitary wasps: *Pseudomasaris maculifrons*, which occupies some of the same ridges frequented by *Hemipepsis ustulata* (Alcock 1985b), and the Australian *Bembix furcata* (Dodson and Yeates 1989). Unlike *H. ustulata*, males of these species show no overt aggression toward one another. The hilltops of Arizona occupied by male tarantula hawks are teeming with a great diversity of hilltopping male insects of other species, including a carpenter bee (Alcock and Smith 1987), several butterflies (Alcock 1983b, 1985a; Alcock and O'Neill 1986), and several species of flies (Alcock 1984a, Nelson 1986). These insects not only share the same ridges, but some also display convergent preferences for territory sites (Alcock 1984a). In addition, receptive females of most species occur in low densities in locations unpredictable to males because their prey, host plants, and nectar sources are widely scattered or superabundant. The high dispersion of females decreases the value of wide-ranging patrolling tactics or defense of nest sites and food resources.

Alternative Male Mating Tactics

Once ethologists started paying attention to male wasps, they found that males within some species use alternative approaches to gain access to females. Alternative mating tactics may be exhibited simultaneously by different males or by the same males at different times in their lives. Males of many populations exhibit condition-dependent strategies, where the expression of a behavior is a response to prevailing conditions that influence the probability of success associated with that behavior (Thornhill and Alcock 1983, Gross 1996). The conditions that favor behavioral flexibility may include intrinsic conditions (qualities of the males themselves) or extrinsic conditions (characteristics of the environment).

SIZE-DEPENDENT TACTICS

The most important intrinsic condition affecting male wasps is body size. Large males are often more successful than smaller rivals at obtaining mates because of their superior ability to defend rendezvous sites or the females themselves. Smaller males that fail at defending territories or individual females have two basic alternatives available. One conceivable option is to continue competing head-on with larger males. However, given the strong influence of size on fighting (O'Neill 1983b) and on the ability to carry females (O'Neill et al. 1989), this approach would almost invariably lead to failure of small males. Nevertheless, small males of some territorial species continue to visit territories, sometimes briefly contesting with the resident but occasionally finding one that has been abandoned recently. For example, intermediate-sized male cicada-killers sometimes mate when they take over territories temporarily abandoned by larger males (Hastings 1989a). The quick reoccupation of vacated territories indicates a similar opportunism among tarantula hawks (Alcock 1979a) and beewolves (Evans and O'Neill 1988). Technically, these "replacers" do not adopt an alternative mating tactic (but rather an alternative means of finding an open territory).

Another option for small males is to adopt a mate-seeking tactic that differs qualitatively from the primary tactic adopted by large males. The secondary tactic provides some likelihood of mating, although it is usually less profitable. These secondary tactics are generally nonaggressive alternatives in which small males search or wait for females at locations with a lower density of females (and of rivals). Five of the examples on Table 8-8 are species in which small males adopt nonaggressive tactics at locations likely to contain fewer receptive females. In the sixth species, *Philanthus zebratus*, it is the large males that adopt the patrolling tactic, while smaller males establish territories (outside of the dense aggregation of nests). The patrolling tactic appears to lead to a greater probability of mating than does territoriality and, perhaps, large males are faster fliers than small males.

Table 8-8 Examples of alternative mating tactics, correlated with body size, of solitary wasp males

Species	Primary mating tactic	Secondary mating tactic	Reference
Pompilidae			
Hemipepsis ustulata	Defend territories on hilltops	Patrol ridge	Alcock 1981a
Sphecidae			
Bembecinus quinquespinosus	Dig for and defend emerging virgin females in emergence area	Patrol periphery of emergence area for females that escape larger males	O'Neill & Evans 1983a, O'Neill et al. 1989
Philanthus zebratus	Patrol air space 3–5 m above nesting area	Defend scent-marked territories on periphery of emergence area	Evans & O'Neill 1978, O'Neill & Evans 1983b
Sphecius grandis	Defend territories in emergence area and await emerging virgin females	Perch in trees on periphery of emergence area	Hastings 1989a
Tachytes tricinctus	Defend territories in emergence area	Patrol emergence area	Elliott & Elliott 1992
Vespidae (Eumeninae)			
Euodynerus foraminatus	Defend emergences sites	Patrol flowers	Cowan 1979, 1981

Some large males of *P. zebratus* have been seen patrolling and holding territories (at different times) (O'Neill and Evans 1983b).

The evolution of size-dependent alternative mating tactics in wasps is probably a response to environmentally induced size variation (Evans and O'Neill 1988). Much of the intraspecific variation in body size of solitary wasps can be explained by variation in the amount of food individuals receive as larvae (Chapter 10). The amount of food is partially under control of the mother, because female offspring tend to receive more food than do male offspring (Chapter 10). However, much of the variation in provisions is undoubtedly out of the control of the female. The potential causes of stochastic variation in food placed in cells are (1) cleptoparasites and fungi that consume part of the provision, (2) temporal and spatial variation in the availability and quality of prey, (3) the influence of temperature on the development of larvae in different cells, (4) variation in the nutritional quality of prey that is not detected by the provisioning female, and (5) intrinsic differences, such as body size, among foraging females (Evans and O'Neill 1988). To this list of problems I could also add bad luck. For example, a nesting female of a progressive-provisioning species may be eaten by a predator before her cell is completely stocked.

Because of the environmentally induced size variation, behavioral flexibility should be a superior strategy compared with obligate adoption of one tactic or another (Alcock et al. 1978, Evans and O'Neill 1988, O'Neill et al. 1989, Gross 1996). Consider a species in which males either patrol for females or defend territories, the latter being more profitable if a male is large enough to win a significant number of fights. Consider further the potential values associated with conditional versus genetically fixed tactics, if body size has low heritability. A male with the flexibility to either patrol or defend can adjust his behavior to his likelihood of success. Success is expected to be high if he is large and establishes a territory, and low (but non-zero) if he is small and patrols. Males without the ability to match their behavior to their size run the risk of adopting a tactic that reduces their potential mating success. Small males whose genes inflexibly direct them to be territorial will spend a lifetime losing fights with larger males, while missing the occasional mating opportunities afforded to small patrollers. Similarly, large males that only patrol will forgo the potential benefits of territoriality.

TEMPORAL VARIATION IN MALE MATING TACTICS

Size variation is not the only factor that favors the evolution of condition-dependent mating tactics. Receptive females may be spatially and temporally distributed in such a way as to present males with multiple approaches to obtaining mates at different times of day or season. At different times, receptive females may be most abundant in an emergence area, at a nest site, on flowers, or at sleeping sites, favoring males that adopt tactics appropriate to female location (Table 8-9). It is likely that more such examples will be uncovered if researchers extend their studies to different times of day and season and observe male behavior in different places.

Female Choice and Male Courtship

Natural selection theory predicts that a female should make the most of her limited time and energy. One way of doing so is to choose a mate who provides her with material benefits or endows her young with superior genes (Andersson 1994). Thus, even if a male manages to run the gauntlet of male competitors and predators to find a female, he will not succeed unless the female permits him to mate. As long as females have opportunities to choose among alternative males, and can reject potential suitors at a cost less than that of accepting the wrong mate, then females are expected to be choosy. In order to overcome a female's reluctance to mate, a male may have to undertake courtship of varying length and form. Information available on solitary wasps allows us to address three questions concerning female mate choice: (1) Can females reject males? (2) By what courtship behaviors do males induce selective females to mate? (3) What do females gain by choosing among potential mates?

MATE REJECTION

Female choice is possible for solitary wasps because females of many species can reject suitors, often because they can physically dominate smaller males. Although mating attempts may be terminated by the male rather than the female, rejection behaviors of females when they are approached or mounted by a male are often so conspicuous that they can be unequivocally interpreted as

Table 8-9 Examples of spatial and temporal variation in male mating tactics in the Sphecidae (unrelated to body size)

Species	Tactic 1	Tactic 2	References
Daily Variation			
Ammophila dysmica	Patrol nesting area in morning	Patrol flowers in afternoon	Rosenheim 1987b
Euodynerus foraminatus	Defend emergence sites in morning	Patrol flowers in afternoon	Cowan 1979, 1981
Stictia heros	Patrol emergence/nesting area early in day when temperatures are low	Defend territories later in day when temperatures are high	Larsson 1989b
Steniolia obliqua	Patrol emergence/nesting area during middle of day	Patrol sleeping clusters in evening	Evans 1966c
Seasonal Variation			
Bembix rostrata	Defend individual emerging females early in season	Patrol for digging or nectar-feeding females later in season	Schöne & Tengö 1981
Oxybelus bipunctatus	Patrol for emerging females early in season	Patrol nest sites and flowers later in season	Peckham et al. 1973
Tachytes distinctus	Defend territories with emergence sites early in season	Defend territories in nesting area later in season	Lin & Michener 1972

attempts to repel the male. A female's means of rejection vary among species and include (1) fleeing from the male, (2) flying at the male to drive him off, (3) raising her wings to prevent mounting, and (4) breaking free from his grasp by writhing, pushing with her legs, biting him, maneuvering her abdomen away from his genitalia, or attempting to sting him. Although males and females often contact one another at flowers, females often seem unreceptive at this time. Rejections of suitors by females at flowers and at nest sites have been observed in Sphecidae (Evans 1957b, Simonthomas and Veenendaal 1974, O'Neill 1979, Gwynne 1980, Schöne and Tengö 1981, Evans et al. 1986), Bethylidae (Gordh 1976, Gordh and Evans 1976, Gordh et al. 1983), Mutillidae (Spangler and Manley 1978), Sapygidae (Torchio 1972a), Tiphiidae (Rivers et al. 1979, Alcock 1981b, Alcock and Gwynne 1987), and Vespidae (Longair 1985a,b, Cowan 1986b).

Male wasps may also abort some copulation attempts, perhaps because of subtle rejection cues from unreceptive females (Longair 1984, 1985b). It is known that the response of males of some species depends on the reproductive status of the female. *Dasymutilla bioculata* males break off attempts to couple if the female is not a virgin, and the attractiveness of a female *Pseudomethoca frigida* declines with time following her only copulation (Brothers 1972). Likewise, *Bembix rostrata* males are more likely to attempt copulation with virgin females than with digging or provisioning females, which are likely to have already mated (Schöne and Tengö 1981); males display the same differential response to the severed abdomens of virgin and mated females.

FORMS OF COURTSHIP

Mating and fertilization in animals are often preceded by courtship, a form of communication whereby a male tries to persuade a female to use his sperm. Some solitary wasps have prolonged courtship that begins before copulation and may continue after intromission and even after the male has withdrawn his genitalia. Females of many wasp species will not mate without courtship, which varies considerably in duration and complexity.

Case Study: *Goniozus legneri*

After chewing his way into a sister's cocoon a male *Goniozus legneri* orients himself head-to-head and dorsum-to-venter with the female before mating commences (Gordh et al. 1983). As this and the next example show, obvious visible courtship is often lacking among solitary wasps.

Case Study: *Philanthus* spp.

When a receptive female beewolf (*Philanthus* spp.) lands within or flies near a territory, a male intercepts her and they quickly couple. The male immediately dismounts so that the pair are oriented end-to-end for the remainder of the brief copulation (Evans and O'Neill 1988). Thus, after the female arrives, the precopulatory phase of mating is over quickly, and, after coupling, there is little opportunity for tactile interactions except via the genitalia. Such a brief, and apparently simple, interaction before copulation is typical of many sphecid wasps.

Case Study: *Sapyga pumila*

After pouncing on a female, the male *Sapyga pumila* exposes his genitalia and brushes them back and forth across the tip of her abdomen for 8–12 seconds. Later, during copulation, the female lowers her antennae, and the male twitches his every 2–3 seconds. However, both the male and female keep their antennae elevated and out of contact (Torchio 1972a).

Case Study: *Aphelopus melaleucus*

A male *Aphelopus melaleucus* walks up to a female from behind, while vibrating his wings in pulses lasting several seconds. After contacting the female with his antennae, the male grasps her with his fore- and midlegs before mounting. During a brief mating, the male remains still (Jervis 1979). This is similar to the courtship of other Dryinidae (e.g., *Anteon brachycerum* and *Dicondylus bicolor*) but differs from that of *Pseudogonatopus distinctus*, in which pairs approach each other head to head and "criss-cross their rapidly vibrating antennae" (Waloff 1974).

Case Study: *Tiphia berbereti*

The *Tiphia berbereti* male approaches the much larger female from behind, mounts, and moves forward until his mandibles rest on her head (Rivers et al. 1979). He then projects his antennae between hers and initiates "antennal play," which lasts 0.5 to 3.0 minutes. During antennal play, the female's hind tibiae stroke the dorsolateral surface of the male's abdomen. Antennal play ceases when the male curls his antennae around those of the female. If receptive, she throws her head back and waves it from side to side, and he moves backward to couple. If unsuccessful, he returns to his former anterior position and reinitiates the sequence. During mating, the female strokes the male's head and thorax with her legs, while he strokes her thorax with his antennae.

Case Study: *Pseudomasaris vespoides*

To initiate mating, a male *Pseudomasaris vespoides* mounts the female dorsally and places his clubbed antennae on her face. He then strokes her with his fully extended antennae, moving them farther from the female's head with each stroke. Coupling is "accompanied by violent buzzing, at which time the female often release[s] her grip" (Longair 1987).

Case Study: *Dasymutilla foxi*

As the male *Dasymutilla foxi* approaches a female he "honks," producing a low-pitched sound by vibrating his wings and thorax. After mounting the female, the male continues to honk, but, during intromission, honking is replaced by stridulation caused by frictional rubbing of the abdominal segments. The male and female then stridulate alternately in bouts lasting approximately 1 second, the male's sound being lower-pitched than the female's (Spangler and Manley 1978). This relatively complex courtship contrasts with that of several other mutillids, whose mating is preceded by behaviors no more complex than the male's prodding the tip of the female's abdomen with his genitalia (Linsley et al. 1955; W. Ferguson 1962; Brothers 1972, 1978; Manley 1975).

Case Study: *Crabro* spp.

Among the sphecids, the most fascinating and visible courtship devices occur in the genus *Crabro*, whose males have thin, translucent, shield-shaped plates on their foretibiae (Fig. 8-10). These plates have patterns of spots and stripes distinctive enough in each species to be used as taxonomic characters (R. Bohart 1976), although some intraspecific variation is evident (Low and Wcislo 1992). Darwin (1871) hypothesized that the elaborate legs of male *Crabro* are used to firmly grasp females during mating, but recent observations show otherwise (Matthews et al. 1979, Low and Wcislo 1992). When a male *Crabro cribrellifer* pounces upon the back of a female, he places his foretibial plates over his partner's eyes (Low and Wcislo 1992). B. Low

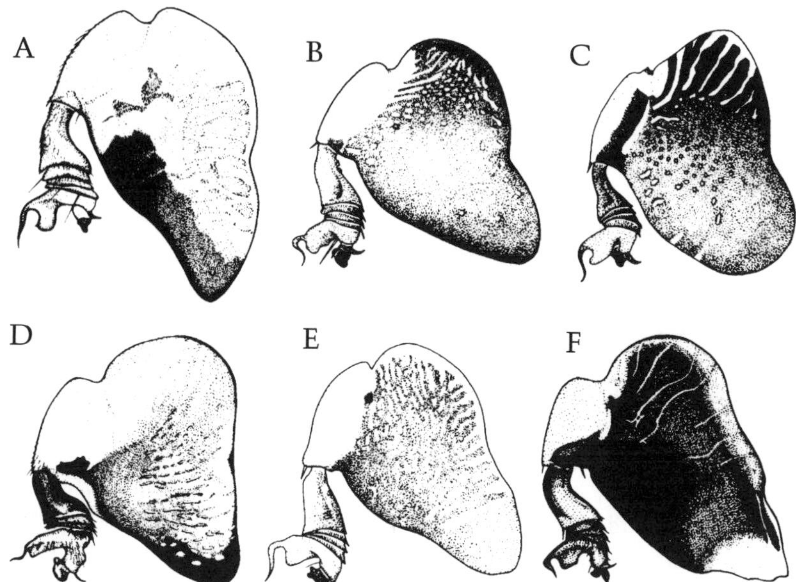

Figure 8-10. Inter- and intraspecific variation in patterns on the foretibial shields of male *Crabro*. (A) *C. pallidus*. (B) *C. cribrellifer*. (C) *C. tenuis*. (D) *C. spinuliferus*. (E) *C. digitatus*. (F) *C. deserticola*. (A–F from Bohart 1976.)

and W. Wcislo and others hypothesize that females detect specific patterns of transmitted light when the plates are placed over their eyes and that the patterns are used to evaluate males. Immediately after leg positioning, the male uses his antennae alternately to stroke those of the female. The three segments of the antennae used for this are modified: the basal segment is flattened, whereas the second and third bear unusually long hairs. Before intromission, the male strokes the female's abdomen with his everted genitalia, but, after intromission, he remains motionless. The complex precopulatory courtship of many male *Crabro* is relatively unusual among sphecids. Although males of some Australian *Bembix* have expansions on the basitarsi of the forelegs similar to those of male *Crabro* (Evans 1982), nothing is known of their function.

Case Study: *Ancistrocerus antilope*

After mounting, a male *Ancistrocerus antilope* uses his antennae to rapidly stroke and tap those of the female, which are held erect. At the same time, he rubs the side and venter of her abdomen with his genitalia, alternately stroking different sides until probing results in intromission. Males that are unsuccessful in achieving intromission remain mounted and reinitiate courtship. In some copulations, the female sits motionless while the male waves or twitches his antennae in her face. In others, the females appear quite agitated, wriggling their abdomens and kicking the males with both their forelegs and hindlegs. The struggling phase follows no stereotypical pattern and varies in intensity. Some pairs repeat the courtship and copulation sequence once or twice, with successive copulations being 24 minutes apart on average (Cowan 1986b). Cowan also examined courtship and copulation in four other eumenines and found interspecific differences in the timing of events and the number of successive matings by pairs. Mating in eumenines sometimes involves the use of novel anatomical structures that may have evolved solely for courtship. Males of *Euodynerus foraminatus* use hooks on the end of their antennae to pull on the antennae of females, and *A. antilope* males stroke females with brushes on their genitalia (as do males of the sphecid *Trypoxylon politum*; Brockmann and Grafen 1989).

FORMS OF COURTSHIP: A SUMMARY

Thus, male solitary wasps tap, stroke, buzz, stridulate, and display visually in courtships that variously involve their antennae, mandibles, legs, wings, stridulatory devices, and genitalia. These structures sometimes exhibit sexual dimorphism that can only be readily explained with reference to courtship. For example, there is no evidence that the tibial shields of male *Crabro* and the hooked antennae of male *Euodynerus* are used by males in fighting.

Nevertheless, despite the complexity and long duration of the courtship of many species, the copulation of other species initiates so abruptly and unceremoniously that the species appear to lack overt precopulatory courtship. The lack of obvious precopulatory courtship suggests that the female choice that determines which males achieve intromission is either lacking in those species or it is rapid and cryptic. However, a brief interaction is not necessarily a simple one (Thornhill and Alcock 1983). Much important information could be transferred quickly (e.g., by a rapid tap of the male's antennae in the milliseconds before mating). Other signals may be transferred in ways not obvious to a human observer, either because they occur internally via genitalic contact or because we cannot easily detect them

(Eberhard 1991, 1996). For example, we cannot usually smell the chemical that males of *Philanthus* deposit on their territories, so we may underestimate the complexity of courtship when we simply see a male intercept a female and quickly couple. The initiation of *Hemipepsis ustulata* copulation is abrupt, and long-distance chemical courtship signals are not used to attract females (Alcock 1979a). Rather, males may use the hills and trees on which they establish their territories as courtship signals.

COPULATORY AND POSTCOPULATORY COURTSHIP?

In some of the courtship sequences described, male behaviors such as antennal stroking and stridulation continue after intromission. Other species display no "courtship" behaviors until *after* mating has started. For example, after achieving intromission without much ado, male *Hemithynnus* species "lash" the dorsum of the females' abdomen with their hindlegs for about 3 minutes (Alcock 1981b). Similarly, male *Oxybelus sericeus* (like small tomcats) bite the female's neck during copulation (Hook and Matthews 1980); male *Glenostictia satan* tap the female's head with their antennae (Longair et al. 1987); and male *Bembix rostrata* stroke the female with their forelegs (Schöne and Tengö 1981). Possible signaling by the male of other species continues even after he has copulated and withdrawn his genitalia. After mating, a male *Abispa splendida* remains mounted on the female and begins to make "agitated movements . . . apparently pulling and biting at his recent mate for a period of about 30 s" (A. Smith and Alcock 1980). *Parancistrocerus pensylvanicus* males follow up mating by repeatedly bouncing up and down on the female and tapping her antennae with their own (Cowan 1986b). Such behaviors are apparently not attempts to guard the female or to induce her to remate, so perhaps they induce the female to use the male's sperm.

THE FUNCTION OF COURTSHIP

The function traditionally attributed to courtship is that it prevents heterospecific matings by providing females with information needed to identify the species of her suitor. Although it is possible that selection against hybrid matings has played a role in the evolution of courtship, courtship may also allow a male to advertise his worth as a mate (Thornhill and Alcock 1983, West-Eberhard 1984, Eberhard 1996).

When considering the function of courtship, we must determine which "virtues" are being touted by males and how the signals can be used by females to judge them. Some forms of courtship may involve arbitrary signals that have spread via Fisherian runaway sexual selection (Andersson 1994), rather than because they provide accurate information on a male's strength, foraging ability, or other tangible qualities. However, there are two forms of nongenetic material benefits that male wasps of some species offer to females and that, in principle, could form the basis for mate choice: food and aid in nesting. I have already noted that male thynnine wasps use gifts of nectar to court females. However, the aid in nesting provided by males of *Trypoxylon* and *Oxybelus* has probably evolved as an incidental consequence of male-male competition rather than as a consequence of female choice (Brockmann and Grafen 1989).

It is probable that some females refuse potential sperm donors for reasons other than the perception of inferior male quality. Female wasps store sperm in their spermatheca that may suffice for a lifetime of egg laying. Having already mated, a female may reject further males simply to save time (Alcock et al. 1978). In addition, if she is disturbed while working at her nest, taking time out to copulate could leave the nest open and her brood susceptible to predators and cleptoparasites. On the other hand, it is possible that females of *Ancistrocerus adiabatus* (Cowan and Waldbauer 1984, Cowan 1986b) mate repeatedly in order to minimize harassment from males who persist in the mating attempts (Alcock et al. 1978).

Conclusion

I managed to get through seven chapters of this book with barely a mention of male wasps (except to note that they are dull-witted enough to be fooled into sexual arousal by orchids). This chapter has rectified the omission, but studies of males lag behind those of females by at least a half century (and studies of females still appear at a faster rate). Fortunately, for those interested in male behavior, the situation has improved greatly in the last 30 years. It is safe to say that an extensive review of male behavior could not have been attempted in 1970.

For a male solitary wasp, the pathway to successful copulation is paved with obstacles, including competitors, predators, and unreceptive females. A male's ability to overcome these problems may be limited by intrinsic shortcomings, such as small body size and limited sensory capabilities, or by extrinsic factors, such as high densities of competitors and predators. It is also constrained when the solution to one problem exacerbates another (e.g., highly conspicuous mate-searching tactics may also increase exposure to predators). Thus, male solitary aculeates have evolved a wide variety of adaptations to aid in their quest for mates. Courtship rituals of varying duration trigger female receptivity and may provide a female with tangible goods (e.g., nectar) in return for sexual favors. Emergence patterns, highly efficient searching tactics, and fighting behaviors may help a male gain and maintain access to receptive females. Males of many species have morphological devices and other body characteristics that complement behavioral tactics, such as the expanded foretarsi of male *Crabro*, the stridulatory devices of mutillid males, the mandibular horns of male *Synagris cornuta*, and the unusually large body size of male *Bembecinus quinquespinosus*. Despite the impressive number of traits that have been documented, however, wasp researchers have barely scratched the surface of the natural diversity of sexually selected adaptations.

9/ Thermoregulation, Sleeping, and Overwintering

Solitary wasps face a number of problems that affect success in reproductive competition. Some problems confront a wasp in sequence: for example, a female must first find prey, then subdue it, and finally transport it to a nest. Other problems need to be faced simultaneously, as when a female must forage while being wary enough to avoid her own predators. Two related constraints that overshadow most activities of wasps are environmental temperature and available sunlight, which, through their effect on body temperature, determine the times and places where efficient (or any) activity is possible. Furthermore, because most solitary wasps are visually oriented, light levels also affect whether they can move about and find prey, flowers, mates, and nests. Thus, the physical environment presents constant selection pressures that affect foraging, nesting, predator avoidance, and mating. Temperature and sunlight are, in fact, such strong constraints that adult wasps typically spend most of their time in an immobile state of "sleep" during the night while awaiting the return of favorable temperature and light conditions.

Thermoregulation

Thermoregulation is a matter of balancing heat gain and heat loss to come as close as possible to a preferred body temperature. Although humans maintain body temperature within a narrow range primarily by physiological mechanisms, we obviously do not take temperature for granted. We experience discomfort on hot, humid days or on cold, windy days as our bodies struggle to stay at 37°C. At such times, we may supplement our physiological mechanisms by seeking different environments or by changing our activity levels and clothing. Our dependence on precise temperature regulation is most evident when we feel debilitated from a fever that raises our body temperature just a degree or two.

In contrast, insects (particularly in the temperate zones and deserts) typically experience wide daily swings in body temperature. In fact, it would not be unusual for body temperature to vary by 30°C over the course of a day, a shift that would kill a human. Nevertheless, insects must avoid both low temperature extremes, in which they risk freezing, and high temperature extremes, in which they first risk dehydration and then irreversible tissue damage. However, thermoregulation is not just a matter of avoiding deadly extremes. The metabolic rates of insects are temperature-dependent, with maximum physiological performance within some narrow range of temperatures. For a wasp, variation in physiological performance would manifest itself as variation in muscle performance, which affects flight and digging speed.

The solitary wasp literature is replete with anecdotal remarks on the relationship between activity and temperature, but body temperatures have rarely been measured, and even the environmental conditions that influence body temperature are usually reported only in general terms. Nevertheless, enough progress has been made recently that we are beginning to a get a broad picture of the range of thermoregulatory constraints faced by solitary wasps and the kinds of adaptations they have evolved to deal with the thermal environment.

ENDOTHERMY

When body temperatures are suboptimal, heat generated metabolically can raise body temperature, if it can be retained long enough. Animals that elevate body temperature primarily by using metabolic heat are referred to as endotherms to contrast them with ectotherms, which depend on heat from external sources (Heinrich 1993). Endotherms, such as birds and mammals, maintain stable and relatively high body temperatures even when the air is cool and direct solar radiation is weak or absent. However, endothermy is energetically costly, particularly when ambient temperature is well below optimal body temperature (even though insects produce metabolic heat using their flight muscles). Compared with large and feathered or furred vertebrates, insects lose heat rapidly: being naked, they are poorly insulated, and being small, they have a relatively high area of body surface over which to lose heat. The net result for insects is that it is too costly to be endothermic 24 hours a day (Heinrich 1993).

Nonetheless, many insects use short-term endothermy (heterothermy) to raise body temperature at critical times. The advantages of short-term endothermy are clear. Early or late in the day, when it is cool and the sun is low on the horizon, direct sunlight may do little to raise body temperature. Thus, a heterothermic insect can extend its daily activity period by using endothermic warm-up to raise its body temperature to that required for initiation of flight. If endothermy is also maintained during flight, a wasp may be able to fly more efficiently. In contrast, a strict ectotherm must await the arrival of favorable conditions so that ectothermic warm-up can commence (e.g., by basking in the sun).

Bees, which (phylogenetically) are really just hairy wasps, are among the best-studied heterotherms (Heinrich 1993, Stone and Willmer 1989), so a brief overview of their endothermic abilities will prepare us to discuss the less extensive literature on wasps. The degree to which a bee can raise its body temperature above ambient in the morning is a function of its ability to generate and retain heat while warm-up is in progress. Wing muscles concentrated in its thorax are both the primary source of heat and the region that needs warming before flight is possible. A heterothermic bee begins warm-up by contracting antagnostic sets of wing muscles in such a way that heat is generated, but no wing movement occurs. Much of the heat produced by this shivering is retained in the thorax via a combination of insulating hairs and a countercurrent heat exchanger in the bee's narrow "waist" that reduces diffusion of heat from the thorax to the abdomen (Heinrich 1993). Nonetheless, because they are small animals, much of the heat leaks to the bee's surroundings. The rate at which bees warm up using metabolic heat increases with body size; warm-up rates in bee species varying in body size from 10 to 1,300 mg range from about 1° to 12°C per minute (Stone and Willmer 1989). Thus, within minutes on cool mornings, many bees can raise thoracic temperature to the minimum necessary to fly and sustain the typical optimum of about 32° to 40°C. A bee's ability to warm up is also partially a function of its species' evolutionary history. Bees that have evolved in cooler climates tend to have faster warm-up rates than similar-sized species from warm environments (Stone and Willmer 1989).

Case Study: *Bembix rostrata* As an Endotherm

Unlike some bees that fly at ambient temperatures as low as 2°C, *Bembix rostrata* does not fly at ambient temperatures below 22°C. Even then, however, flight is typically not initiated until thoracic temperature is about 36°C (i.e., just below normal human body temperature of 37°C). J. Ghazoul and P. Willmer (1994) provided conclusive evidence that *B. rostrata* use preflight endothermic warm-up to bridge the thermal gap between 22° and 36°C.

To determine whether the wasps could warm to the minimum flight temperature endothermically, Ghazoul and Willmer brought *B. rostrata* into the laboratory, inserted minute temperature probes (thermocouples) into their thoraces, and then cooled the wasps to about 10°C. The wasps were then moved to a 21°–22°C environment where their thoracic temperatures were recorded. Despite being in a relatively cool location, with no access to an external heat source, the wasps quickly raised their thoracic temperatures to 33°–38°C and initiated flight (Fig. 9-1A). In the lab, the average rates of warm-up were 1.8°C per minute for males and 3.4°C per minute for females, the latter value being slightly lower than for bees of similar size. Thus, once a female *B. rostrata* passively warms to 22°C (using sunlight and external heat), it can then accelerate warm-up physiologically and attain the minimum flight temperature of 36°C in less than 5 minutes. The reason for the sexual difference in warm-up rates is unknown, but it is apparently not due to differences in gross morphology, because females and males of this species are equal in size and both sexes have dense pubescence on their thoraces.

Like a car idling on a cold winter morning, a stationary *B. rostrata* shivering to warm itself expends energy without going anywhere. Once it flies, however, it can quickly replenish its energy stores, recouping the approximately 1.7 calories expended in a typical warming bout in little more than 2 minutes of foraging on flowers. Therefore, warm-up is not only rapid but also energetically efficient. Once in flight, patrolling male *B. rostrata* usually have thoracic temperatures 15° to 20°C in excess of ambient temperature, suggesting that a tremendous amount of heat is produced during flight (Larsson 1991). Endothermy during flight may increase the flight efficiency

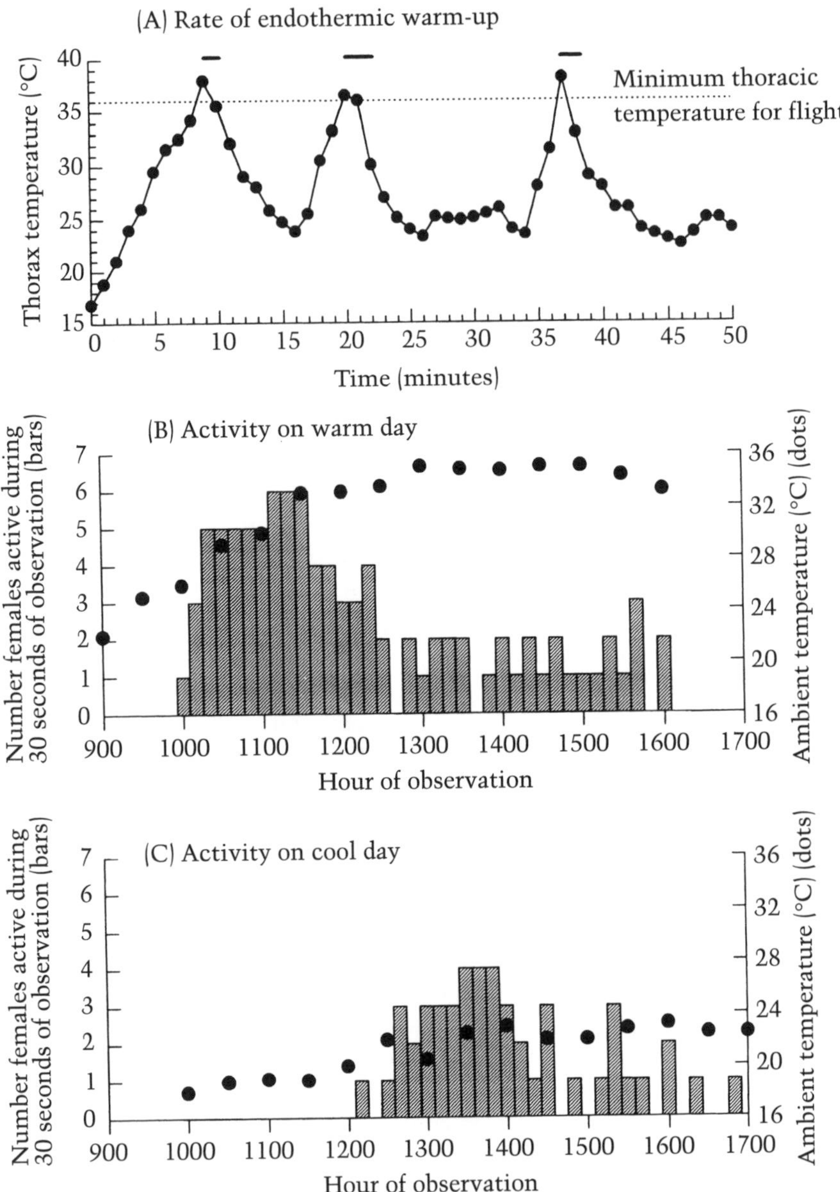

Figure 9-1. (A) Thoracic temperature of a single *Bembix rostrata* female, showing several sequences of endothermic warm-up under ambient conditions of 22°C air temperature and no direct sunlight; solid bars indicate periods of flight. (B, C) Diurnal changes in the number of *B. rostrata* active (bars) and ambient temperature (dots) on two days with different temperature conditions. (A–C redrawn from Ghazoul and Willmer 1993.)

and load-carrying ability of females carrying prey and of males carrying females (see Chapter 8). Any advantage gained in flight speed may be of critical importance to a foraging female *Bembix*, which chase down flies on the wing.

Endothermy in Other Solitary Wasps

Endothermy has been documented for a few other species of solitary aculeate wasps. In a study in Papua, New Guinea, the chrysidid *Stilbum cyanurum*, warmed from 24° to 30°C at a mean rate of 1.7°C per minute and maintained body temperatures 3°C above ambient while in flight (Stone 1989). Females of a *Scolia* species also exhibited preflight endothermic warm-up, and those caught in flight had body temperatures several degrees in excess of ambient.

Indirect evidence for endothermy during flight has been found in several other species. P. Willmer (1985a,b) found that the thoracic temperatures of female *Cerceris arenaria* caught in flight while returning to their nests with prey fell within the narrow range of about 34° to 36°C, even though ambient temperature varied from 18° to 28°C. Because thoracic temperatures were up to 18°C above ambient, it seems unlikely that ectothermy alone could account for the observed body temperatures. Even females just leaving the cool confines of their nests had body temperatures that varied independently of ambient, sometimes exceeding it by as much as 13°C. These observations suggest that *C. arenaria* warms up before flying, either using endothermy or ectothermic basking. Temperatures of flying male *Stictia heros* also vary independently of ambient temperature over an air temperature range of 32° to 38°C and sometimes exceed ambient by a wide margin (Villalobos and Shelly 1994).

Endothermy is undoubtedly common in large solitary wasps (Stone 1989, Ghazoul and Willmer 1994). However, no direct evidence exists to ascertain whether it is widespread or whether wasps can achieve the same high warm-up rates as some of the larger bumble bees and anthophorid bees (Stone and Willmer 1989). Some wasps, such as the cicada killer (*Sphecius speciosus*), are as large as some of the largest endothermic bees. Furthermore, scoliids of the genus *Campsomeris* are both large and hairy (and so perhaps well insulated), although endothermy does not occur just in large, hairy insects (Stone 1989). Endothermic *Stilbum cyanurum* are hairless wasps weighing just 90 mg (compared with the 500–1,300 mg masses of many endothermic bees), and the endothermic halictid bee *Lasioglossum smeath-*

manellum weighs a mere 10 mg (Stone and Willmer 1989). The warm-up rates of these species are low compared with those of large bees, but they may still provide a thermal boost at low temperatures. Nevertheless, endothermy is a relatively costly luxury that many solitary wasps may be unable to afford or are able to use only under restricted conditions. Certainly, it would seem that most dryinids, bethylids, and small sphecids would be incapable of rapid endothermic warm-up, because they are as small or smaller than the small bees with warm-up rates of only about 1°C/min.

ECTOTHERMY

Nonendothermic wasps are not completely at the mercy of their thermal environments, fluctuating in temperature through a day like a similar-sized rock. They differ from a rock, of course, because they can move about and take advantage of environmental heterogeneity in temperature, wind speed, and available sunlight. After the sun has risen sufficiently, most vegetated environments present a mixture of sunny and shady areas and windy and calm areas that provide a wide range of microclimatic conditions. Thus, as a general rule, a wasp trying to reduce the danger of overheating can move to where wind speed is higher and air temperature and incident sunlight are lower (O'Neill and O'Neill 1988). Conversely, a wasp can bask in a sunny, relatively warm and calm location to increase radiative heat gain and decrease convective heat loss. When basking, wasps typically press their bodies against warm substrates (e.g., Olberg 1959), so they may also be gaining heat conductively.

Case Study: *Bembix rostrata* As an Ectotherm

Bees, which do not bask (Heinrich 1993), nevertheless often fly at ambient temperatures much below 20°C (Stone and Willmer 1989) by relying on endothermic warm-up. In contrast, *B. rostrata* is a borderline heterotherm that must combine ectothermic and endothermic warm-up, employing the latter only when ambient temperature reaches 22°C. Before flight in the morning, *B. rostrata* sit on the ground basking in the sun, absorbing heat both convectively (from the warm, relatively still air near the soil surface) and conductively (from the soil). This is a cheap source of heat, but because of their dependence on direct sunlight, *B. rostrata* are inactive on cloudy days. Even on warm days, females return to their nests if clouds block direct sunlight (Schöne and Tengö 1992, Ghazoul and Willmer 1994). Retreating to the nest appears to be a direct response to the disappearance of the sun, because the wasps respond after the sun is obscured but before air temperature declines (Schöne and Tengö 1992). Because their body temperatures are directly linked to solar radiation and ambient temperatures, female *B. rostrata* exhibit full activity during a relatively narrow portion of each sunny day (Fig. 9-1B). On cooler days, activity is reduced and commences later (Fig. 9-1C). Male *B. rostrata* are also constrained by sunlight, delaying their patrolling each morning until the sun hits the soil surface in the nesting area (Larsson 1991).

Ectothermy in Other Solitary Wasps

Although basking is energetically cheaper than endothermy, optimal thermal and radiative conditions are available only at certain times and places. As a result, many wasps make their appearance only when direct sunlight is available for warming the body directly (Corbet and Backhouse 1975). Thereafter, they must avoid areas that are either too cool or too hot. Low temperatures limit the ability of wasps to fly early and late in the day, and high midday temperatures often constrain wasps to be active in late morning and late afternoon.

High temperature risks may be of particular importance to solitary wasps active near fully insolated soil surfaces at times when soil surface temperatures exceed 50°C. R. Chapman and colleagues (1926) examined heat stress and mortality in six species of *Anoplius*, *Bembix*, *Chlorion*, *Dasymutilla*, *Microbembex*, and *Sphex* active on sand dunes in Minnesota, where temperatures well above 50°C were common in the summer. After bringing the wasps into the lab, they first cooled the wasps and then raised the temperature gradually. None of the insects survived at temperatures exceeding 55°C and some, especially the *Bembix* and *Microbembex*, showed signs of stress at temperatures as low as 45°C.

Because of these constraints, temperature has a strong effect on the nesting activity of many female ground-nesting wasps. For example, *Sphex ichneumoneus* females are more likely to dig nests when cloud cover is less than 30% and when soil surface temperature is 35°–55°C (Brockmann 1979). The activity patterns of female wasps often exhibit a bimodal distribution of activity on hot sunny days, with peaks of activity in mid- to late morning and late afternoon that allow females to avoid stressful midday temperatures (Kurczewski et al. 1969, Cane and Miyamoto 1979, Alexander 1985, Rosenheim 1987b, Karsai 1989, Larsson 1990, Schöne and Tengö 1992, Punzo 1994b). Even during these times, females may have to make minor adjustments in body temperatures to maintain efficient metabolic rates and to avoid stressful temperatures. J. Rosenheim (1987b) noted that female *Ammophila dysmica* bask on the soil surface during cool periods, whereas later in the day they alternate digging on the hot soil surface with trips to vegetation, where they temporarily cool off. (Olberg [1959] provides some excellent photographs showing the posture of basking sphecids.) When soil temperatures exceed 45°C, female *Oxybelus bipunctatus* often forgo making temporary closures during hunting trips, perhaps as a means of reducing temperature stress at the hot nest entrance (Peckham et al. 1973) or perhaps because their enemies are not abroad at this time.

Male wasps are also affected by the availability and intensity of direct sunlight. For example, male *Philanthus* do not occupy mating territories on overcast days, and they abandon territories when clouds appear or when shade from trees encroaches on territories late in the afternoon (Evans and O'Neill 1988); the activity pattern of female wasps may also be constrained by changing patterns of sun and shade (Linsley and MacSwain 1956, Móczár et al. 1973). On the other hand, when it is too hot, males that remain in unshaded locations risk overheating. Male *Philanthus psyche* establish territories in sparsely vegetated sandy areas, commonly remaining active when the sand surface temperatures approach 60°C. If prevented from leaving the sand at surface temperatures greater than 50°C, they die within minutes at temperatures from 50° to 52°C and within 10 seconds at higher temperatures (O'Neill and O'Neill 1988). Male *P. psyche* avoid this fate at high temperatures by perching on low plants where convective cooling probably offsets radiative heat gain. Similarly, male *Bembecinus nanus* and *B. quinquespinosus*, which patrol for females in unshaded emergence areas, spend very little time perching on the hot soil surface as temperatures exceed 45°C (Evans and O'Neill 1986, O'Neill et al. 1989). When it becomes too hot in such areas, male wasps often head for flowers to feed on nectar or they become completely inactive and wait for the return of better conditions on the following morning. For example, male *Tachysphex albocinctus* excavate and retreat to short (1.5–2.3 cm) burrows when it is hot (Asis et al. 1989).

BODY SIZE AND COLOR

For insects in full sunlight, the rate of heating and the extent to which body temperature exceeds ambient increase with body size

(Willmer and Unwin 1981). For example, larger *Cerceris arenaria* females warm faster when basking and are better able to maintain high body temperatures while in flight (Willmer 1985a,b). As a result, they become active earlier each morning, make shorter foraging trips, and bring in more prey each day than do smaller females. On very hot days, however, larger females may be at a disadvantage, because they are more likely to reach the maximum tolerated body temperature of 38°C. Small and large *Microbembex* males, in contrast, display no differences in activity attributable to differential abilities to cope with high temperatures (Fraizer 1997).

Body coloration also affects thermoregulation because darker pigments absorb a greater proportion of incident solar radiation than do lighter colors (Willmer and Unwin 1981). Evidence consistent with the hypothesis that body color plays a role in the thermoregulation of solitary wasps can be found in correlations between the type of environment that a wasp inhabits and its coloration. W. Ferguson (1962) notes that females of most species of diurnally active mutillid wasps in hot deserts of North America are light in color, whereas nocturnally active species and those living at higher altitudes tend to be darker. Similarly, *Philanthus albopilosus*, which inhabits open sand dunes, are much darker in Alberta, Canada, than in Texas and Colorado (Hilchie 1982). Presumably, the darker northern wasps absorb solar radiation more efficiently to enhance ectothermic warming, whereas the lighter southern wasps use their greater reflectances to help avoid overheating.

In species of at least one genus, intrapopulational color variation may have evolved in response to differential thermal constraints on different members of the population. Small male *Bembecinus quinquespinosus* patrol relatively cool microhabitats in vegetation in the periphery of emergence areas. Large males, on the other hand, frequent the unshaded soil surface of emergence areas where they compete for females (O'Neill et al. 1989). Thus, large males experience potential thermal stresses exacerbated by the fact the female emergences peak at soil surface temperatures of 36°–45°C and continue at even higher temperatures. The key to the larger males' ability to remain on hot, fully insolated soil surfaces may lie in their body coloration. Males above mean body size exhibit an increasing amount of yellow (rather than black) coloration on their bodies as body size increases (Fig. 9-2). Males below average in size are nearly completely black, whereas the largest males are mostly yellow. Measurements

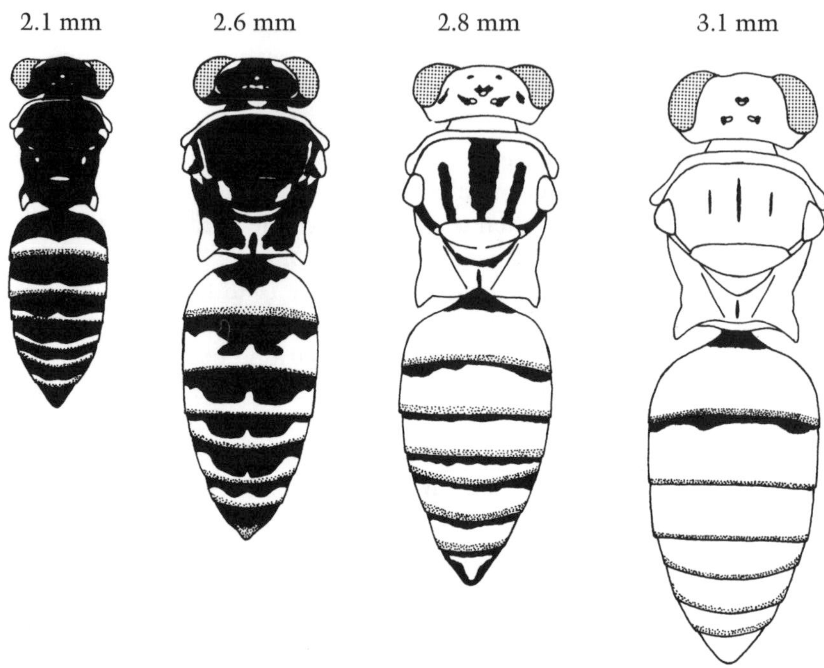

Figure 9-2. Variation in color patterns of male *Bembecinus quinquespinosus*; headwidth of each male is given (from O'Neill and Evans 1983a).

of the reflectances of the integument of males indicate that yellow cuticle reflects 29% of incident solar radiation (in the 290 to 2,600 nm range), whereas black cuticle reflects only 9%. Thus, the yellow coloration, which is an evolutionarily derived condition in the genus *Bembecinus*, may have evolved to allow larger males to remain on the soil surface longer in their quest for emerging females. However, body temperature measurements necessary to test this hypothesis have not been made.

VARIATION IN THRESHOLD TEMPERATURES FOR ACTIVITY

H. Evans and M. West-Eberhard (1970) commented that "wasps are decidedly sun-loving animals." As a general statement, this is probably true, because many wasps seem to require external heat sources to initiate and maintain efficient flight. Nevertheless, many wasps live and fly in cool, shaded habitats, apparently without the benefit of endothermy. *Mellinus arvensis* and *Crabro cribrarius* are active at temperatures lower than 15°C, although neither is endothermic (Ghazoul and Willmer 1994). Many North American species of *Crabro* are also active at lower temperatures than other sphecids (Kurczewski and Acciavatti 1968, Miller and Kurczewski 1976, Evans et al. 1980b). Similarly, nesting female *Alysson conicus* live in shady, wooded habitats where ambient temperatures do not exceed 22°C and where the wasps remain active "during periods of unfavorable weather" (O'Brien and Kurczewski 1982b). Females of at least one population of *Bembix americana* also can forage on overcast (or foggy) days when air temperatures are as low as 12°C (Lane et al. 1986); recall that *B. rostrata* do not even initiate flight until ambient temperatures reach 22°C, and after that they fly only after a period of endothermic warm-up.

Other species are quite active during the early morning hours or in the evening. Much nest digging by *Sericophorus viridis* occurs during predawn hours (Matthews and Evans 1970). (Note also that, although wasps may require direct sunlight for activity outside of their nests, they are not completely inactive in the dark, because females excavate burrows deep within the cool soil.) The pompilid *Batozonellus annulatus* hunts during the day but waits until night to dig its burrows (Tsuneki 1968). Finally, although *Bembix u-scripta* females dig nests during the daylight hours, they do most of their hunting from 5:00 to 7:00 P.M., often finishing after the sun sets, and they are difficult to observe (Evans 1960).

Sleeping

Most people, biologists included, would probably consider the study of wasp sleeping to be pretty uncompelling. Perhaps it deserves study if for no other reason than that adult wasps engage in sleeping longer than they do any other single activity. The study of sleep also underscores the overriding influence of thermal constraints on solitary wasps, many of which are adapted to conditions available only during warm sunny days. Thus, diurnally active wasps face severe thermal constraints at night, when they are incapable of efficient activity (especially flight) because of the lack of direct sunlight and external heat sources and because of the high cost of endothermy. Besides, for insects with eyes evolved to deal with daytime light levels, it is certainly too dark at night for most activities.

The behavior of adult wasps (and other insects) during their nighttime period of inactivity has long been referred to as sleep (e.g., E. Schwarz 1901, N. Banks 1902, Bradley 1908, Rau and Rau 1916, Evans 1966c). Evans (1966c) notes that the adoption of the term *sleep* is a matter of convenience and does not mean that wasp sleep should be considered neurophysiologically or functionally equivalent to vertebrate sleep. However, there are some similarities. Although sleep has been hard to define, it can be characterized by a "list of symptoms" shared by many animals during their major daily rest period (Meddis 1987). Thus, sleep is a prolonged period of immobility that occurs each day, generally at the same location (or same type of location) and at the same time, during which an animal often adopts a species-specific posture and is typically less responsive to external stimuli than at other times. Sleep is distinct from longer periods of inactivity such as diapause (overwintering) and from shorter periods of inactivity forced on the animal such as occur when a passing cloud temporarily cools the environment. Although this characterization of sleep comes mainly from consideration of vertebrates, it rather neatly describes the sleeping behavior of species of Tiphiidae, Scoliidae, Pompilidae, Vespidae (Eumeninae and Masarinae), and at least three subfamilies of Sphecidae.

Stereotypical sleeping behavior has been well documented in many species of solitary nest-building wasps, parasitoids, and cleptoparasites, revealing quite a bit of interspecific variation in (1) where sleeping occurs, (2) postures adopted during sleep, and (3) whether a wasp sleeps in the company of other members of its species, members of the opposite sex, and members of other species (E. Schwarz 1901, N. Banks 1902, Bradley 1908, Rau and Rau 1916, Evans 1966c, Freeman and Johnston 1978b, Evans et al. 1986, Rosenheim 1987b, S. Gess 1996).

Like *Steniolia obliqua*, which sleep in the tips of conifer branches (Fig. 9-3), many species of aculeate wasps sleep on vegetation. Most often, wasps sleep on plants either alone or in an isolated cluster. For example, I have seen small clusters of sleeping *Myzinum quinquecinctum* in Colorado and loose, mixed clusters of *Prionyx* and *Stizoides* in Montana. The number of species gathering in a small area can be large and can include both wasps and bees. In Arizona, up to 76 bees and wasps were found in discrete sleeping aggregations whose species membership changed between nights (Evans and

Figure 9-3. A sleeping cluster of *Steniolia obliqua* on the tip of a pine branch (from Evans 1966c).

Linsley 1960, Linsley 1962). During the study, sleeping aggregations variously included 3 species of Scoliidae, 13 Sphecidae, 14 Eumeninae, and 14 bees, many of which adopted specific sleeping postures and sites within aggregations. For example, from 1 to 30 individuals of four species of *Ammophila* were found each night, resting at heights of 0.6–1.2 m on vertical stems in the center of the sleeping cluster, each wasp grasping the stem with its mandibles and all six legs, while holding its body at a 50°–70° angle to the stem. Sleeping *Scolia dubia*, on the other hand, rested at heights of 1.2–1.7 m on the tops of bushes with their bodies curled around flowers. Most of the species observed in Arizona by Evans and Linsley slept individually or within loose clusters so that they were not in bodily contact. Other species, such as *Steniolia obliqua*, sleep packed close together in compact clusters. Usually, clusters occur on plants (Evans 1966c), but sleeping clusters of *Chalybion caerulum* have been observed under a rock overhang and under the roof of an open cow shed (Rau and Rau 1916). *Bembecinus quinquespinosus* sleep in dense clusters beneath rocks within emergence areas (Evans et al. 1986).

Some solitary wasps sleep in the ground. For example, females of *Bembix*, *Clitemnestra gayi*, *Glenostictia scitula* (Evans 1966c), *Philanthus* (Evans and O'Neill 1988), and some masarine wasps (S. Gess 1996) spend nights within their nests. Female *C. gayi* may be joined by males, whereas *G. scitula* males sleep apart from females within loose clusters on plants. *Bembix* and *Philanthus* males dig their own short sleeping burrows, often near the females' nesting area. Both sexes of *Microbembex* spend the night individually in sleeping burrows (Evans 1966c). As are nest burrows, sleeping burrows are often closed from the inside when the wasp enters for the night.

Overwintering

In temperate areas, most wasps survive cold winters as prepupae within nest cells, emerging as adults only with the return of favorable conditions during the following spring or summer. However, adults are known to pass the winter in a small number of species of solitary wasps, including several species of *Anoplius*, *Priocnemis*, and *Dipogon* in the Pompilidae and *Liris* and *Podalonia* in the Sphecidae. Females of both *Liris argentata* and *Podalonia luctuosa* overwinter in burrows constructed in late summer and early fall, with up to 10 females of *P. luctuosa* sharing a single burrow. In the mountains of Europe, *Podalonia hirsuta* overwinter in aggregations sheltered in rock crevices or similar microhabitats. By overwintering as adults, these wasps may be able to use prey available only in early spring or avoid parasitoids whose peak of activity occurs later in the summer (O'Brien and Kurczewski 1982a,c, and references therein).

Conclusion

Thermoregulation is an overriding concern to wasps because temperature-dependent physiological performance has a strong influence on the duration and efficiency of daily activity and on survival in stressful environments. Because of the relative paucity of information on thermoregulation by solitary wasps, this chapter is as much an appeal for more research as it is a review of what we know. Biologists interested in wasp thermoregulation must now move from anecdote to detailed analysis, as several researchers have already done. A model study might proceed as follows. One would determine the body temperatures achieved by wasps in the field at different times of day and in different microhabitats while assessing the relative contributions of physiological, morphological, and behavioral thermoregulatory adaptations. Physiological mechanisms include endothermy and countercurrent heat exchangers; morphological mechanisms include coloration and insulating hairs; and behavioral mechanisms include basking, posturing, and microhabitat selection. The next step might be to determine the effectiveness of thermoregulatory adaptations by comparing the observed body temperatures with (1) body temperatures that would be achieved if the wasp behaved randomly with respect to its thermal environment and (2) preferred body temperatures determined under controlled conditions (Hertz et al. 1993).

Perhaps the easiest response to the question "Why do wasps sleep?" is "Why not?" What else can they do when it is cool and dark? However, given the diversity and species specificity of sleeping behaviors that have been described, the question as to why they choose particular locations and postures deserves a less facile answer. During the night, activity is impossible because of low temperatures, and a wasp adapted to daytime conditions may be vulnerable to nocturnal predators. Thus, at night the best that a wasp can do is await the return of favorable conditions, perhaps in a location that warms quickly the following morning (e.g., in the east-facing edge of a forest clearing or at the tops of plants). Wasps may also choose a spot that minimizes predation risk, dehydration, or physical damage caused by wind. The choice of a particular sleeping site may not solve all of these problems simultaneously. Those species that sleep in the ground can probably avoid predation easily while occupying a windless and relatively humid microenvironment. However, they may warm less readily in the morning than do wasps that sit high atop plants, where the morning sun will first strike. Further research, it is hoped, will provide definitive answers. For now, we can only speculate.

10/ Parental Strategies

Most female insects invest little in their offspring beyond producing eggs and placing them in an appropriate habitat. Mosquitoes lay their eggs in water; butterflies oviposit on their host plants; and grasshoppers bury their eggs in the soil. But these and most other insects do not build shelters for their young, bring them food, or stand by to protect them from predators. A few insects, however, do one or more of those things. For example, certain female bugs guard their eggs and young from parasitoid wasps and marauding ants (Tallamy and Schaeffer 1997), whereas dung beetles provide their young a discrete package of food in a burrow (Halffter 1997).

The aculeate Hymenoptera exhibit the full range of levels of parental care. On one end of the spectrum, most parasitoid wasps merely oviposit on a host in situ and then abandon their eggs to search for another host. At the other extreme lie the eusocial wasps, bees, and ants, whose young are pampered throughout their preadult lives by a group of specialized caregivers. Most bees (whether social or not) provision nests, and, although there are a few cleptoparasitic species, no bees are parasitoids. All ants are highly social and even the so-called parasitic species live by co-opting the system of caregiving in other ant colonies (Hölldobler and Wilson 1990). The aculeate wasps, however, include not only eusocial and cleptoparasitic species but also parasitoids and nest provisioners, and these vary considerably in the amount of parental investment provided. In fact, the single (paraphyletic) family Sphecidae contains the entire range of behaviors, although the sphecids have barely crossed the threshold of eusociality (Matthews 1991). Thus, solitary wasps are an excellent focal group for studies of parental investment that explore reproductive trade-offs, sex allocation, and patterns in the evolution of parental behavior.

Reproductive Trade-Offs

Female wasps have a finite amount of time and energy to provide for their young, so when increasing the average investment in each offspring they should have to decrease lifetime offspring production (Trivers 1972). Data are available to test this prediction, along with two related predictions concerning egg number and egg size. First, females of species with low levels of parental investment should carry larger numbers of mature eggs (reflecting a potentially greater egg-laying rate). Second, there should be an inverse correlation between the number of eggs carried and the size of each egg.

PARENTAL INVESTMENT AND LIFETIME FECUNDITY

Estimates available from various lab and field studies reveal that lifetime reproductive output is indeed lowest among the solitary aculeate wasps with the highest levels of parental investment (Table 10-1). A clear dichotomy can be seen in the data: the parasitoids, which invest relatively little in each offspring, have higher average lifetime fecundities than the nest provisioners. Although the number of species for which data are available is small, increased parental care is clearly paid for with lower potential lifetime reproductive output among the solitary wasps.

PARENTAL INVESTMENT AND EGG PRODUCTION

Using data from K. Iwata (1955, 1960, 1964, 1965), we can also look at how *potential* fecundity may vary with levels of parental investment. For each genus in some of the major families examined by Iwata, I took the highest published value for the number of mature oocytes in a female's ovaries and then calculated the mean of those values for each family; in some cases data were available for only one genus (Table 10-2). This (admittedly crude) analysis indicates that females of species with less extensive parental care (i.e., parasitoids) generally carry larger numbers of mature oocytes than do nest provisioners. Female Ichneumonidae and Braconidae often carry huge numbers of eggs, giving them the ability to lay multiple egg clutches in (or on) a rapid succession of hosts. The higher egg-laying rates of some Ichneumonidae, Braconidae, and Chrysididae are made possible by the presence of dozens of ovarioles that allow multiple eggs to mature at the same time. Most aculeate parasitoids and cleptoparasites have intermediate numbers of eggs, whereas solitary nest-provisioning wasps with extensive parental care tend to have one to three mature eggs and just six ovarioles. The extreme condition is

Table 10-1 Estimates from laboratory and field studies of lifetime reproductive success of female solitary wasps

Species	Mean number of eggs laid during lifetime[1]	Reference
Parasitoids		
Dryinidae		
Dicondylus indianus	587 (max)	Sahragard et al. 1991
Pseudogonatopus flavifemur	143	Chua et al. 1986
Tetradontochelys unicus	79	Barrett et al. 1965
Bethylidae		
Cephalonomia waterstoni[2]	42, 69	Finlayson 1950a
Goniozus gallicola	154	Gordh 1976
G. emigratus	233	Gordh & Hawkins 1981
G. legneri[3]	149, 170	Gordh et al. 1983
G. triangulifer	81	Legaspi et al. 1987
Laelius pedatus	71	Mertins 1980
Chrysididae		
Praestochrysis shanghaiensis	29	Yamada 1987b
Tiphiidae		
Tiphia popilliavora[4]	36, 55	Clausen 1940
T. parallela	70	Clausen 1940
Nest Provisioners		
Sphecidae		
Ammophila sabulosa	6	Field 1992d
Bembix rostrata	5 (max)	Larsson & Tengö 1989

Table 10-1 *Continued*

Species	Mean number of eggs laid during lifetime[1]	Reference
Sceliphron assimile	10	Freeman 1981b
Sphex ichneumoneus	10	Brockmann 1985a
Trypoxylon politum[5]	12, 13	Molumby 1997
T. texense	10	Freeman 1981a
Pompilidae		
Episyron arrogans	6	Endo 1981
Vespidae (Eumeninae)		
Ancistrocerus adiabatus	11	Cowan 1981
Eumenes alluaudi	5	Brooke 1981
Euodynerus foraminatus	11	Cowan 1981
Pachodynerus nasidens	14	Freeman & Jayasingh 1975
Zeta abdominale	7	Taffe 1979

[1]Values are mean, except where noted as potential maximum (max) values. [2]Laboratory data (from wasps reared at 25° and 30°C, respectively). [3]Two different host species. [4]Two strains of different geographic origin. [5]Two different years.

represented by progressively provisioning bembicines and eumenines, which tend to have lower egg numbers than even their mass-provisioning relatives (Evans 1966c, Cowan 1991). Compare the extremely low egg numbers for nest-provisioning aculeates with the approximately 15,000 mature eggs that Iwata found in a single female *Stilbula cynioformis* (a eucharitid parasitoid); this number certainly exceeds the total number of mature eggs present within entire aggregations of most nest-provisioning wasps.

A comparison can also be made between nest-provisioning Sphecidae and the few members of the family that are parasitoids (Table 10-2). Females of two species of *Larra* (parasitoids of mole crickets) examined by Iwata carried 10 and 21 mature oocytes, many more than the range of one to four oocytes found in the nest-provisioning Larrinae. Similarly, *Irenangelus pernix* (a pompilid cleptoparasite of

Table 10-2 Mean maximum and range of the number of mature oocytes found in ovaries of different genera and families of wasps

Family and lifestyle	Number of genera	Maximum number of mature oocytes: Mean of all genera[1]	Maximum number of mature oocytes: Range of values among genera	Range: number of ovarioles per ovary
Apocritan Parasitoids				
Ichneumonidae[2]	81	49	2–366	6–120
Braconidae	19	204	4–1500	4–60
Aculeate Parasitoids and Cleptoparasites				
Chrysididae	3	39	3–104	6–108
Bethylidae (*Goniozus* spp.)	1	26	—	6
Sphecidae (*Chlorion* spp.)	1	21	—	6
Sphecidae (*Larra* spp.)	1	10.5	—	6
Scoliidae	2	4.5	2–7	6
Mutillidae	5	4.4	6–10	1–8
Pompilidae *(Irenangelus)*	1	10	—	6
Aculeate Solitary Nest Provisioners				
Sphecidae (Sphecinae)	4	2.5	2–4	6
Sphecidae (Pemphredoninae)	5	1.6	1–2	6
Sphecidae (Larrinae)	9	1.6	1–4	6
Sphecidae (Crabroninae)	4	1.8	1–2	6
Sphecidae (Bembicinae)	6	1.5	1–2	6
Sphecidae (Philanthinae)	2	2.0	—	6
Pompilidae	15	2.5	1–6	6
Vespidae (Eumeninae)	11	1.3	1–3	6
Eusocial Aculeates				
Vespidae	6	4.7	3–7	6–22

Sources: Iwata 1955, 1960, 1964, 1965; O'Neill 1985.
[1]Where there is information for more than one genus, calculation is based on highest value for each genus, where N = number of genera. [2]Does not include genus *Euceros*, in which females may carry up to ~5000 mature eggs.

other spider wasps) and *Chlorion lobatum* (a sphecine parasitoid of field crickets) have higher mature oocyte numbers than their nest-provisioning relatives. Intermediate between the relatively highly fecund parasitoid Sphecidae and the low-fecundity nest provisioners is *Podalonia valida,* whose females may carry as many as seven mature oocytes. Although *P. valida* is a nest provisioner, several aspects of its behavior allow females to have higher egg-laying rates than most sphecids: each female builds a series of shallow nests in a localized area, and each larva receives a single large prey item (O'Neill and Evans 1999).

FECUNDITY AND EGG SIZE

Another means of quantifying investment is to compare the sizes of eggs produced by species with contrasting parental strategies. The major constituent of insect eggs is yolk, which contains the proteins and other nutrients necessary for development of the embryo to a first-instar larva. Adult females of many species of parasitoids feed on host hemolymph, taking in significant amounts of protein after reaching the adult stage (Jervis and Kidd 1991). However, adult females of such groups of nest provisioners as the Vespidae (Eumeninae and Masarinae) and the Sphecidae (Bembicinae and Philanthinae) consume very little protein or fat in their diets of nectar and honeydew. Therefore, most of the nutrients in these eggs are probably derived from those that the female took in as a larva. Thus, each female has a limited reservoir of materials with which to make eggs. When producing larger eggs, she must necessarily produce fewer eggs. Females are further constrained by the inflexibility of their chitinous exoskeletons, which limits abdominal space for carrying eggs. In fact, the large oocytes of some female aculeates occupy such constricted spaces within the abdomen that they are often deformed from being appressed against the inner surface of the exoskeleton (O'Neill 1985), although this deformation probably does not affect egg viability. Queens of some ant and stingless bee species can expand to accommodate higher numbers of eggs by stretching the intersegmental membranes of their abdomens (E. Wilson 1971). However, queens remain in their nests most of their lives, never having to carry the added weight of eggs in flight, as do female solitary wasps.

Y. Itô (1978) tested the hypothesis that mature oocyte size and number are inversely correlated in the Hymenoptera using the extensive data on egg size gathered by Iwata (Iwata 1955, 1960, 1964; Iwata

Table 10-3 The egg index (length of egg/length of thorax) and the number of mature ovarian oocytes in species of Parasitica and Aculeata

Family	Species	Egg index	Maximum number of mature eggs
Parasitica			
Ichneumonidae	*Habronyx insidiator*	0.11	130
	Trogus mactator	0.31	60
	Diplazon laetatorius	0.43	28
	Polysphincta tuberculata	0.44	17
Parasitoid Aculeata			
Chrysididae	*Chrysis ignata*	0.48	21
Tiphiidae	*Tiphia popilliavora*	0.67	20
Sphecidae (Larrinae)	*Larra amplipennis*	0.54	21
Solitary Nest-Provisioning Aculeata			
Sphecidae (Sphecinae)	*Sphex nigellus*	0.60	2
Sphecidae (Pemphredoninae)	*Psenulus lukricus*	0.90	1
	Microstigmus comes	1.12	1
Sphecidae (Bembicinae)	*Bembecinus japonicus*	1.01	1
Pompilidae	*Batozonellus maculifrons*	0.96	2
Vespidae (Eumeninae)	*Eumenes micado*	0.81	1
	Ancistrocerus fukaianus	0.88	1
Eusocial Aculeata			
Vespidae	*Polistes yokohamae* (queen)	0.69	6

Source: Modified from Itô 1978.

and Sakagami 1966). To make the comparisons, he controlled for differences in egg size among different-sized species by transforming absolute egg size to an egg index, the ratio of the length of the thorax to the length of the largest mature oocyte. Itô found that female apocritans (Ichneumonidae) and parasitoid aculeates generally carry large numbers of relatively small mature oocytes (Table 10-3,

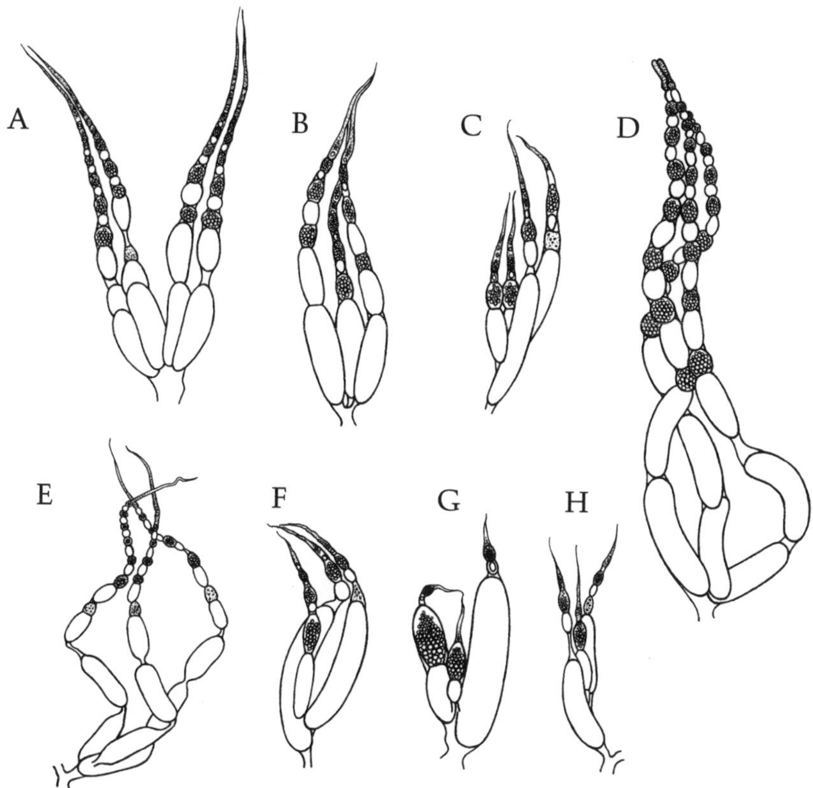

Figure 10-1. Ovaries of some aculeate wasps; each drawing shows one (of two) ovaries of each female. (A) *Mutilla europea* (Mutillidae). (B) *Scolia erythrosoma* (Scoliidae). (C) *Stilbum cyanurum* (Chrysididae). (D) *Larra erebus* (Larrinae). (E) *Chlorion lobatum* (Sphecidae). (F) *Sphex crudelis* (Sphecinae). (G) *Bembecinus japonicus* (Bembecinae). (H) *Eumenes campaniformis* (Eumeninae). (A–H redrawn from Iwata 1955, 1965.)

Fig. 10-1A–E), whereas nest-provisioning wasps carried fewer large eggs (Table 10-3, Fig. 10-1F–H). The progressively provisioning Bembicinae and Eumeninae, which take several days to provision a cell, tend to carry just one mature oocyte (Iwata 1955, Evans 1966c).

Large egg size, low fecundity, and advanced parental care are phylogenetically derived conditions within the Hymenoptera. However, the trend may be reversed, not only in aculeates that revert to a parasitoid lifestyle (e.g., *Chlorion* and *Larra*) but also in eusocial species. Because queens have been emancipated from parental care by their workers, there has undoubtedly been selection for higher egg-laying rates. In addition, the highly advanced cooperative brood care of eusocial species may lessen the need to start an embryo with a huge investment in yolk. Thus, compared with their solitary eumenine relatives, queens of eusocial Vespidae typically carry a greater number of mature eggs in a greater number of ovarioles (Table 10-2). On the basis of data from *Polistes yokohamae*, Itô (1978) suggested that queens of eusocial wasps also carry relatively smaller eggs (Table 10-3). However, females of *Microstigmus comes*, a eusocial sphecid, are like their solitary relatives in carrying a maximum of just one mature egg (Matthews 1968). This value may reflect their low reproductive potential, since they live in relatively small colonies.

Parental Investment Decisions

Even among parasitoids, reproduction is not just a matter of finding a host and ovipositing. A mother's fitness may be strongly affected by several types of reproductive decisions made during the course of egg laying. The first (and this applies mainly to the Bethylidae) is clutch size: how many eggs to lay on each host. The second decision (really a set of related decisions) falls under the category of sex allocation: how many offspring of each sex to produce and how much investment to provide sons relative to daughters. The sex of the offspring of other animals (such as vertebrates) is not controlled by the mother, but the question of sex allocation is highly relevant to Hymenoptera because they have haplodiploid sex determination (Whiting 1935, Kerr 1962). Female wasps can determine the sex of each offspring simply by controlling the release of stored sperm from the spermatheca as the egg passes down the oviduct: unfertilized eggs become haploid males, whereas fertilized eggs usually become diploid females, or occasionally sterile diploid males (Cook 1992).

CLUTCH SIZE

Nonaculeate parasitoid wasps commonly lay multiple eggs on hosts, but within the Aculeata, multiple egg clutches occur only in the Bethylidae and a few Dryinidae and Chrysididae. Within the Bethylidae, mean clutch sizes of one to five eggs are typical, but values from 25 to 105 have been reported, the higher values being typical of the genus *Sclerodermus* (Fig. 10-2A) (Hardy and Mayhew

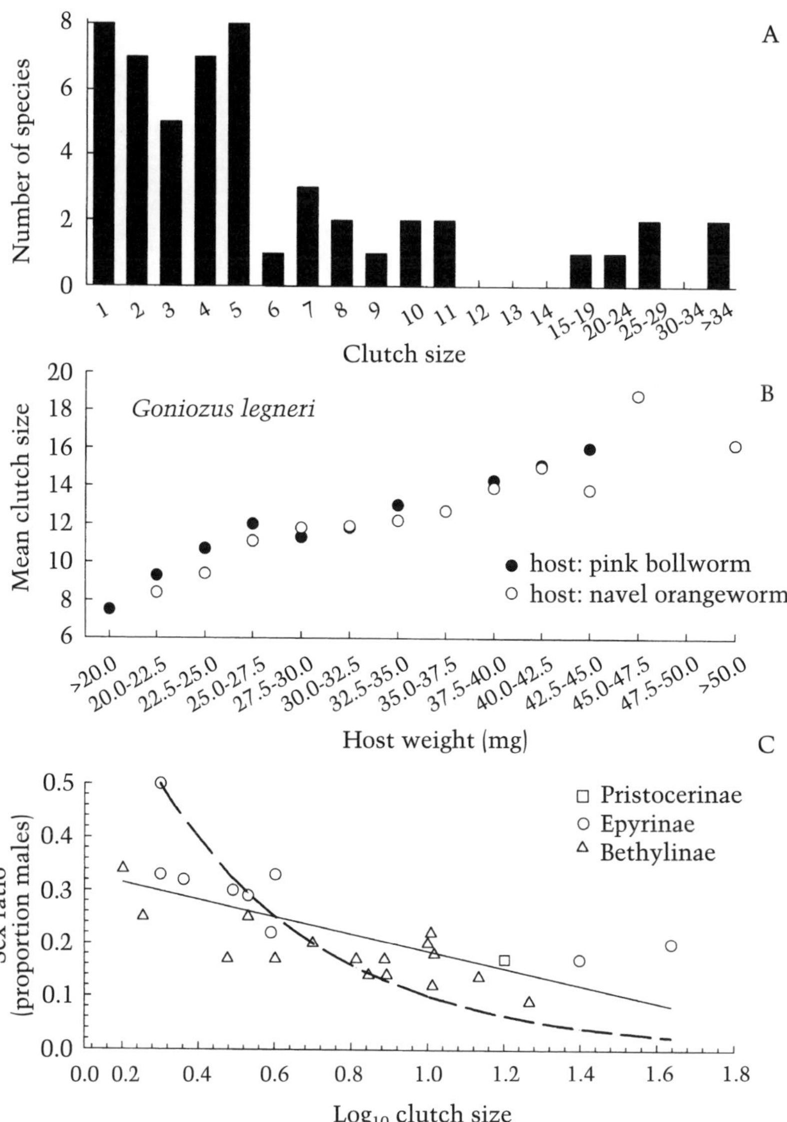

Figure 10-2. (A) Frequency distribution of clutch sizes among 52 species of Bethylidae (modified from Mayhew and Hardy 1998). (B) The relationship between host weight and clutch size for two species of host Lepidoptera used by *Goniozus legneri* (data from Gordh et al. 1983). (C) The relationship of sex ratio (proportion males) to $\log_{10}$ clutch size for 26 gregarious species in three subfamilies of Bethylidae; solid line is the linear regression of sex ratio on clutch size; dashed line is the sex ratio predicted from Hamilton's local mate competition theory (i.e., the reciprocal of clutch size) (modified from Hardy and Mayhew 1998).

1998, Mayhew and Hardy 1998). As expected, clutch size is positively correlated with host size within species of Bethylidae (Fig. 10-2B; see also Hardy et al. 1992, Mayhew 1998), indicating that females add more eggs to a host when it is capable of supporting more larvae. In principle, overly large clutches could increase competition among siblings for host resources, resulting in starvation or reduced body size of survivors. Reduced body size could, in turn, translate into lower fecundity, because smaller bethylid females tend to carry fewer and smaller eggs (O'Neill and Skinner 1990). When researchers experimentally increased clutch size in bethylids by transferring eggs among hosts, offspring size decreased (Hardy et al. 1992, Mayhew 1997). However, under normal conditions, when *Laelius pedatus* lay more eggs on larger hosts, they allow proportionally greater resources for each offspring (Mayhew 1998). As a result, adult *L. pedatus* emerging from larger broods tend to be *larger* than those from smaller clutches. In addition, developmental mortality does not increase with clutch size, again suggesting that females take advantage of the greater resources available on larger hosts by increasing clutch size without increasing competition for resources.

SEX ALLOCATION

Given that a female wasp is ready to lay an egg, which sex should it be? This question lies at the heart of studies of the evolution of parental strategies and eusocial behavior in the Hymenoptera (Godfray 1994, S. Frank 1995, Crozier and Pamilo 1996). In biology, the standard answer to the question of sex ratios is derived from R. Fisher's (1930) sex ratio theory, which states that frequency-dependent natural selection should drive the population sex ratio toward an equal number of males and females. At a 1 : 1 (male : female) sex ratio, an evolutionary equilibrium is reached because males and females in a population have the same average reproductive success. However, because sons and daughters are equally valuable to their mother as grandchild-producing entities, it should not matter which sex of offspring is produced in any particular case (Crozier and Pamilo 1996). Thus, according to Fisher's basic theory, although the population sex ratio should be 1 : 1, sex determination during reproduction need only be random: analogous to a fair coin toss. Such a result is achieved in vertebrates (including humans) via heterogametic sex determination.

However, Fisher's basic model makes several implicit assumptions that, if violated, lead to different predictions for sex allocation decisions by individual mothers; these altered decisions may lead, in turn, to different predictions for sex ratios of the entire population. First, Fisher's basic model assumes that the relationship between the amount of parental investment that an offspring receives and the offspring's eventual reproductive success is the same for both sons and daughters (S. Frank 1995). Second, it assumes that the sons of a given female have the potential of mating with any female in the population (Hamilton 1967). Third, it assumes that the relative reproductive value of the different sexes does not change from generation to generation (Werren and Charnov 1978). Violations of these assumptions generate novel predictions about sex allocation decisions of solitary wasps.

Sex Allocation and the Amount of Investment Given Each Son and Daughter

Even in species in which the population sex ratio is evolutionarily stable at 1:1, females may have something to gain from assigning a specific sex to a specific offspring (rather than assigning sex randomly, for example). Nonrandom sex assignment is predicted when (1) the amount of investment available to be given to offspring is variable, (2) the amount of investment that an offspring receives affects its future reproductive success, and (3) variation in the amount of investment affects sons and daughters differentially. R. Trivers and D. Willard (1973) predicted that, when these circumstances hold, species that can readily control the sex of offspring should evolve a condition-dependent sex allocation strategy. In particular, if small body size is likely to have a greater detrimental effect on female offspring than on male offspring, then a mother should produce a daughter when she can provide a high level of investment but a son when she is constrained to give a lower level.

The three assumptions of the Trivers-Willard model apply to solitary wasps. First, the ability of females to invest in young does indeed vary with both extrinsic and intrinsic factors. Extrinsic (environmental) factors that could affect the ability of a female to invest in specific offspring include temporal and spatial variation in prey availability caused by such factors as weather or seasonal changes in prey availability (see Chapter 3). For species that nest in existing cavities, the diameter of the nest hole is important, because it may constrain the size of offspring that can be produced (Krombein 1967). The major intrinsic factor that affects females' ability to invest in offspring is body size: larger females can invest more heavily in each offspring (as we shall document shortly).

Second, the amount of food that a larva receives is correlated with its eventual adult size in various solitary wasps, including the mutillid *Chrestomutilla glossinae* (Heaversedge 1970; see also Mickel 1928), the eumenine *Euodynerus foraminatus* (Cowan 1981), and the sphecids *Ammophila sabulosa* (Field 1992d), *Sceliphron assimile* (Freeman 1981b), *Sceliphron spirifex* (E. White 1962), *Sphecius speciosus* (Dow 1942), and *Trypoxylon politum* (Cross et al. 1975, Brockmann and Grafen 1989, Molumby 1997). Moreover, larger size often confers a reproductive advantage on offspring when they become adults. As we saw in Chapter 8, large males are often more successful in mate competition than are small males. Similarly, larger females produce more or larger eggs, or both (O'Neill 1985, Larsson 1989a, O'Neill and Skinner 1990), and they provision nests with more and larger prey (Linsley and MacSwain 1956, Kurczewski et al. 1969, Byers 1978, Kurczewski and Elliott 1978, Laing 1979, Cowan 1981, Gwynne and Dodson 1983, O'Neill 1985, Willmer 1985b, Hastings 1986, Molumby 1997, Strohm and Linsenmair 1997a). Larger female wasps may also be better able to usurp and defend nests from conspecifics (Mueller et al. 1992), and they may be more efficient nest builders (Freeman and Johnston 1978a). In contrast, smaller individuals of a wasp species rarely have consistently higher success than larger conspecifics (but see Larsson and Tengö 1989).

Third, although size often affects the fitness of both sexes of solitary wasps, small females generally suffer more than do small males. The simple fact that females of most species of solitary wasps are larger than males argues for the greater importance of large body size for females. More direct evidence is available from A. Molumby (1997), who showed that the reproductive success of *Trypoxylon politum* females increased with body size but that the mating success of males did not. Much of the selection for larger female body size probably comes from natural selection favoring females in a species that produces greater numbers of eggs and larger eggs (O'Neill 1985, Alcock and Gwynne 1987, Kurczewski 1987a, Larsson 1989a, O'Neill and Skinner 1990). In examining the relationship between female size and egg size in five species of Sphecidae in which females were larger than males, I (O'Neill 1985) noted that some of the eggs of the largest females were almost as long as the abdomens of smaller conspecifics. Thus, if the smallest females of these species were as small as the smallest males they probably could not produce even one egg of viable size. In addition, larger female wasps can carry larger loads during provisioning (O'Neill 1985, Hastings 1986, Strohm and Linsenmair 1997a). That load-carrying ability favors larger body size can be seen in those relatively rare cases among wasps in which mean male size equals or exceeds that of females. These exceptions to the rule involve species whose males carry females during courtship and mating (Chapter 8).

It seems likely, then, that the three conditions of the Trivers-Willard model are met in at least some species of nest-provisioning solitary wasps. Thus, it is not surprising that females often provide more generous provisions to daughters. J. Fabre (1921) observed this differential provisioning early on, and more recent quantitative studies confirm his conclusion (Table 10-4). In other studies, indirect evidence of the same provisioning bias can be inferred from the fact that daughters are placed in larger cells, which generally are stocked with more provisions (e.g., Krombein 1967, Danks 1970, Evans 1971). But is sex allocation condition-dependent, as the Trivers-Willard model predicts? Do mothers tend to produce daughters when high levels of investment are possible and sons when conditions are poor? Several studies report that daughters are more likely to be produced by larger mothers, which enjoy greater provisioning success. In *Sceliphron assimile* (Freeman 1981b) and *Trypoxylon politum* (Molumby 1997), larger females build larger cells, stock them with a greater mass of provisions, and produce larger offspring, a greater percentage of which are females. Moreover, larger mothers of both species tend to produce larger daughters, but not larger sons. P. Willmer (1985b) inferred a similar pattern for *Cerceris arenaria*.

In a study of the beewolf *Philanthus triangulum*, sex allocation was a function of both the size of the mother and prey availability (Strohm and Linsenmair 1997a,b). Body size affected provisioning because larger female *P. triangulum* have a higher proportion of successful hunting trips, engage in hunting trips of shorter duration, and take heavier honey bees. Thus, as predicted, large females produced a greater proportion of daughters than did small females. In contrast, because mothers in the lower 33% of the size range are less efficient foragers, they produced only sons, which require 40% fewer prey on average. Most larger females, on the other hand, produced a mixture of sons and daughters, and none of the females in the largest 20% of the size range produced only sons.

Sex allocation in *P. triangulum* could also be influenced by the changes in honey bee availability, which are known to be caused by

Table 10-4 Evidence for sex-specific provisioning by solitary wasps

Species	Indication that female offspring receive greater amount of investment	References
Ammophila sabulosa	Females receive a 37% greater prey mass on average.	Field 1992d
Ancistrocerus adiabatus	Females receive a 65% greater prey mass on average.	Cowan 1981
A. antilope	Females receive a mean of 7.4 prey caterpillars; males receive 6.1.	Krombein 1967
Cerceris arenaria	Females receive a mean of 7–13 prey weevils; males receive 4–8.	Willmer 1985b
Euodynerus foraminatus	Females receive a mean of 14 prey caterpillars; males receive 8. Females receive 69% greater prey mass on average.	Krombein 1967, Cowan 1981
Pachodynerus nasidens	Females receive 123% more prey caterpillars. Eggs that become females are 23% longer than those that become males.	Jayasingh 1980, Jayasingh & Taffe 1982
Philanthus triangulum	Females receive a mean of 3.8 prey bees; males receive 2.2.	Strohm & Linsenmair 1997a
Sceliphron assimile	Females are placed in larger cells, and larger cells are stocked with a greater mass of prey.	Freeman 1981b
S. spirifex	Females receive an estimated 49% greater prey mass on average.	E. White 1962
Sphecius speciosus	Females receive 2.2–3.9 g of provisions; males receive 0.9–2.1 g. All females receive two prey items, but most males receive only one.	Dow 1942, Lin 1979a
Tiphia popilliavora	On larger third-instar hosts, 67% of offspring are females; on smaller second-instar hosts, 94% of offspring are males.	Brunson 1938
Trypoxylon politum	Proportion of males decreases in cells with greater mass of provisions.	Molumby 1997

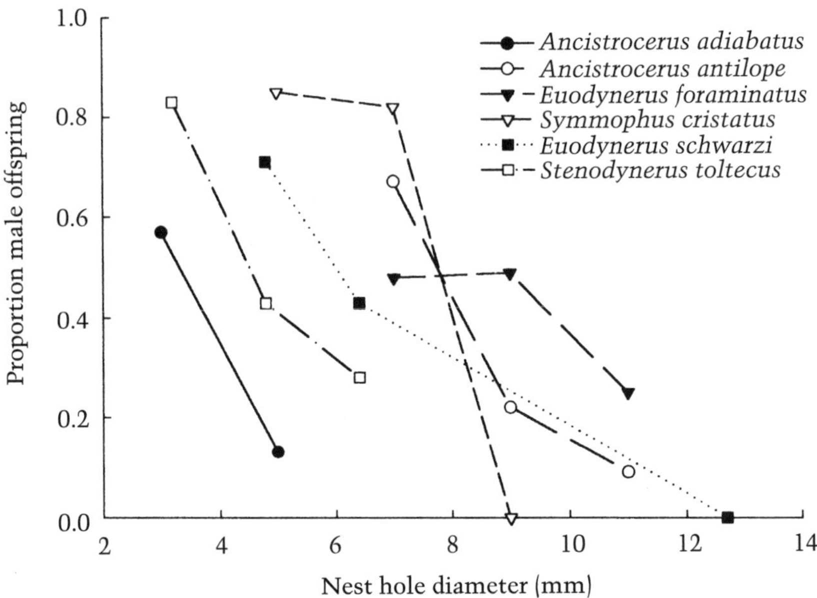

Figure 10-3. The relationship between the proportion of males emerging from trap nests and the diameter of the nest hole. (Data for top four species from Longair 1981; for bottom two, from Krombein 1967.)

weather conditions (Simonthomas and Simonthomas 1972). By experimentally controlling prey availability to simulate changes in prey density, E. Strohm and K. Linsenmair (1997b) found that females given access to 0.5–2.0 bees/day allocated all of them to sons. However, if given 4.0 bees/day or 4.0 bees every fourth day, more than half the prey were allocated to daughters; it takes at least three prey items to feed a daughter, but a son can develop on just one honey bee. Thus, limitations on prey availability, superimposed on size constraints, explain why *P. triangulum* males (as a group) receive nearly 75% of provisions and why the population sex ratio is male-biased.

Perhaps the largest body of indirect evidence that females adjust sex to prevailing conditions comes from observations of females of cavity-nesting species. These females tend to provision fewer female cells in nest holes with smaller-bore diameters (Fig. 10-3) (Krombein 1967, Coville and Coville 1980, Longair 1981, Coville and Griswold 1983). Two related factors could be at work. First, larger daughters may not fit well in small-diameter holes. Second, large mothers may avoid smaller holes, and small mothers that do use them may be unable to find enough prey to provision for a large daughter.

Sex Allocation and the Mating Structure of a Population

It is probable that a large amount of mixing occurs before mating in populations of many species of solitary wasp. If it does, then a given male has an equal likelihood of mating with any given female in the population. For example, in the many bembicines and philanthines, males emerge in large numbers before females and appear to mix freely over a large area before the appearance of receptive females (Evans 1957b, Evans and O'Neill 1978, Longair et al. 1987, O'Neill et al. 1989). Similarly, *Hemipepsis ustulata* (Alcock 1981a) and *Philanthus basilaris* (O'Neill 1983b) males gather on leks at some distances from their scattered emergence sites and probably draw dispersed females from equally great distances.

However, such panmictic mating structures do not occur in those solitary wasp species in which (1) females produce isolated clusters of offspring, (2) the male offspring have little opportunity to disperse and mate with females from other broods (because the distance to the next brood is large relative to the males' dispersal abilities), and (3) male offspring can mate with their sisters. In these species, a female that accepts a brother as a mate reduces the risk of not obtaining sperm. Because close inbreeding is possible in Hymenoptera, a mother that produced sons in excess of the number needed to inseminate her daughters would be squandering her limited parental resources. She would produce sons who would compete unnecessarily with one another for access to sisters and who would not have much chance of fathering offspring with females from other broods. Such competition among brothers to mate with their sisters was referred as *local mate competition* (LMC) by W.D. Hamilton (1967), who also pointed out that under LMC, natural selection should favor a parental sex allocation strategy that maximizes the number of mated females produced in each brood (Green et al. 1982). In other words, the adaptive strategy for mothers in spatially structured populations with LMC is to produce broods with the minimum number of sons required to inseminate all their sisters. For example, a mother producing one son that mates all nine of her daughters in a brood of 10 offspring ends up with nine grandchild-producing daughters. In contrast, a mother with the same number of offspring in a brood with a 1:1 sex ratio produces five sons who compete with one another for access to their sisters and just five daughters, which produce a proportionately lower number of grandchildren. This argument led Hamilton to predict that, if males can mate only with their sisters and if males are not limited in the number of females they can inseminate, then the optimal sex ratio of broods (from the mothers' standpoint) is the reciprocal of clutch size (i.e., one son per *N* sisters).

Bethylid wasps are exactly the kind of animals to which Hamilton's LMC theory applies. Females of many bethylids lay batches of eggs on beetle or moth larvae (Chapter 2), and sons from these broods mate with their sisters (Chapter 8). In qualitative agreement with Hamilton's predictions, bethylid broods contain a low proportion of males, and the proportion decreases with increasing clutch size (Griffiths and Godfray 1988, Hardy and Mayhew 1998) (Fig. 10-2C).

Although a small number of males in a clutch could maximize the number of inseminated females produced by their mother, anything that causes the loss of the few sons leaves the daughters without mates. G. Heimpel (1994) referred to the cost of producing male-less broods under LMC as the "virginity load." One problem that could leave a brood without males (or with superfluous males for that matter) would be imprecise sex determination. If sex determination in a population with a 50:50 sex ratio is considered analogous to a toss of a fair coin, what would happen if the equivalent process in bethylids with a female-biased sex ratio was the toss of a weighted coin (Hardy 1992)? Such a mechanism could result in broods with the specific mean proportion of males that maximizes production of inseminated daughters. However, some females in the population would produce broods with zero sons, while others had broods with superfluous sons. For example, in a species with a clutch size of 10, an optimal value of one male/clutch, and a binomial distribution of clutch sizes, fully 35% of broods would lack males and 26% would have in excess of the optimum.

Thus, bethylid females could gain a substantial fitness advantage from allocating sex to eggs *precisely* rather than binomially, thereby minimizing the number of male-less broods and maximizing the number of daughters (Green et al. 1982, Hardy 1992). Precise ("less than binomial") control of the sex of offspring has been demonstrated (to a greater or lesser extent) for at least seven bethylids of the genera *Goniozus* and *Laelius* (Green et al. 1982, Legaspi et al. 1987, Cook 1992, Hardy et al. 1992, Morgan and Cook 1994, Hardy and

Cook 1995, Mayhew and Godfray 1997). For example, for brood sizes of *Goniozus gordhi* ranging from 2 to 9, the proportion of single male broods (0.89) was much higher than would be expected by chance if sex control were binomial; for brood sizes from 10 to 15, 61% of the broods contained more than one male, perhaps as a means of ensuring an adequate supply of sperm for the additional females (Green et al. 1982). Similarly, the proportion of single-male broods in *Laelius pedatus* clutches of 2–6 offspring was 0.87, many more than a process analogous to a weighted coin toss would generate (Morgan and Cook 1994). Thus, precise sex allocation reduces the virginity load.

Another problem that could result in broods without males is mortality among males before adult emergence. The effect of developmental mortality was clearly seen in *L. pedatus*, where the proportion of male-less broods (0.10) was close to the proportion of males that died during development (0.09) (Morgan and Cook 1994), suggesting that the precision of sex allocation is even higher than indicated above. Male developmental mortality increases the proportion of all-female broods in *Goniozus nephantidis* to as high as 32% (Hardy and Cook 1995). Presumably, selection for optimal sex ratio should result in some balance between the cost of producing all-female broods and the cost of producing too many males as an insurance against mortality. However, the effect of such selection may vary if mortality is different on the alternative hosts of a bethylid species (Morgan and Cook 1994).

The application of LMC theory to bethylid sex allocation studies clearly holds much promise, particularly as data on variation in population mating structure, developmental mortality, host quality, and the sperm production capacity of males are integrated into Hamilton's basic theory (Hardy and Cook 1995, Mayhew and Godfray 1997, Hardy and Mayhew 1998).

As an aside, I should note that the mating systems of wasps may affect sex allocation in another way. The nest cells of most species of cavity-nesters studied with trap nests are provisioned in a sex-biased sequence, with the female cells deeper within the nest tunnel than the male cells (Krombein 1967). Male cells are provisioned last because these species are protandrous and brothers have to be able to emerge first without having to chew through their sisters' cells to get outside. The sequence of sexes in nests provides further evidence that females control the sex of their offspring precisely.

Sex Allocation in Bivoltine Species with Overlapping Generations

Local mate competition models deal with the situation in which males of one brood have little opportunity to mate with females of other broods. The models of sex allocation discussed above also apply to species in which males of one generation do not have the opportunity to mate with females of the next generation. In species with a single generation each year (univoltine species), adults of different generations never meet. However, in bivoltine species, males of the first summer generation may live long enough to mate with females of the second generation (but males of the second generation cannot mate with females of the first generation, which are no longer receptive by this time). As a result, males of the first generation can have greater average reproductive success than females of their own generation. Therefore, females that produce the *first* generation should produce an excess of sons. Moreover, males of the second generation have lower reproductive value than their sisters because they have to compete with males from the first generation (but they die before females of the next generation appear). Thus, females producing the *second* generation should produce a greater proportion of daughters than those of the first (Werren and Charnov 1978, Seger 1983). The result would be male-biased sex ratios in the first generation, but female-biased ratios in the second.

Alternating sex ratios in bivoltine solitary wasps, with a male-biased sex ratio in the first generation, have been found in both eumenine (Krombein 1967, Longair 1981) and sphecid wasps (Danks 1983, Brockmann and Grafen 1992). For example, the percentage of males dropped from 91% to 33% between generations of *Ancistrocerus antilope* and from 50% to 41% between generations of *Euodynerus foraminatus* (Longair 1981). Similarly, in a 7-year study of *Trypoxylon politum*, sex ratios were biased in favor of sons in the first generation but biased toward daughters (or were 1:1) in the second. In contrast, the sex ratio was 1:1 in a univoltine population of *T. politum* in the northern United States (Brockmann and Grafen 1992).

Although these temporal patterns give qualitative support to the sex allocation model of J. Seger (1983), there are a number of alternative hypotheses for male-biased sex ratios in the first (overwintering) generation of bivoltine species. Some of the alternatives are adaptive hypotheses, some are not. The alternative hypotheses include the following suggestions: (1) Male-biased ratios compensate for greater male mortality in the overwintering generation (Longair 1981). (2) Male-biased ratios provide for greater genetic variability among mates in inbreeding species (Cowan 1979). (3) Male-biased sex ratios in the first generation of the year are an artifact of failing to count those females that actually emerged before winter (Danks 1983). (4) Although numerical sex ratios vary among generations, the amount invested in each sex is equal in the two generations (as would be predicted by Fisher's [1930] sex ratio model). H. Danks (1983) and H. Brockmann and A. Grafen (1992) suggest further reasons why sex ratios could alternate (or only apparently alternate) between generations. The dizzying array of possible explanations for alternating sex ratios can be sorted out only in carefully designed, long-term studies of bivoltine species (Brockmann and Grafen 1992).

Sex Allocation and Biased Sex Ratios: Other Causes

If the complexities of the models discussed above do not give those planning sex ratio studies enough cause to worry, still other factors might affect patterns of sex allocation. First, depletion of their stored sperm may force older female bethylid wasps to produce males (if they cannot remate) (Gordh and Hawkins 1981, Gordh et al. 1983). Second, sex-biased developmental mortality can be a problem, because it obscures the primary sex ratio and therefore the mother's sex allocation decisions. Relative rates of male and female mortality during development cannot be measured because immature wasps are difficult to sex. Third, despite the fact that diploid Hymenoptera are usually females, diploid male wasps occasionally appear (and are often sterile) (Cook 1992, T. Chapman and Stewart 1996). Although these males may usually do no more than add noise to sex allocation estimates (because the mother "meant" some of her sons to be daughters), high levels of inbreeding can sometimes result in a high proportion of diploid males in a population (nearly 25% in the case of one inbred population of *Ancistrocerus antilope* [T. Chapman and Stewart 1996]).

Worse still are the practical difficulties of obtaining an unbiased sex ratio estimate for a population. At first glance, it would seem that species that nest in trap nests would be ideal for sex ratio studies because one can obtain large samples of broods. However, data on sex ratios in trap nests are difficult to interpret because the percentage of male offspring in trap nests typically decreases with increasing nest hole diameter (Fig. 10-3). Unless the observer provides a distribution of trap nest hole diameters that matches that typical of the recent evolutionary history of the species (an unlikely coincidence), the observed sex ratios may be atypical and therefore difficult to interpret in an adaptive sense (Cowan 1981, Helms 1994). To compound matters, sex ratio may also vary with trap nest depth (Danks 1983). The problem of inadvertently biasing sex ratio when using artificial nests is avoided when studying natural nests. Relatively unbiased samples of some naturally occurring nest cells are sometimes easy to gather, for example from the mud nests of *Sceliphron* (Freeman 1981b) and *Trypoxylon* (Brockmann and Grafen 1992). However, random samples of nest cells of ground-nesting species are more difficult to obtain. My co-workers and I (O'Neill et al. 1989) partially overcame this problem by using an emergence trap for the ground-nesting species *Bembecinus quinquespinosus*, but we still could not rule out an effect of preemergence mortality on the observed sex ratio (which did not differ from 1:1 in either of two years).

Future studies may tell us why one can obtain such sex ratio estimates in the field as 100% males in *Sphaeropthalma pensylvanica* (Matthews 1997) and 0% males in *Elampus viridicyaneus* (Rosenheim and Grace 1987) and *Gonatopus clavipes* (Waloff and Jervis 1987). Other studies have reported seasonally variable sex ratios (Heaversedge 1970) and an unexpectedly high proportion of single-sex broods (Rosenheim 1993). Such results are not likely to be just artifacts of field collecting or lab-rearing techniques, so they may lead us to further evolutionary insights into the weirdness of Hymenopteran sex ratios.

Sex allocation decisions by individual females and overall sex ratios of populations are difficult to predict because of the numerous selective factors acting on each species of wasp. Consider that sex allocation in a bivoltine population of hole-nesting eumenine wasps could be a mixed function of decisions related to (1) condition-dependent investment in males and females, (2) some level of local mate competition, (3) bivoltinism, (4) the emergence sequence of the sexes, and (5) nest hole diameter and length. Actually, it is surprising that we know as much as we do about sex allocation in the solitary Hymenoptera.

The Evolution of Parental Behavior

THE FOSSIL RECORD OF ACULEATE PARENTAL CARE

In principle, there are two ways to reconstruct the evolutionary history of a trait: the fossil record and the comparative method. Except for occasional fossil nests and footprints, fossils that record behavior are rare, and fossils that record parental behavior in particular are nearly nonexistent. The most direct evidence we could expect to have of the existence of parental care in extinct wasps would be fossil nests. The mud-dauber nests that have been found in Eocene and Oligocene deposits (Bown and Ratcliffe 1988) are exceptions. A less direct approach would be to search the fossil record for forms of rake spines, pygidial plates, mandibles, and psammophores known to be associated with particular forms of nests in extant species. However, the fossil record for solitary wasps is scanty. F. Carpenter (1992) lists just 21 sphecids, 7 pompilids, and 6 eumenids in his comprehensive treatise on insect fossils. Thus, fossils can at best give us hints as to the minimum age of certain types of nesting behaviors (which they can only underestimate), and they certainly cannot provide enough data to reconstruct the evolution of wasp nesting behavior. Fortunately, the evolutionary history of this behavior can be deduced from comparative studies of living wasps.

HYPOTHETICAL SEQUENCES IN THE EVOLUTION OF NESTING BEHAVIOR

Sophisticated parental and social behaviors of aculeate Hymenoptera did not appear full-blown overnight, with parasitoid species making a rapid evolutionary transition to eusociality. Instead, the ancestral parasitoid state (with minimal parental care and high fecundity) gradually gave rise to forms of parental care in which females spend much of their lives providing prey, nest, and protection for a few offspring. In order to understand the sequence of steps in the evolution of parental care in solitary wasps, we need to define discrete categories of nesting behavior. Next, we need to arrange the categories in some general hypothetical evolutionary order. Such historical reconstructions require one to make decisions as to which behaviors are likely to be ancestral and which derived (descendant) in each lineage. In Table 10-5, I offer a probable consensus on which traits related to oviposition, nesting, and provisioning are ancestral and which are derived in solitary wasps. Some have been discussed in previous chapters. Others form the basis of the schemes developed

Table 10-5 Adaptive changes required for particular transitions in the evolution of parental strategies of solitary aculeate wasps

Evolutionary transition		Required adaptations and refinements associated with derived trait
Ancestral female trait	Derived female trait	
Adaptations Related to Oviposition and Nesting		
Leaves host in situ after capture and oviposition	Moves host to refuge before oviposition	Prey carriage; means of paralyzing to facilitate prey translocation and storage; site evaluation
Leaves refuge (nest) open	Closes refuge (nest) before departure	Means for selection, carriage, and placement of materials for closure
Modifies an existing cavity (e.g., a crack in the soil)	Constructs cell or nest from scratch (e.g., in undisturbed soil)	Nesting behaviors (e.g., digging with legs and mouthparts); associated structural modifications (e.g., rake spines, flattened pygidial plates, psammophores)
Prepares nest after capturing prey	Prepares nest before foraging	Reversal of sequence of behavior from "prey-nest" to "nest-prey"; improved homing and orientation
Builds single-celled nest	Builds multicelled nest	Long-term burrow maintenance; partition building
Deposits egg on provisions in cell	Deposits egg in empty cell before provisioning	Reversal of "prey-egg" sequence; materials (soil pedestal or silken thread) for attachment of egg to cell wall (most species)
Solitary nesting	Communal nesting	Tolerance of conspecific females
Adaptations Related to Provisioning		
Carries prey directly to refuge (nest)	Deposits prey in temporary cache en	Rudimentary homing and orientation

Table 10-5 *Continued*

Evolutionary transition		Required adaptations and refinements associated with derived trait
Ancestral female trait	Derived female trait	
immediately after capture	route to potential nest	capabilities that allow female to learn location of and travel between prey and nest
Carries prey with mandibles	Carries prey with legs only	Behavioral change
Carries prey with legs only	Carries prey with aid of abdominal structures	Modification of sting (e.g., *Sericophorus*) or of terminal abdominal segment (e.g., *Clypeadon*)
Uses single large host or prey	Uses prey (or pollen load) so small that more than one foraging trip is required for each cell	Estimation of total amount of prey (or pollen) placed in cell as provisioning proceeds
Prepares prey simply by stinging it	Prepares prey beyond simple stinging and placement in cell	Female macerates prey (some Sphecidae and Eumeninae) or severs legs before placing it in cell (some Pompilidae)
Lays many small eggs	Lays a few large eggs	Extended oocyte development
Mass provisions	Progressively provisions	Monitoring total amount of prey provided while the prey are being fed upon
Provisions with arthropods	Provisions with nectar and pollen (carried in crop)	Novel forms of food location, gathering, carriage, and manipulation in cell; pollen-gathering morphology
Progressively provisions one nest at a time	Progressively provisions two nests at a time	Remembering position and contents of two separate nests

Note: The transitions are not necessarily in the order in which they appeared in any given lineage.

by Iwata (1942) and Evans (1953, 1958a), who classified the parental strategies of females wasps in terms of the form and temporal sequence of the following behaviors: hunting, provisioning, oviposition, use of refuges (niches) or nests, and degree of modification and elaboration of nests.

Evans (1958a) used this classification to propose a potential sequence of stages for the evolution of parental strategies. The stages, in slightly modified form, are listed in Table 10-6, along with notes on the taxa in which each stage is represented. Below, I discuss the stages and their proposed order of occurrence, but first a few cautionary notes. First, not all stages need be represented in the history of each lineage (i.e., the list in the table is not an inflexible, linear evolutionary sequence). Second, the fact that we have identified a likely set of evolutionary stages does not imply that a lineage represented today by the "early" stages in the sequence is on a deterministic trajectory toward "later" stages. "Stages," of course, are a human construct, developed after a process has occurred. Finally, the use of the term *stages* does not imply a one-way evolutionary pathway. As we shall see, several lineages have reverted to forms of behavior that are undoubtedly ancestral in the aculeate wasps as a whole.

Disposition of Hosts in Nonnesting Species

The first taxa to consider are those that have remained on the lowest "rungs of the social ladder" (Evans 1958a), the parasitoid aculeates. Most parasitoid and cleptoparasitic aculeate wasps attack their hosts and then leave them where they were found, with one or more eggs attached and often at least temporarily paralyzed. Wasps using this strategy (stage 1 in Table 10-6) include all Dryinidae, Mutillidae, and Sapygidae, most Bethylidae, Tiphiidae, and Chrysididae, and some Pompilidae and Sphecidae (see Chapters 2, 4, and 5). In some cases (e.g., the Dryinidae and Chrysididae that attack walking-stick eggs), the host and the wasp egg are left exposed to the elements and other organisms. Many species, however, attack the host in its own refuge and so it remains in a sheltered location. For example, many bethylids find hosts in such cryptic microhabitats as leaf rolls (hosts of *Goniozus japonicus*), sugarcane stems (*G. indicus*), almond hulls (*G. legneri*), seeds (*Cephalonomia gallicola* and *Laelius pedatus*), and tunnels in wood (*Sclerodermus immigrans*) (see Table 2-2 for references). Similarly, Tiphiidae (Table 2-5) and Scoliidae (Table 2-6) capture beetle larvae in their subterranean feeding cells

Table 10-6 Sequences of parental behaviors in the solitary aculeate wasps

Stage/provisioning sequence[1]	Occurrence
1. Parasitoids that leave prey in situ: **Prey-egg**. Paralysis often incomplete and/or temporary. Some prey in refuges (cryptic locations) when found.	The ancestral condition of the Aculeata; occurs in all Dryinidae, all noncleptoparasitic Chrysididae and Mutillidae, most Bethylidae, Tiphiidae, and Scoliidae, and some Pompilidae and Sphecidae
2. Parasitoids that drag prey to nearby niche that is sometimes modified slightly, usually by covering the prey: **Prey-niche-egg**.	Some Bethylidae, Tiphiidae, Scoliidae, Pompilidae, and Sphecidae
3. Single prey in a single-celled nest built *after* hunting. **Prey-nest-egg-nest closure**.	Most Pompilidae and a few Sphecidae *(Podalonia, Prionyx)*
4. Single prey in a nest built *before* wasp hunts:	
A. Single-celled nest: **Nest-prey-egg-nest closure**.	Pompilidae *(Pompilus cinereus)*, Sphecidae *(Ammophila procera*, a few *Tachysphex)*
B. Multicelled nest: **Nest-prey-egg-cell (multiple cells in same nest)-nest closure**.	Pompilidae (*Auplopus*, *Dipogon*, *Macromeris*, several *Priocnemis* spp.)
5. Mass provisioning	
A. Single-celled nest: **Nest-prey-egg-more prey-nest closure**.	*Ammophila, Bicyrtes*
B. Multicelled nest (egg laid on first prey): **Nest-(prey-egg-more prey-cell closure)-(multiple cells in same nest)-nest closure**.	*Bicyrtes, Sphex, Stictiella*
C. Multicelled nest (egg laid on last prey): **Nest-(many prey-egg-prey-closure)-(multiple cells in same nest)-nest closure**.	Common in Sphecidae (e.g., *Sphex*, *Gorytes*, *Crabro*, *Cerceris*, *Philanthus*, *Tachysphex*)

Behavior[1]	Examples
6. Progressive provisioning (egg laid after first prey brought in):	
A. Single-celled nest: **Nest-(prey-egg-more prey over several days)-nest closure**.	*Bembix americana, B. sayi, Glenostictia, Steniolia*
B. Multicelled nest: **Nest-(prey-egg-more prey over several days-cell closure)-(multiple cells)-nest closure**.	*Bembix amoena* (and other *Bembix*), *Pareumenes quadrispinosus*
C. Several single-celled nests provisioned progressively at the same time	*Ammophila azteca, A. harti, A. pubescens*
7. Mass or progressive provisioning (egg laid in empty cell before provisioning)	
A. Progressively provisioned single-celled nest: **Nest-(egg-many prey over several days)-nest closure**.	A few Sphecidae (*Bembix occidentalis, B. pallidipicta, Bembecinus* spp.)
B. Mass-provisioned multicelled nest: **Nest-(egg-many prey [or pollen mass]-cell closure)-(multiple cells)-nest closure**.	Most Eumeninae, all Masarinae
C. Progressively provisioned multicelled nest: **Nest-(egg-many prey over several days-cell closure)-(multiple cells)-nest closure**.	Eumeninae *(Orancistrocerus, Paraleptomenes, Synagris)*

Sources: Modified from Evans 1953 and Evans & West-Eberhard 1970, with additional information from reviews in Bohart & Menke 1976, Iwata 1976, Cowan 1991, and Shimizu 1994.

[1]Text in bold gives actual sequence of behaviors. The list of genera for each category gives only a few well-studied examples.

(scarab hosts) or burrows (tiger beetle hosts), whereas parasitoid pompilids find and leave their spider hosts in their webs or tunnels (Table 2-7). Perhaps these taxa did not evolve to build their own nests because they gained no net benefit from doing so, since their choice of hosts already places their young in a protected situation. Many of these parasitoids do nothing to modify the refuge before departing, but others cover the host or block the refuge with debris found nearby. For example, after *Methoca* and *Pterombrus* sting and oviposit on a tiger beetle larva (Table 2-5), they pack the burrow with soil, stones, and twigs. *Methoca californica* (Burdick and Wasbauer 1959) and *M. ichneumonoides* (F. Williams 1916) go a step further; they make the entrance to the burrow indistinguishable from the surroundings.

Other parasitoid aculeates move the host a short distance and place it in a natural refuge (stage 2 in Table 10-6). Female *Campsomeris annulata* and *Scolia japonica* move their scarab host deeper into the soil, where a new cell is prepared before an egg is laid (Table 2-6). Similarly, some Bethylidae (Table 2-2) and ampulicine Sphecidae (Table 2-8) drag or lead their hosts to refuges, where they are covered with debris. Thus, although most parasitoids display little postoviposition parental behavior (by definition), some exhibit rudimentary forms of "nesting" behavior: prey carriage over short distances, crude modifications of natural refuges, and construction of simple cells in which to house the host and egg (Rubink and Evans 1979).

Evans (1958a) is undoubtedly correct in concluding that "the movement of prey into a more suitable niche" was a critical early stage in the evolution of parental strategies and social behavior in the Hymenoptera. The exact sequence of the stages taken after this in each lineage is difficult to determine for various reasons. However, we can make a start in reconstructing the evolution of parental strategies by considering a few lineages.

Evolutionary Trends in Parental Strategies of the Pompilidae

The Pompilidae are a good group to start with, because the family has representatives of the first four stages in Table 10-6 (Evans 1953, 1958a; Iwata 1976; Shimizu 1994). All pompilids of the subfamilies Notocyphinae, Ctenocerinae, and Epipompilinae (as well as some Pepsinae and Pompilinae) are parasitoids (stages 1 or 2 in Table 10-6). From stage 2 (prey-niche-egg) the next likely stage (stage 3) in the evolution of pompilid behavior occurred when females began digging their own nests, rather than finding an existing niche; most pompilids fit into this category (Evans 1953). This step gave females greater flexibility when a preexisting niche was hard to find and probably allowed them to build more secure hiding places. Perhaps, intermediate between stages 2 and 3 were species (represented by *Agenioideus cinctellus*) that use preexisting niches but that cover the prey with debris before departing (Richards and Hamm 1939). The habit of caching prey probably occurred sometime during this transition, because females that dig nests from scratch must leave prey unattended (and unprotected) for longer periods.

The transition to stage 4 involved what Evans (1953) referred to as a "fundamental transposition" of behaviors: building the nest *before*, rather than *after*, hunting. The reversal of steps in the sequence is a radical change, because as Fabre first showed over 100 years ago, wasps can have quite inflexible sequences of innate behaviors. As Evans and West-Eberhard (1970) note: "a wasp in a 'nest-building mood' is largely incapable of either departing grossly from its species-specific digging or masonry behavior or suddenly switching to some other phase of its behavioral cycle." However, something of the sort must have occurred sometime during the evolution of the Pompilidae (as well as the Sphecidae and Vespidae). The reversal of the order of behaviors was probably accompanied by improved abilities at homing and orientation because the wasp would leave the nest behind to forage.

Pompilids have not gone beyond stage 4 of the proposed evolutionary sequence. By not evolving an ability to use more than one prey item per cell, they have restricted themselves to larger prey. However, a number of variations on the theme represented by stage 4A have appeared in the Pompilidae. Some groups have evolved the use of multicellular nests (stage 4B), whereas others have switched from ground-nesting to using preexisting cavities (*e.g., Dipogon*) or building mud nests (e.g., *Auplopus*) (Chapter 6). Finally, females of a few pompilid species nest communally (Evans and Shimizu 1996) (see Communal Nesting, below).

Reversals from more to less complex parental strategies have also occurred in the Pompilidae. Cleptoparasitism has evolved independently at least twice in the Pompilidae. *Ceropales* and *Irenangelus* of the Ceropalinae may have evolved cleptoparasitism directly from a parasitoid ancestor (Shimizu 1994), a pathway also followed by the chrysidine cleptoparasites. However, *Evagetes* probably evolved from pompilids that dug their own nests after hunting spiders (Evans 1953, Shimizu 1994). Similarly, pompilids in certain subgenera of *Arachnospila, Anoplius,* and *Agenioideus* have reverted to parasitoid lifestyles from ancestors that were nest-provisioners (Shimizu 1994). Evans (1953) suggested that such reversals may have occurred when a nesting species that preyed on a burrowing spider began to take advantage of the ready-made niche provided by the spider, enabling the wasp to drop costly nest-building behavior.

Evolutionary Trends in Parental Strategies of the Sphecidae

The Sphecidae has representatives of all seven stages of nesting behavior seen in solitary wasps (Table 10-6). Parasitoid lifestyles (stages 1 and 2) are practiced by all Ampulicinae, as well as by some *Chlorion* (Sphecinae) and *Larra* (Larrinae) (Table 2-8). However, the parasitoid species of *Chlorion* and *Larra* probably evolved from species that dug their own nests and so represent cases of evolutionary reversals to an ancestral trait. Stage 3 (building a nest after hunting), though the norm in the Pompilidae, is rare in the Sphecidae. In the subfamily Sphecinae, species of *Podalonia* and *Prionyx* capture a single large prey item and then dig a shallow single-celled nest. It is uncertain whether *Podalonia* and *Prionyx* represent primitive forms in the evolution of the sphecid nesting behavior or whether they have reverted to stage 3 secondarily. In addition, some *Podalonia* species are capable of switching from a *prey-nest* to a *nest-prey* sequence, depending on prey availability (Evans 1987).

Stage 4, which requires the reversal of the ancestral prey-nest sequence, can be considered a critical preadaptation for the evolution of multicelled nests and the use of multiple prey items in each cell. Stage 4A is represented now by a relatively few species of sphecids, including some *Ammophila* and *Tachysphex* (Kurczewski and Elliott 1978), and was probably an important transitional stage in the evolution of nesting behavior in the Sphecidae. Otherwise, we would have to make the dubious assumption that certain lineages jumped directly from stage 3 to multicellular nests or multiple prey items per cell or both (stage 5). Stage 4B (multicellular nests, with a single prey item per cell) is not represented by contemporary sphecids.

Stages 5–7 are distinguished mainly by the use of more than one prey item per cell, with subcategories based on the number of cells in each nest and the rate at which cells are provisioned. In the Sphecidae, prey are commonly mass provisioned: all prey are provided over a short period of time, usually before the egg hatches. In all mass-provisioning sphecids, the egg is laid either after the first prey item is brought in (stages 5A and 5B) or after the last (stage 5C). When compared with mass provisioners that lay the egg on the first prey in the cell, those that lay the egg on the last of many prey in a cell can probably better determine which sex of egg should be deposited because by then the female has information on how much food is available.

Among the mass-provisioning sphecids that lay the egg on the first prey in the cell, it sometimes happens that the egg hatches before the female brings in the last prey. Such *delayed provisioning* occurs (sometimes) in certain species of Bembicinae (e.g., *Bicyrtes quadrifasciata*). Delayed provisioners may only occasionally provision cells long enough to contact their larvae in a cell. However, the laying of the egg early in the provisioning cycle and the delaying of provisioning represent important prerequisites to the evolution of progressive provisioning (Evans 1966c, Evans and West-Eberhard 1970, Cowan 1991). True progressive provisioners extend the period of provisioning so that it always overlaps the larval phase of the offspring. Progressive provisioners lay the egg either after the first prey item is brought in (stage 6) or in the empty cell before provisioning (stage 7). In some species, progressive provisioning may be truncated: the female begins provisioning when the egg is about to hatch, but she then fills the cell up quickly (when prey is available) and closes it before the larva finishes feeding. Truncated progressive provisioning occurs in certain species of Bembicinae, including *Bembecinus neglectus*, *Bembix pruinosa*, and *Microbembex monodonta* (Evans 1966c).

Fully progressive provisioning is prolonged even further, until the time when the larva is ready (or nearly ready) to spin the cocoon. In the Sphecidae, a few bembicines (e.g., *Glenostictia scitula*, *Steniolia obliqua*, *S. duplicata*, and *Bembix u-scripta*) exhibit this form of provisioning (Evans 1966c). By lengthening the period of provisioning, progressive provisioners place themselves in a position to carefully adjust the amount of provision to the developmental progress of the larvae and (in a few *Rubrica* and *Bembix*) to clean refuse from the cell as the larva finishes feeding on each prey item (Evans 1966c). Fully progressive provisioners represent the most complex form of parental care in the solitary Sphecidae: low-fecundity females that invest tremendous amounts of time and energy on individual offspring. Some progressive provisioners of the genus *Ammophila* lessen the cost of this strategy by simultaneously provisioning two or perhaps more single-celled nests at one time (Evans 1965, Hager and Kurczewski 1986).

Evolutionary Trends in Parental Strategies of the Solitary Vespidae

Like the Pompilidae, the solitary Vespidae span only a small range of nesting strategies. With perhaps only a few exceptions, all of the species of solitary Eumeninae and Masarinae fall within stage 7 (Table 10-6): they have multicellular nests, and they mass or progressively provision cells in which the egg has been deposited before provisioning. Fully progressive provisioning occurs in species of *Synagris* (Roubaud 1911) and *Paraleptomenes* (Jayakar and Spurway 1966). A possible analog to ancestral species that made the transition to progressive provisioning can be seen in *Synagris spiniventris*, which mass provisions if prey is abundant but progressively provisions during prey shortages (Roubaud 1916, in Cowan 1991).

No contemporary eumenines and masarines exhibit stages 1–5. Thus, a study of the solitary Vespidae cannot help a researcher understand the earlier stages in the evolution of parental care in solitary wasps. Nevertheless, taken in the context of the Vespidae as a whole, the comparative study of eumenine nesting strategies will certainly shed light on the evolution of eusociality (Cowan 1991). A few eumenines exhibit traits thought to be critical preadaptations for the evolution of more "advanced" stages of social behavior: less reliance on the sting for subduing prey (leaving stings and venoms free to evolve as purely defensive weapons), feeding offspring with macerated prey, extended periods of parental care, long adult life spans, and communal nesting (Evans and West-Eberhard 1970, Cowan 1991).

Communal Nesting: A Brief Survey

The subfamily Eumeninae includes at least one quasi-social species (one in which groups of females cooperate in preparing and provisioning cells but in which all females oviposit) (Cowan 1991), and there are a few eusocial Sphecidae in the genera *Arpactophilus* and

Microstigmus (Matthews 1991). There is also some evidence of incipient eusociality in a few pompilids (Evans and Shimizu 1996). Intermediate between truly solitary species and quasi-social or eusocial species are the communal nesters, whose females share a nest, each one provisioning her own cells (Table 10-7). Co-occupation of nests can take various forms, with varying degrees of cooperation and antagonism and varying durations of association between females. Short-term nest sharing may also occur when females provision the same nest (and often the same cell) until one decides to leave or is forced out (Brockmann et al. 1979, G. Bohart et al. 1982, Field and Foster 1995). Temporary associations occur in *Stenodynerus claremontensis*; emerging sisters remain within their natal nest for several days, after which all but one disperses (Markin and Gittins 1967). *Philanthus gibbosus* also co-occupy nests soon after emergence, simultaneously provisioning different cells for several days before all but one of the females is forced out (Evans 1973a).

In many cases, however, females remain somewhat amicably at communal nests for prolonged periods; each female constructs and provisions her own cells (Table 10-6). Several advantages to joining a communal nest have been proposed. One is increased efficiency of nest building, when a main tunnel is shared or when females reuse old cells and nests. Another possible benefit is reduced parasitism, a side effect of the fact that one or more adults is likely to be in the nest to confront natural enemies (Evans 1977, Evans and Hook 1986a, Matthews 1991). Both factors appear to be important in *Cerceris antipodes*, which share nests dug in extremely hard soil (McCorquodale 1989c). Once within shared nests, females also gain from improved nest defense against mutillids (McCorquodale 1989c). Female *Auplopus semialatus* actually cooperate in attacking cleptoparasitic *Irenangelus eberhardi* that intrude upon their communal nests (Wcislo et al. 1988).

Given these potential advantages, why is communal nesting not more common? First, the benefits documented for some species may not be available for all species, if for example nests cannot be reused or if shared nests are not much easier to build than solitary nests. Second, nest sharing and reuse of old nests have costs as well. Fights often erupt between females in shared nests. Such interactions could result in subordinate females being expelled or having lower fitness if they stay (e.g., Hook 1987, McCorquodale 1989c). These costs could be ameliorated if females in nests are close relatives that can have a

Table 10-7 Selected examples of communal nesting or nest sharing in the Sphecidae, Pompilidae, and Vespidae

Species	Type of nest[1]	Number of females	Maximum number of cells per nest[2]	References
Pompilidae				
Auplopus argentifrons	Mud nest (R)	8	22	F. Williams 1919c
A. esmerelda	Mud nest	4	95	Kimsey 1980
A. semialatus	Mud nest (R)	2–8	21	Wcislo et al. 1988
Macromeris violacea	Mud nest (R)	3	15	F. Williams 1919c
Sphecidae				
Cerceris antipodes	Ground nest (R)	1–8	196	Evans & Hook 1986a; McCorquodale 1988b, 1989a,b,c
C. australis	Ground nest (R)	1–11	179	Evans & Hook 1982
C. californica	Ground nest (R)	1–5	66	Hook 1987
C. goddardi	Ground nest (R)	1–7	231	Evans & Hook 1986a
Moniaecera asperata	Ground nest	2–3	—	Evans 1964
Spilomena subterranea	Ground nest	2–4	—	McCorquodale & Naumann 1988
Vespidae: Eumeninae				
Montezumia cortesia	Mud nest	3	17	Evans 1973b
Xenorhynchium nitidulum	Mud nest (R)	2–3	14	West-Eberhard 1987
Vespidae: Masarinae				
Trimeria howardi	Ground nest (R)	1–7	≥24	Zucchi et al. 1976

[1]R, cells or nests reused.

[2]Often includes large percentage of cells left from previous generation in reused nests.

positive effect on each others' inclusive fitness (McCorquodale 1988b, Pfennig and Reeve 1993).

Conclusion

The biology of solitary wasps is as well known as it is mainly because their parental strategies have fascinated amateur and professional entomologists for over a century. However, as the cliché in science goes, the more we know, the more questions arise, particularly as we develop new ways to study wasps and new theories to understand them. In the last several decades, for example, increased data on sex ratios, combined with new conceptual tools from evolutionary theory, have led to a new approach to the problem of sex allocation in wasps. At least in a general sense, females seem to be making sex allocation decisions consistent with the notion that they act to maximize their number of grandchildren.

Despite a large amount of comparative information that could be used to address the evolution of parental care in solitary wasps, no major steps have been taken since Evans's publications on the matter (Evans 1958a, 1966b; Evans and West-Eberhard 1970). Now, however, the time is ripe to apply quantitative tools of phylogenetic systematics (Martins 1996) to analyses of the evolution of parental strategies. A cladistic approach will allow us to examine patterns of evolution of nesting strategies carefully, identifying cases of evolutionary reversals and distinguishing convergence from common ancestry as explanations for similar nesting strategies in different taxa. The first step will be to conduct formal phylogenetic analyses of taxa that exhibit a diversity of nesting strategies (e.g., the Pompilidae, the sphecid subfamilies Sphecinae and Bembicinae). In the long run, any comprehensive comparative analysis of nesting strategies will have to include bees as well as wasps (Alexander 1992) and eusocial as well as solitary species. As Evans (1977) pointed out, solitary aculeates hold the key to understanding the early stages of the evolution of eusociality.

Appendix A
Superfamilies (-oidea), Families (-idae), and Subfamilies (-inae) of Solitary Aculeate Hymenoptera

Chrysidoidea
- Bethylidae (B)
  - Bethylinae (B-Be)
  - Epyrinae (B-Ep)
  - Galodoxiinae
  - Mesitiinae
  - Pristocerinae (B-Pr)
  - Protopristocerinae
- Chrysididae (C)
  - Amiseginae (C-Am)
  - Chrysidinae (C-Ch)
  - Cleptinae (C-Cl)
  - Loboscelidiinae (C-Lo)
- Dryinidae (D)
  - Anteoninae (D-An)
  - Aphelopinae (D-Ap)
  - Apodryininae
  - Biaphelopinae
  - Bocchinae (D-Bo)
  - Conganteoninae
  - Dryininae (D-Dr)
  - Gonatopodinae (D-Go)
  - Plesiodryininae
  - Thaumatodryininae
  - Transdryininae
- Embolemidae (Emb)
- Plumariidae
- Sclerogibbidae
- Scolebythidae

Apoidea
- Heterogynaidae (H)
- Sphecidae (Sph)
  - Ampulicinae (Sph-Am)
  - Astatinae (Sph-As)
  - Bembicinae (Sph-Be)
  - Crabroninae (Sph-Cr)
  - Entomosericinae (Sph-En)
  - Laphyragoginae (Sph-Lap)
  - Larrinae (Sph-La)
  - Pemphredoninae (Sph-Pe)
  - Philanthinae (Sph-Ph)
  - Sphecinae (Sph-Sp)
  - Xenosphecinae (Sph-Xe)

Vespoidea
- Bradynobaenidae (Bra)
  - Apterogyninae
  - Bradynobaeninae
  - Chyphotinae
  - Typhoctinae (Bra-Ty)
- Mutillidae (Mu)
  - Mutillinae (Mu-Mu)
  - Myrmillinae (Mu-Myr)
  - Myrmosinae (Mu-Mys)

Pseudophotopsidinae (Mu-Pse)
Rhopalomutillinae
Sphaeropthalminae (Mu-Sp)
Ticoplinae
Pompilidae (P)
Ceropalinae (P-Ce)
Ctenocerinae
Epipompilinae
Notocyphinae (P-No)
Pepsinae (P-Pe)
Pompilinae (P-Po)
Sapygidae (Sap)
Fedtschenkiinae (Sap-Fe)
Sapyginae (Sap-Sa)
Scoliidae (Sco)
Campsomerinae (Sco-Ca)
Proscoliinae (Sco-Pr)
Scoliinae (Sco-Sc)
Sierolomorphidae (Sie)
Tiphiidae (T)
Anthoboscinae (T-An)
Brachycistidinae
Diamminae (T-Di)
Methocinae (T-Me)
Myzininae (T-My)
Thynninae (T-Th)
Tiphiinae (T-Ti)
Vespidae (V)
Eumeninae (V-Eum)
Euparagiinae (V-Eup)
Masarinae (V-Ma)

Sources: Bethylidae: Gordh & Móczár 1990; Chrysididae: Kimsey & Bohart 1990; Dryinidae: Olmi 1984a,b; Pompilidae: Shimizu 1996; Sphecidae: Bohart & Menke 1976; Tiphiidae: Kimsey 1991; Vespidae, Eumeninae: Brothers & Carpenter 1993; Vespidae, Masarinae: S. Gess 1996; all groups: Krombein et al. 1979b.

Appendix B
Solitary Wasp Genera Mentioned in This Book

See Appendix I for abbreviations, which designate the genus's family and subfamily

Abispa (V-Eum)
Acanthostethus (Sph-Be)
Adelphe (C-Am)
Ageniella (P-Pe)
Agenioideus (P-Po)
Allepyris (B-Ep)
Allochares (P-Po)
Allocoelia (C-Ch)
Alysson (Sph-Be)
Amisega (C-Am)
Ammophila (Sph-Sp)
Ampulex (Sph-Am)
Ampulicomorpha (Emb)
Anacrabro (Sph-Cr)
Ancistrocerus (V-Eum)
Ancistromma (Sph-La)
Anisepyris (B-Ep)
Anoplius (P-Po)
Anospilus (P-Po)
Anteon (D-An)
Antepipona (V-Eum)
Anterhynchium (V-Eum)
Anthobosca (T-An)
Apenesia (B-Pr)
Aphelopus (D-Ap)
Aphilanthops (Sph-Ph)
Aporinellus (P-Po)
Aporus (P-Po)
Arachnospila (P-Po)
Argochrysis (C-Ch)
Argogorytes (Sph-Be)
Arpactophilus (Sph-Pe)
Astata (Sph-As)
Auplopus (P-Pe)
Austrochares (P-Po)
Austrogorytes (Sph-Be)
Batozonellus (P-Po)
Belomicrus (Sph-Cr)
Bembecinus (Sph-Be)
Bembix (Sph-Be)
Bethylus (B-Be)
Bicyrtes (Sph-Be)
Bocchus (D-Bo)
Bothynostethus (Sph-La)
Bradynobaena (Bra)
Caenochrysis (C-Ch)
Calligaster (V-Eum)
Campsomeris (Sco-Ca)
Campsoscolia (Sco-Sc)
Carinostigmus (Sph-Pe)
Celonites (V-Ma)
Cephalonomia (B-Ep)
Ceramius (V-Ma)
Cerceris (Sph-Ph)
Ceropales (P-Ce)
Chalybion (Sph-Sp)
Chilosphex (Sph-Sp)
Chirodamus (P-Pe)
Chlorion (Sph-Sp)
Chrestomutilla (Mu-Sp)
Chrysis (C-Ch)
Chrysura (C-Ch)
Chrysurissa (C-Ch)
Cleptes (C-Cl)
Cleptidea (C-Cl)
Clitemnestra (Sph-Be)
Clypeadon (Sph-Ph)
Crabro (Sph-Cr)
Crossocerus (Sph-Cr)
Crovettia (D-Ap)
Cryptocheilus (P-Pe)
Cyphotes (Bra)

Dasymutilla (Mu-Sp)
Dasyproctus (Sph-Cr)
Diamma (T-Di)
Dichragenia (P-Pe)
Dicondylus (D-Go)
Dicranoplius (P-Po)
Dienoplus (Sph-Be)
Dimorphothynnus (T-Th)
Dinetus (Sph-As)
Diodontus (Sph-Pe)
Diploplectron (Sph-As)
Dipogon (P-Pe)
Discoelius (V-Eum)
Dissomphalus (B-Pr)
Dolichurus (Sph-Am)
Dryinus (D-Dr)
Dryudella (Sph-As)
Duckeia (C-Am)
Dynatus (Sph-Sp)
Echthrodelphax (D-Go)
Ectemnius (Sph-Cr)
Elampus (C-Ch)
Enchemicrum (Sph-Cr)
Encopognathus (Sph-Cr)
Entomocrabro (Sph-Cr)
Entomognathus (Sph-Cr)
Entomosericus (Sph-En)
Epactothynnus (T-Th)
Ephuta (Mu-Mu)
Ephutomorpha (Mu-Mys)
Epinysson (Sph-Be)
Episyron (P-Po)
Epsilon (V-Eum)
Epyris (B-Ep)
Eremiasphecium (Sph-Ph)
Eremochares (Sph-Sp)
Eucerceris (Sph-Ph)
Eumenes (V-Eum)
Euodynerus (V-Eum)
Euparagia (V-Eup)
Eusapyga (Sap-Sa)
Evagetes (P-Po)
Exeirus (Sph-Be)
Fabriogenia (P-Pe)
Fedtschenkia (Sap-Fe)
Gastrosericeus (Sph-La)
Gayella (V-Ma)
Glenostictia (Sph-Be)
Gonatopus (D-Go)
Goniozus (B-Be)
Gorytes (Sph-Be)
Hapalomellinus (Sph-Be)
Haplogonatopus (D-Go)
Hedychridium (C-Ch)
Hedychrum (C-Ch)
Hemipepsis (P-Pe)
Hemithynnus (T-Th)
Heterogyna (H)
Homonotus (P-Po)
Hoplisoides (Sph-Be)
Hylomesa (T-My)
Irenangelus (P-Ce)
Iridomimus (P-Pe)
Isodontia (Sph-Sp)
Jugurtia (V-Ma)
Krombeinictus (Sph-Cr)
Laelius (B-Ep)
Laphyragogus (Sph)
Larra (Sph-La)
Larropsis (Sph-La)
Lestica (Sph-Cr)
Lestiphorus (Sph-Be)
Leucodynerus (V-Eum)
Lindenius (Sph-Cr)
Liris (Sph-La)
Listropygia (Sph-Ph)
Loboscelidia (C-Lo)
Lonchodryinus (D-An)
Lyroda (Sph-La)
Macromerella (P-Pe)
Macromeris (P-Pe)
Macrothynnus (T-Th)
Masarina (V-Ma)
Megadryinus (D-Dr)
Megalothynnus (T-Th)
Megascolia (Sco-Sc)
Mellinus (Sph-Be)
Metanysson (Sph-Be)
Methoca (T-Me)
Microbembex (Sph-Be)
Microstigmus (Sph-Pe)
Mimesa (Sph-Pe)
Mimumesa (Sph-Pe)
Minagenia (P-Pe)
Miscophus (Sph-La)
Moniaecera (Sph-Cr)
Monobia (V-Eum)
Montezumia (V-Eum)
Mutilla (Mu-Mu)
Myrmecomimesis (C-Am)
Myrmosa (Mu-Mys)
Myrmosula (Mu-Mys)
Myzinum (T-My)
Neochrysis (C-Ch)
Neozeloboria (T-Th)
Nesomimesa (Sph-Pe)
Nitela (Sph-La)
Nitelopterus (Sph-La)
Notocyphus (P-No)
Nysson (Sph-Be)
Ochleroptera (Sph-Be)
Odontosphex (Sph-Ph)
Odynerus (V-Eum)
Omalus (C-Ch)
Orancistrocerus (V-Eum)
Oryttus (Sph-Be)
Oxybelus (Sph-Cr)
Pachodynerus (V-Eum)
Palarus (Sph-La)
Palmodes (Sph-Sp)
Pappognatha (Mu)
Parachilus (V-Eum)
Paragia (V-Ma)
Paragymnomerus (V-Eum)
Paralastor (V-Eum)
Paraleptomenes (V-Eum)
Parascleroderma (B-Pr)
Parancistrocerus (V-Eum)
Pareumenes (V-Eum)
Parnopes (C-Ch)
Passaloecus (Sph-Pe)
Pemphredon (Sph-Pe)
Penepodium (Sph-Sp)
Pepsis (P-Pe)

Phanagenia (P-Pe)
Philanthus (Sph-Ph)
Pison (Sph-La)
Pisonopsis (Sph-La)
Plenoculus (Sph-La)
Pluto (Sph-Pe)
Podagritus (Sph-Cr)
Podalonia (Sph-Sp)
Podium (Sph-Sp)
Poecilopompilus (P-Po)
Polemistus (Sph-Pe)
Polochrum (Sap-Sa)
Pompilus (P-Po)
Praestochrysis (C-Ch)
Priocnemis (P-Pe)
Priocnessus (P-Pe)
Prionyx (Sph-Sp)
Pristocera (B-Pr)
Prorops (B-Ep)
Proscolia (Sco-Pr)
Prosopigastra (Sph-La)
Psen (Sph-Pe)
Psenulus (Sph-Pe)
Pseudisobrachium (B-Pr)
Pseudogonatopus (D-Go)
Pseudolopyga (C-Ch)
Pseudomalus (C-Ch)
Pseudomasaris (V-Ma)
Pseudomethoca (Mu-Sp)
Pseudophotopsis (Mu-Pse)
Pseudopompilus (P-Po)
Pseudoscolia (Sph-Ph)
Psorthaspis (P-Po)
Pterocheilus (V-Eum)
Pterombrus (T-My)
Pulverro (Sph-Pe)
Quartinia (V-Ma)
Quartinoides (V-Ma)
Raphiglossa (V-Eum)
Rhagigaster (T-Th)
Rhopalum (Sph-Cr)
Rhynchium (V-Eum)
Rolandia (V-Ma)
Rubrica (Sph-Be)
Sapyga (Sap-Sa)
Sceliphron (Sph-Sp)
Scleroderma (B-Ep)
Scolia (Sco-Sc)
Sericophorus (Sph-La)
Sericopompilus (P-Po)
Sierlomorpha (Sie)
Solierella (Sph-La)
Sphaeropthalma (Mu-Sp)
Sphecius (Sph-Be)
Sphex (Sph-Sp)
Sphictostethus (P-Pe)
Spilomena (Sph-Pe)
Stangeella (Sph-Sp)
Steniolia (Sph-Be)
Stenodynerus (V-Eum)
Stictia (Sph-Be)
Stictiella (Sph-Be)
Stigmus (Sph-Pe)
Stilbum (C-Ch)
Stizoides (Sph-Be)
Stizus (Sph-Be)
Symmorphus (V-Eum)
Synagris (V-Eum)
Tachynomyia (T-Th)
Tachypompilus (P-Po)
Tachysphex (Sph-La)
Tachytes (Sph-La)
Tetradontochelys (D-Go)
Thynnoides (T-Th)
Timulla (Mu-Mu)
Tiphia (T-Ti)
Tracheliodes (Sph-Cr)
Trachypus (Sph-Ph)
Tricarinodynerus (V-Eum)
Trichrysis (C-Ch)
Trigonopsis (Sph-Sp)
Trimeria (V-Ma)
Trypoxylon (Sph-La)
Typhoctes (Bra-Ty)
Xenorhynchium (V-Eum)
Xenosphex (Sph-Xe)
Zanysson (Sph-Be)
Zaspilothynnus (T-Th)
Zeta (V-Eum)
Zethus (V-Eum)
Zyzzyx (Sph-Be)

Bibliography

Abdul-Nour, H. 1971. Contribution á l'étude des Parasites d'Homoptére Auchenorrhyches du Sud de la France: Dryinidae (Hymenoptéres) et Strepsiptéres. Thése, Acad. Montpellier, Univ. Sci. et Tech. Languedoc.

Abraham, Y.J., D. Moore, and G. Godwin. 1990. Rearing and aspects of biology of *Cephalonomia stephanoderis* and *Prorops nasuta* (Hymenoptera: Bethylidae) parasitoids of the coffee berry borer, *Hypothenemus hampei* (Coleoptera: Scolytidae). Bull. Entomol. Res. 80: 121–128.

Achterberg, C. van, and B. van Aartsen. 1986. The European Pamphiliidae (Hymenoptera: Symphyta), with special reference to the Netherlands. Zoologische Verhandelingen 234: 1–98.

Ågren, L., and A.-K. Borg-Karlson. 1984. Responses of *Argogorytes* (Hymenoptera: Sphecidae) males to odor signals from *Ophrys insectifera* (Orchidaceae). Preliminary EAG and chemical investigation. Nova Acta Regiae Societatis Scientiarum Upsaliensis, Serie V: C,3: 111–117.

Ågren, L., B. Kullenberg, and T. Sensenbaugh. 1984. Congruences in pilosity between three species of *Ophrys* (Orchidaceae) and their hymenopteran pollinators. Nova Acta Regiae Societatis Scientiarum Upsaliensis, Serie V: C,3: 15–25.

Ainslie, C.N. 1920. Notes on *Gonatopus ombrodes,* a parasite of jassids. Entomol. News 31: 169–173, 187–190.

Alcock, J. 1975a. The behavior of western cicada-killer males, *Sphecius grandis* (Sphecidae, Hymenoptera). J. Nat. Hist. 9: 561–566.

Alcock, J. 1975b. Male mating strategies of some philanthine wasps (Hymenoptera: Sphecidae). J. Kans. Entomol. Soc. 48: 532–545.

Alcock, J. 1975c. The nesting behavior *Philanthus multimaculatus* Cameron (Hymenoptera: Sphecidae). Amer. Midl. Nat. 93: 222–226.

Alcock, J. 1975d. Territorial behaviour by males of *Philanthus multimaculatus* (Hymenoptera: Sphecidae) with a review of territoriality in male sphecids. Anim. Behav. 23: 889–895.

Alcock, J. 1979a. The behavioural consequences of size variation among males of the territorial wasp *Hemipepsis ustulata* (Hymenoptera: Pompilidae). Behaviour 71: 322–335.

Alcock, J. 1979b. The evolution of intraspecific diversity in male reproductive strategies in some bees and wasps. Pages 381–402 in M.S. Blum and M.A. Blum (eds.), Sexual Selection and Reproductive Competition in Insects. New York: Academic Press.

Alcock, J. 1981a. Lek territoriality in the tarantula hawk wasp *Hemipepsis ustulata* (Hymenoptera: Pompilidae). Behav. Ecol. Sociobiol. 8: 309–317.

Alcock, J. 1981b. Reproductive behavior of some Australian thynnine wasps (Hymenoptera: Tiphiidae). J. Kans. Entomol. Soc. 54: 681–693.

Alcock, J. 1983a. Consistency in the relative attractiveness of a set of landmark territorial sites to two generations of male tarantula hawk wasps (Hymenoptera: Pompilidae). Anim. Behav. 31: 74–80.

Alcock, J. 1983b. Hilltopping territoriality by males of the great purple hairstreak *Atlides halesus* (Lepidoptera: Lycaenidae): convergent evolution with pompilid wasps. Behav. Ecol. Sociobiol. 13: 57–62.

Alcock, J. 1984a. Convergent evolution in perching and patrolling site preferences of some hilltopping insects of the Sonoran Desert. Southwest. Nat. 29: 475–480.

Alcock, J. 1984b. Ridgetop rendezvous. Nat. Hist. 93: 43–46.

Alcock, J. 1985a. Hilltopping in the nymphalid butterfly *Chlosyne californica* (Lepidoptera). Amer. Midl. Nat. 113: 69–75.

Alcock, J. 1985b. Hilltopping behavior in the wasp *Pseudomasaris maculifrons* (Fox) (Hymenoptera: Masaridae). J. Kans. Entomol. Soc. 58: 162–166.

Alcock, J. 1988. The Kookaburra's Song: Exploring Animal Behavior in Australia. Tucson: University of Arizona Press.

Alcock, J., and W.J. Bailey. 1997. Success in territorial defence by male tarantula hawk wasps *Hemipepsis ustulata*: the role of residency. Ecol. Entomol. 22: 377–383.

Alcock, J., and M. Carey. 1988. Hilltopping behaviour and mating success of the tarantula hawk wasp, *Hemipepsis ustulata* (Hymenoptera: Pompilidae), at a high elevation peak. J. Nat. Hist. 22: 1173–1178.

Alcock, J., and G.J. Gamboa. 1975. Home ranges of *Cerceris simplex macrosticta* (Hymenoptera, Sphecidae). Psyche (Camb., Mass.) 81: 528–533.

Alcock, J., and D.T. Gwynne. 1987. Courtship feeding and mate choice in thynnine wasps (Hymenoptera: Tiphiidae). Austr. J. Zool. 35: 451–458.

Alcock, J., and M.D. Johnson. 1990. Male behavior in the tarantula-hawk wasp *Pepsis thisbe* Lucas (Hymenoptera: Pompilidae). J. Kans. Entomol. Soc. 63: 399–404.

Alcock, J., and K.M. O'Neill. 1986. Density-dependent mating tactics in the grey hairstreak, *Strymon melinus* (Lepidoptera: Lycaenidae). J. Zool., Lond. 209: 105–113.

Alcock, J., and K.M. O'Neill. 1987. Territory preferences and the intensity of competition in the grey hairstreak, *Strymon melinus* (Lepidoptera: Lycaenidae) and the tarantula hawk wasp *Hemipepsis ustulata* (Hymenoptera: Pompilidae). Amer. Midl. Nat. 118: 128–138.

Alcock, J., and A. Smith. 1987. Hilltopping, leks, and female choice in the carpenter bee *Xylocopa (Neoxylocopa) varipuncta*. J. Zool., Lond. 211: 319–328.

Alcock, J., C.E. Jones, and S.L. Buchmann. 1977. Male mating strategies in the bee *Centris pallida* (Hymenoptera: Anthophoridae). Amer. Nat. 111: 145–155.

Alcock, J., E.M. Barrows, G. Gordh, L.J. Hubbard, L. Kirkendall, D.W. Pyle, T.L. Ponder, and F.G. Zalom. 1978. The ecology and evolution of male reproductive behaviour in bees and wasps. Zool. J. Linn. Soc. 64: 293–326.

Alexander, B.A. 1985. Predator-prey interactions between the digger wasp *Clypeadon laticinctus* and the harvester ant *Pogonomyrmex occidentalis*. J. Nat. Hist. 19: 1139–1154.

Alexander, B.A. 1986. Alternative methods of nest provisioning in the digger wasp *Clypeadon laticinctus* (Hymenoptera: Sphecidae). J. Kans. Entomol. Soc. 59: 59–63.

Alexander, B.A. 1992. An exploratory analysis of cladistic relationships within the superfamily Apoidea, with special reference to sphecid wasps. J. Hym. Res. 1: 25–61.

Alexander, B.A., and C.D. Michener. 1995. Phylogenetic studies of the families of short-tongued bees (Hymenoptera: Apoidea). Univ. Kans. Sci. Bull. 55: 377–424.

Alm, S.R., and F.E. Kurczewski. 1984. Ethology of *Anoplius tenebrosus* (Cresson) (Hymenoptera: Pompilidae). Proc. Entomol. Soc. Wash. 86: 110–119.

Andersson, M. 1994. Sexual Selection. Princeton, N.J.: Princeton University Press.

Antony, J., and C. Kurian. 1960. Studies on the habits and life history of *Perisierola nephantidis* Muesbeck. Indian Coconut J. 13: 145–153.

Ardrey, R. 1966. The Territorial Imperative: A Personal Inquiry into the Animal Origins of Property and Nations. New York: Atheneum.

Armitage, K.B. 1965. Notes on the biology of *Philanthus bicinctus* (Hymenoptera: Sphecidae). J. Kans. Entomol. Soc. 38: 89–100.

Asis, J.D., S.F. Gayubo, and J. Tormos. 1989. Nesting behaviour of three species of *Tachysphex* from Spain, with a description of the mature larva of *Tachysphex tarsinus* (Hymenoptera Sphecidae). Ethol. Ecol. Evol. 1: 233–239.

Asis, J.D., J. Tormos, and S.F. Gayubo. 1994. Biological observations on *Trypoxylon attenuatum* and description of its mature larva and its natural enemy *Trichrysis cyanea* (Hymenoptera: Sphecidae, Chrysididae). J. Kans. Entomol. Soc. 67: 199–207.

Asis, J.D., S.F. Gayubo, and J. Tormos. 1996. Behavior of *Philanthus puchellus* (Hymenoptera: Sphecidae) with a description of its mature larva. Ann. Entomol. Soc. Amer. 89: 452–458.

Askew, R.R. 1971. Parasitic Insects. London: Heinemann.

Baerends, G.P. 1941. Fortpflanzungsverhalten und Orientierung der Grabwespe *Ammophila campestris* Jur. Tijdschr. Entomol. 84: 68–275.

Baldridge, R.S., and H.D. Blocker. 1980. Parasites of leafhoppers (Homoptera: Cicadellidae) from Kansas grasslands. J. Kans. Entomol. Soc. 53: 441–447.

Banks, D. 1995. Male nest defense in the digger wasp *Cerceris binodis* (Hymenoptera: Sphecidae). J. Hym. Res. 4: 77–79.

Banks, N. 1902. Sleeping habits of certain Hymenoptera. J. N.Y. Entomol. Soc. 10: 209–214.

Barber, H.S. 1915. *Macrosiagon flavipennis* in cocoon of *Bembex spinolae*. Proc. Entomol. Soc. Wash. 17: 187–188.

Barnard, C.J., and R.M. Sibley. 1981. Producers and scroungers: a general model and its application to captive flocks of house sparrows. Anim. Behav. 29: 543–550.

Barrett, C.F., P.H. Westdal, and H.P. Richardson. 1965. Biology of *Pachygonatopus minimus* Fenton (Hymenoptera: Dryinidae), a parasite of the six-spotted leafhopper, *Macrosteles fascifrons* (Stål) in Manitoba. Can. Entomol. 97: 216–221.

Barrows, E.M. 1978. Male behavior in *Evagetes subangulatus* (Hymenoptera: Pompilidae). Great Lakes Entomol. 11: 77–80.

Barth, G.P. 1907. On the nesting habits of *Psen barthi*. Bull. Wisc. Nat. Hist. Surv. 5: 251–257.

Batra, S. 1965. Organisms associated with *Lasioglossum zephyrum* (Hymenoptera: Halictidae). J. Kans. Entomol. Soc. 38: 367–389.

Bayliss, P.S., and D.J. Brothers. 1996. Biology of *Tricholabiodes* Radoszkowski in southern Africa, with a new synonymy and review of recent biological literature (Hymenoptera: Mutillidae). J. Hym. Res. 5: 249–258.

Bechtel, R.C., and E.I. Schlinger. 1957. Biological observations on *Ectemnius* with particular reference to their *Ogcodes* prey (Hymenoptera: Sphecidae.—Diptera: Acroceridae). Entomol. News 58: 225–232.

Bellmann, V.H. 1983. Observations on the breeding-behavior of *Celonites abbreviatus* Villers (Hymenoptera: Masaridae). Zool. Anz., Jena. 212: 321–328.

Berry, J.A. 1998. The bethyline species (Hymenoptera: Bethylidae: Bethylinae) imported into New Zealand for biological control of pest leafrollers. New Zealand J. Zool. 25: 329–333.

Betbeder-Matibet, M. 1990. Rearing of some species of the genus *Chilo* and of some of their parasites for the biological control of boring insects of the Gramineae in Africa. Ins. Sci. Appl. 11: 617–623.

Beusekom, G. van. 1948. Some experiments on the optical orientation in *Philanthus triangulum* Fabr. Behaviour 1: 195–225.

Bohart, G.E., F.D. Parker, and V.J. Tepedino. 1982. Notes on the biology of *Odynerus dilectus* (Hymenoptera: Eumenidae), a predator of the alfalfa weevil, *Hypera postica* (Coleoptera: Curculionidae). Entomophaga 27: 23–31.

Bohart, R.M. 1941. A revision of the Strepsiptera with special reference to the species of North America. Univ. Calif. Pub. Entomol. 7: 91–159.

Bohart, R.M. 1976. A review of the nearctic species of *Crabro* (Hymenoptera: Sphecidae). Trans. Amer. Entomol. Soc. 102: 229–287.

Bohart, R.M., and L.S. Kimsey. 1982. A synopsis of the Chrysididae in America north of Mexico. Mem. Amer. Entomol. Inst. 33: 1–266.

Bohart, R.M., and J.W. MacSwain. 1940. Notes on two chysidids parasitic on western bembicid wasps. Pan-Pac. Entomol. 16: 92–93.

Bohart, R.M., and P.M. Marsh. 1960. Observations on the habits of *Oxybelus sericeus* Robertson (Hymenoptera: Sphecidae). Pan-Pac. Entomol. 36: 115–118.

Bohart, R.M., and A.S. Menke. 1976. Sphecid Wasps of the World: A Generic Revision. Berkeley: University of California Press.

Bohart, R.M., and R.O. Schuster. 1972. A host record for *Fedtschenkia* (Hymenoptera: Sapygidae). Pan-Pac. Entomol. 48: 149.

Bohart, R.M., C.S. Lin, and J.F. Holland. 1966. Bionomics of *Oxybelus sparideus* at Lake Texoma, Oklahoma (Hymenoptera: Sphecidae: Crabroninae). Ann. Entomol. Soc. Amer. 59: 818–820.

Bordage, E. 1913. Notes biologiques recueilles a l'île de la Reunion. Bull. Sci. France Belg. 47: 377–412.

Borg-Karlson, A.-K. 1990. Chemical and ethological studies of pollination in the genus *Ophrys* (Orchidaceae). Phytochemistry 29: 1359–1387.

Bowden, J. 1964. Notes on the biology of two species of *Dasyproctus* Lep. and Br. in Uganda (Hymenoptera). J. Entomol. Soc. Sth. Africa 26: 425–437.

Bown, T.M., and B.C. Ratcliffe. 1988. The origin of *Chubutolithes* Ihering, ichnofossils from the Eocene and Oligocene of Chubut province, Argentina. J. Paleont. 62: 163–167.

Bradley, J.C. 1908. A case of gregarious sleeping habits among aculeate Hymenoptera. Ann. Entomol. Soc. Amer. 1: 127–130.

Bridwell, J.C. 1919. Some notes on Hawaiian and other Bethylidae (Hymenoptera) with descriptions of new species. Proc. Haw. Entomol. Soc. 4: 21–38.

Bridwell, J.C. 1920. Some notes on Hawaiian and other Bethylidae (Hymenoptera) with the description of a new genus and species. Proc. Haw. Entomol. Soc. 4: 291–315.

Bridwell, J.C. 1958. Biological notes on *Ampulicomorpha confusa* Ashmead and its fulgorid host (Hymenoptera: Dryinidae and Homoptera: Achilidae). Proc. Entomol. Soc. Wash. 60: 3–26.

Brockmann, H.J. 1979. Nest-site selection in the great golden digger wasp, *Sphex ichneumoneus* L. (Sphecidae). Ecol. Entomol. 4: 211–224.

Brockmann, H.J. 1980. House sparrows kleptoparasitize digger wasps. Wilson Bull. 92: 394–398.

Brockmann, H.J. 1985a. Provisioning behavior of the great golden digger wasp, *Sphex ichneumoneus* (L.) (Sphecidae). J. Kans. Entomol. Soc. 58: 631–655.

Brockmann, H.J. 1985b. Tool use in digger wasps (Hymenoptera: Sphecidae). Psyche 92: 309–329.

Brockmann, H.J. 1988. Father of the brood. Nat. Hist. 97: 33–37.

Brockmann, H.J. 1992. Male behavior, courtship, and nesting in *Trypoxylon (Trypargilum) monteverdae* (Hymenoptera: Sphecidae). J. Kans. Entomol. Soc. 65: 66–84.

Brockmann, H.J., and A. Grafen. 1989. Mate conflict and male behaviour in a solitary wasp, *Trypoxylon (Trypargilum) politum* (Hymenoptera: Sphecidae). Anim. Behav. 37: 232–255.

Brockmann, H.J., and A. Grafen. 1992. Sex ratios and life-history patterns of a solitary wasp, *Trypoxylon (Trypargilum) politum* (Hymenoptera: Sphecidae). Behav. Ecol. Sociobiol. 30: 7–27.

Brockmann, H.J., R. Dawkins, and A. Grafen. 1979. Evolutionarily stable nesting strategies in a digger wasp. J. Theor. Biol. 77: 473–496.

Brooke, M. De L. 1981. The nesting biology and population dynamics of the Seychelles potter wasp *Eumenes alluaudi* Perez. Ecol. Entomol. 6: 365–377.

Brothers, D.J. 1972. Biology and immature stages of *Pseudomethoca F. frigida*, with notes on other species (Hymenoptera: Mutillidae). Kans. Univ. Sci. Bull. 50: 1–38.

Brothers, D.J. 1978. Biology and immature stages of *Myrmosula parvula* (Hymenoptera: Mutillidae). J. Kans. Entomol. Soc. 51: 698–710.

Brothers, D.J. 1981. Note on the biology of *Ycaploca evansi* (Hymenoptera: Scolebythidae). J. Entomol. Soc. Sth. Africa 44: 107–108.

Brothers, D.J. 1984. Gregarious parasitoidism in Australian Mutillidae (Hymenoptera). Aust. Entomol. Mag. 11: 8–10.

Brothers, D.J. 1992. The first Mesozoic Vespidae (Hymenoptera) from the Southern Hemisphere, Botswana. J. Hym. Res. 1: 119–124.

Brothers, D.J., and J.M. Carpenter. 1993. Phylogeny of Aculeata: Chrysidoidea and Vespoidea (Hymenoptera). J. Hym. Res. 2: 227–302.

Brothers, D.J., and A.T. Finnamore. 1993. Superfamily Vespoidea. Pages 161–278 in H. Goulet and J.T. Huber (eds.), Hymenoptera of the World: An Identification Guide to Families. Publication 1894/E, Research Branch, Agriculture Canada.

Brunson, M.H. 1934. The fluctuation of the population of *Tiphia popilliavora* Rohwer in the field and its possible causes. J. Econ. Entomol. 27: 515–518.

Brunson, M.H. 1938. Influence of Japanese beetle instar on the sex and population of the parasite *Tiphia popilliavora*. J. Agric. Res. 57: 379–386.

Buckell, E.R. 1928. Notes on the life-history and habits of *Melittobia chalybii* Ashmead. (Chalcidoidea: Elachertidae). Pan-Pac. Entomol. 5: 14–22.

Burdick, D.J., and M.S. Wasbauer. 1959. Biology of *Methoca californica* Westwood (Hymenoptera: Tiphiidae). Wasmann J. Biol. 17: 75–89.

Burk, T. 1982. The evolutionary significance of predation on sexual signalling males. Fla. Entomol. 65: 90–104.
Burrell, R.W. 1935. Notes on the habits of certain Australian Thynnidae. J. N.Y. Entomol. Soc. 43: 19–29.
Buysson, R du. 1898. La *Chrysis shanghaiensis* Sm. Ann. Soc. Entomol. Fr. 67: 80.
Byers, G.W. 1978. Nests, prey, behavior and development of *Cerceris halone* (Hymenoptera: Sphecidae). J. Kans. Entomol. Soc. 51: 818–831.
Callan, E. McC. 1939. A note on the breeding of *Protobethylus callani* Richards (Hymenopt., Bethylidae), an embiopteran parasite. Proc. R. Entomol. Soc. Lond., Series B, 8: 223–224.
Calmbacher, C.W. 1977. The nest of *Zethus otomitus* (Hymenoptera: Eumenidae). Fla. Entomol. 60: 135–137.
Cane, J.H., and M.M. Miyamoto. 1979. Nest defense and foraging ethology of a neotropical sand wasp, *Bembix multipicta* (Hymenoptera: Sphecidae). J. Kans. Entomol. Soc. 52: 6667–6672.
Carpenter, F.M. 1992. Treatise on Invertebrate Paleontology. Part R. Arthropoda 4, Vol. 4: Superclass Hexapoda, pp. 279–655. Lawrence, Kans., and Boulder, Colo.: Geological Society of America and University of Kansas.
Carpenter, J.M. 1986. Cladistics of the Chrysidoidea (Hymenoptera). J. N.Y. Entomol. Soc. 94: 303–330.
Carrillo, J.L., and L.E. Caltagirone. 1970. Observations on the biology of *Solierella peckhami, S. blaisdelli* (Sphecidae), and two species of Chrysididae (Hymenoptera). Ann. Entomol. Soc. Amer. 63: 672–681.
Castner, J.L. 1988. Biology of the mole cricket parasitoid *Larra bicolor* (Hymenoptera: Sphecidae). Adv. Parasitic Hym. Res. 1988: 423–432.
Castner, J.L., and H.G. Fowler. 1987. Diel patterns of *Larra bicolor* (Hymenoptera: Sphecidae) in Puerto Rico. J. Entomol. Sci. 22: 77–83.
Cazier, M.A., and E.G. Linsley. 1974. Foraging behavior of some bees and wasps at *Kallstroemia grandiflora* flowers in southern Arizona and New Mexico. Amer. Mus. Novitates 2546: 1–20.
Chambers, V.H. 1955. Some hosts of *Anteon* spp. (Hym., Dryinidae) and a hyperparasite *Ismarus* (Hymenoptera: Belytidae). Entomol. Mon. Mag. 91: 114–115.
Chapman, R.F. 1959. Some observations on *Pachyopthalmus africa* Curran (Diptera: Calliphoridae), a parasite of *Eumenes maxillosus* De Geer (Hymenoptera: Eumenidae). Proc. R. Entomol. Soc. Lond. 34: 1–6.
Chapman, R.N, C.E. Mickel, J.R. Parker, G.E. Miller, and E.G. Kelly. 1926. Studies in the ecology of sand dune insects. Ecology 7: 416–426.
Chapman, T.W., and S.C. Stewart. 1996. Extremely high levels of inbreeding in a natural population of the free-living wasp *Ancistrocerus antilope* (Hymenoptera: Vespidae: Eumeninae). Heredity 76: 65–69.
Charnov, E.L. 1982. The Theory of Sex Allocation. Princeton, N.J.: Princeton University Press.

Cheng, W.Y. 1991. Importation of natural enemies for the control of sugarcane insect pests in Taiwan in 1955 to 1989. Taiwan Sugar 38: 11–17.

Chilcutt, C.F., and D.P. Cowan. 1992. Carnivory in adult female eumenid wasps (Hymenoptera: Vespidae: Eumeninae) and its effect on egg production. Great Lakes Entomol. 25: 297–301.

Chmurzynski, J.A. 1964. Studies on the stages of spatial orientation in female *Bembex rostrata* (Linne 1758) returning to their nests (Hymenoptera, Sphegidae). Acta Biol. Exper. 24: 103–132.

Chmurzynski, J.A. 1967. On the role of relations between landmarks and nest-hole in the proximate orientation of female *Bembex rostrata* (Linne 1758) returning to their nests (Hymenoptera, Sphegidae). Acta Biol. Exper. 27: 221–254.

Chua, T.H., V.A. Dyck, and N.B. Peña. 1986. Mass rearing of the parasitoid *Pseudogonatopus flavifemur* (Hymenoptera: Dryinidae) for field-introduction trials. Pages 209–220 in M.Y. Hussein and A.G. Ibrahim (eds.), Biological Control in the Tropics: Proceedings of the First Regional Symposium on Biological Control. Pertanian, Malaysia: Penerbit University Press.

Clausen, C.P. 1928. *Hyperalonia oenomus* Rond., a parasite of *Tiphia* larvae (Dip., Bombyliidae). Ann. Entomol. Soc. Amer. 21: 643–659.

Clausen, C.P. 1940. Entomophagous Insects. New York: McGraw-Hill.

Clausen, C.P. 1956. Biological control of pests in the continental United States. USDA Tech. Bull. 1139.

Clausen, C.P., T.R. Gardner, and K. Sato. 1932. Biology of some Japanese and Chosenese grub parasites (Scoliidae). USDA Tech. Bull. 308.

Clauss, B. 1985. The status of the banded bee pirate, *Palarus latifrons*, as a honeybee predator in southern Africa. Proc. 3rd Int. Conf. Apic. Trop. Climates, Nairobi, pp. 157–159.

Coleman, E. 1928. Pollination of an Australian orchid by the male ichneumonid, *Lissopimpla semipunctata* Kirby. Trans. Entomol. Soc. Lond. 76: 533–539.

Collins, J.A., and D.T. Jennings. 1987a. Nesting height preferences of eumenid wasps (Hymenoptera: Eumenidae) that prey on spruce budworm (Lepidoptera: Tortricidae). Ann. Entomol. Soc. Amer. 80: 435–438.

Collins, J.A., and D.T. Jennings. 1987b. Spruce budworm and other lepidopterous prey of eumenid wasps (Hymenoptera: Eumenidae) in spruce-fir forests of Maine. Great Lakes Entomol. 20: 127–133.

Conlong, D.E. 1990. A study of pest-parasitoid relationships in natural habitats: an aid towards the biological control of *Eldana saccharina* (Lepidoptera: Pyralidae) in sugarcane. Proc. Sth. Afr. Sug. Technol. Ass. 64: 111–115.

Conlong, D.E., and D.Y. Graham. 1988a. A comparison of the known life-histories of the bethylids *Parasierola* sp. and *Goniozus natalensis* Gordh collected from *Eldana saccharina* Walker (Lepidoptera: Pyralidae) in Uganda and southern Africa, respectively. J. Entomol. Soc. Sth. Afr. 51: 143–144.

Conlong, D.E., and D.Y. Graham. 1988b. Notes on the natural host surveys and laboratory rearing of *Goniozus natalensis* Gordh (Hymenoptera: Bethylidae), a parasitoid of *Eldana saccharina* Walker (Lepidoptera: Pyralidae) larvae from *Cyperus papyrus* L. in southern Africa. J. Entomol. Soc. Sth. Afr. 51: 115–127.

Cook, J.M. 1992. Experimental tests of sex determination in *Goniozus nephantidis* (Hymenoptera: Bethylidae). Heredity 71: 130–137.

Cooper, K.W. 1952. Records and flower preferences of masarid wasps. II. Polytropy or oligotropy in *Pseudomasaris*? (Hymenoptera: Vespidae). Amer. Midl. Nat. 48: 103–110.

Cooper, K.W. 1953. Biology of eumenine wasps, I. The ecology, predation, and competition of *Ancistrocerus antilope* (Panzer). Trans. Amer. Entomol. Soc. 79: 13–35.

Cooper, K.W. 1955. Venereal transmission of mites by wasps, and some evolutionary problems arising from the remarkable association of *Ensliniella trisetosa* with the wasp *Ancistrocerus antilope*. Biology of eumenine wasps II. Trans. Amer. Entomol. Soc. 80: 119–174.

Cooper, K.W., and J. Bequaert. 1950. Records and flower preferences of masarid wasps (Hymenoptera: Vespidae). Psyche 57: 137–142.

Corbet, S.A., and M. Backhouse. 1975. Aphid-hunting wasps: a field study of *Passaloecus*. Trans. R. Entomol. Soc. Lond. 127: 11–30.

Costa Lima, A. da. 1936. Sur un noveau chryside: *Duckeia cyanea*, parasite des oeufs de phasmide. Pages 173–175 in Livre Jublaire E.L. Bouvier. Paris: Firmin-Didot et Cie.

Cottrell, R.G. 1936. The biology of *Dasymutilla bioculata* (Cresson). M.S. thesis, University of Minnesota.

Court, H.K. 1961. Taxonomy and biology of the genus *Lindenius* in North America. M.S. thesis, University of California.

Coville, R.E. 1976. Predatory behavior of the spider wasp *Chalybion californicum* (Hymenoptera: Sphecidae). Pan-Pac. Entomol. 52: 229–233.

Coville, R.E. 1981. Biological observations on three *Trypoxylon* wasps in the subgenus *Trypargilum* from Costa Rica: *T. nitidum schulthessi*, *T. saussurei*, and *T. lactitarse* (Hymenoptera: Sphecidae). Pan-Pac. Entomol. 57: 332–340.

Coville, R.E., and P.L. Coville. 1980. Nesting biology and male behavior of *Trypoxylon (Trypargilum) tenoctitlan* in Costa Rica (Hymenoptera: Sphecidae). Ann. Entomol. Soc. Amer. 73: 110–119.

Coville, R.E., and C. Griswold. 1983. Nesting biology of *Trypoxylon xanthandrum* in Costa Rica with observations on its spider/prey (Hymenoptera: Sphecidae; Araneae: Senoculidae). J. Kans. Entomol. Soc. 56: 205–216.

Coville, R.E., and C. Griswold. 1984. Biology of *Trypoxylon (Trypargilum) superbum* (Hymenoptera: Sphecidae), a spider-hunting wasp with extended guarding of the brood by males. J. Kans. Entomol. Soc. 57: 365–376.

Cowan, D.P. 1979. Sibling matings in a solitary hunting wasp: adaptive inbreeding? Science 205: 1403–1405.

Cowan, D.P. 1981. Parental investment in two solitary wasps, *Ancistrocerus adiabatus* and *Euodynerus foraminatus* (Eumenidae: Hymenoptera). Behav. Ecol. Sociobiol. 9: 95–102.

Cowan, D.P. 1986a. Parasitism of *Ancistrocerus antilope* (Hymenoptera: Eumenidae) by *Leucospis affinis* (Hymenoptera: Leucospididae). Great Lakes Entomol. 19: 177–179.

Cowan, D.P. 1986b. Sexual behavior of eumenid wasps (Hymenoptera: Eumenidae). Proc. Entomol. Soc. Wash. 88: 531–541.

Cowan, D.P. 1991. The solitary and presocial Vespidae. Pages 33–73 in K.G. Ross and R.W. Matthews (eds.), The Social Biology of Wasps. Ithaca, N.Y.: Cornell University Press.

Cowan, D.P., and G.P. Waldbauer. 1984. Seasonal occurrence and mating at flowers by *Ancistrocerus antilope* (Hymenoptera: Eumenidae). Proc. Entomol. Soc. Wash. 86: 930–934.

Coward, S.J., and R.W. Matthews. 1995. Tufted titmouse *(Parus bicolor)* predation on mud-dauber wasp prepupae *(Trypoxylon politum)*. J. Kans. Entomol. Soc. 68: 371–373.

Crawford, R.L. 1982. Tufted titmouse raids mud-dauber's nest. Oriole 47: 42.

Crompton, J. 1955. The Hunting Wasp. Boston: Houghton-Mifflin.

Cronin, J.T., and D.R. Strong. 1994. Parasitoid interactions and their contribution to the stabilization of Auchenorrhyncha populations. Pages 400–428 in R.E. Denno and T.J. Perfect (eds.), Planthoppers: Their Ecology and Management. New York: Chapman and Hall.

Cross, E.A., M.G. Stith, and T.R. Baumann. 1975. Bionomics of the organ-pipe mud-dauber, *Trypoxylon politum* (Hymenoptera: Sphecidae), a spider-hunting wasp with extended guarding of the brood by males. Ann. Entomol. Soc. Amer. 68: 901–916.

Crozier, R.H., and P. Pamilo. 1996. Evolution of Social Insect Colonies: Sex Allocation and Kin Selection. Oxford: Oxford University Press.

Currado, I., and M. Olmi. 1972. Dryinidae italiani: conoscenze attuali e nuovi reperti (Hymenoptera, Bethyloidea). Boll. Mus. Zool. Univ. Torino 7: 137–176.

Dahlsten, D.L. 1961. Life history of a pine sawfly, *Neodiprion* sp., at Willits, California (Hymenoptera: Diprionidae). Can. Entomol. 93: 182–195.

Dahlsten, D.L. 1967. Preliminary life tables for pine sawflies in the *Neodipron fulviceps* complex (Hymenoptera: Diprionidae). Ecol. 48: 275–289.

Dahms, E.C. 1984. A review of the biology of species in the genus *Melittobia* (Hymenoptera: Eulophidae) with interpretations and additions using observations on *Melittobia australica*. Mem. Qd. Mus. 21: 337–360.

Dambach, C.A., and E. Good. 1943. Life-history and habits of the cicada-killer *(Sphecius speciosus)* in Ohio. Ohio J. Sci. 43: 32–41.

Danks, H.V. 1970. Biology of some stem-nesting aculeate Hymenoptera. Trans. R. Entomol. Soc. Lond. 122: 323–399.

Danks, H.V. 1983. Differences between generations in the sex ratio of aculeate Hymenoptera. Evolution 37: 414–416.

Danthanarayana, W. 1980. Parasitism of the light brown apple moth, *Epiphyas postvittana* (Walker), by its larval ectoparasite, *Goniozus jacintae* Farrugia (Hymenoptera: Bethylidae), in natural populations in Victoria. Aust. J. Zool. 28: 685–692.

Darling, D.C., and D.R. Smith. 1985. Description and life history of a new species of *Nematus* (Hymenoptera: Tenthredinidae) on *Robinia hispida* (Fabaceae) in New York. Proc. Entomol. Soc. Wash. 87: 225–230.

Darwin, C. 1871. The Descent of Man and Evolution in Relation to Sex. London: Murray.

Davidson, A. 1905. An enemy of the trap door spider. Entomol. News 14: 233–234.

Davidson, R.H., and B.J. Landis. 1938. *Crabro davidsoni* Sandh., a wasp predacious on adult leafhoppers. Ann. Entomol. Soc. Amer. 31: 5–8.

Davis, J.J. 1919. Contributions to the knowledge of the natural enemies of *Phyllophaga.* Bull. Ill. Nat. Hist. Surv. 13: 53–138.

Davis, W.T. 1920. Mating habits of *Sphecius speciosus*, the cicada-killing wasps. Bull. Brooklyn Entomol. Soc. 15: 128–129.

Day, M.C. 1977. A new genus of Plumariidae from southern Africa, with notes on Scolebythidae (Hymenoptera, Chrysidoidea). Cimbebasia (A) 4: 171–177.

Day, M.C. 1988. Spider wasps, Hymenoptera, Pompilidae. Handb. Identif. Br. Insects 6: 1–60.

Day, M.C., and K.G.V. Smith. 1981. Insect eggs on adult *Rhopalum clavipes* (L.) (Hymenoptera: Sphecidae): a problem solved. Entomologist's Gaz. 31: 173–176.

DeBach, P., and D. Rosen. 1991. Biological Control of Natural Enemies, 2d ed. Cambridge: Cambridge University Press.

DeMichelis, S., and A. Manino. 1998. Electrophoretic detection of parasitism by Dryinidae in typhlocybinae leafhoppers (Homoptera: Auchenorrhyncha). Can. Entomol. 130: 407–414.

Deyrup, M. 1988. Review of adaptations of velvet ants (Hymenoptera: Mutillidae). Great Lakes Entomol. 21: 1–4.

Deyrup, M., J.T. Cronin, and F.E. Kurczewski. 1988. *Allochares azureus*: an unusual wasp exploits unusual prey (Hymenoptera: Pompilidae; Arachnida: Filistatidae). Psyche 95: 265–281.

Disney, R.H.L. 1994. Scuttle Flies: The Phoridae. London: Chapman and Hall.

Dodson, G.N., and D.T. Gwynne. 1984. A digger wasp preying on a Jerusalem cricket. Pan-Pac. Entomol. 60: 297–299.

Dodson, G.N., and D.K. Yeates. 1989. Male *Bembix furcata* Erichson (Hymenoptera: Sphecidae) behavior on a hilltop in Queensland. Pan-Pac. Entomol. 65: 172–175.

Donisthorpe, H. St. J.K. 1927. The guests of British ants: their habits and life histories. London: Routledge and Sons.

Dorr, L.J., and J.L. Neff. 1982. *Pseudomasaris marginalis* nesting in logs in Colorado (Hymenoptera: Masaridae). Pan-Pac. Entomol. 58: 124–128.

Doutt, R.L. 1973. Maternal care of immature progeny by parasitoids. Ann. Entomol. Soc. Amer. 66: 486–487.

Dow, R. 1942. The relation of the prey of *Sphecius speciosus* to the size and sex of the adult wasp (Hym.: Sphecidae). Ann. Entomol. Soc. Amer. 35: 310–317.

Eberhard, W.G. 1974. The natural history and behaviour of the wasp *Trigonopsis cameronii* Kohl (Sphecidae). Trans. R. Entomol. Soc. Lond. 125: 295–328.

Eberhard, W.G. 1991. Copulatory courtship and cryptic female choice in insects. Biol. Rev. 66: 1–31.

Eberhard, W.G. 1996. Female Choice: Sexual Selection by Cryptic Female Choice. Princeton, NJ: Princeton University Press.

Elliott, N.B. 1984. Behavior of males of *Cerceris watlingensis* (Hymenoptera: Sphecidae, Philanthinae). Amer. Midl. Nat. 112: 85–90.

Elliott, N.B., and W.M. Elliott. 1992. Alternative mating tactics in *Tachytes tricinctus* (F.) (Hymenoptera: Sphecidae, Larrinae). J. Kans. Entomol. Soc. 65: 261–266.

Emlen, S.T., and L.W. Oring. 1977. Ecology, sexual selection, and the evolution of mating systems. Science 197: 215–223.

Endo, A. 1976. Factors influencing prey selection of a spider wasp, *Episyron arrogans* (Smith) (Hymenoptera: Pompilidae). Physiol. Ecol. Japan 17: 335–350.

Endo, A. 1980. Behaviour of the miltogrammine fly *Metopia sauteri* (Townsend) (Diptera, Sarcophagidae) cleptoparasitizing on the spider wasp *Episyron arrogans* (Smith) (Hymenoptera, Pompilidae). Kontyû 48: 445–457.

Endo, A. 1981. Nesting success of the spider wasp, *Episyron arrogans* (Smith) (Hymenoptera, Pompilidae), and the effect of interactions with other insects around its nesting site. Physiol. Ecol. Japan 18: 39–75.

Evans, H.E. 1949. The strange habits of *Anoplius despressipes* Banks: a mystery solved. Proc. Entomol. Soc. Wash. 51: 206–208.

Evans, H.E. 1953. Comparative ethology and the systematics of spider wasps. Syst. Zool. 2: 155–172.

Evans, H.E. 1956. Notes on the biology of four species of ground-nesting Vespidae. Proc. Entomol. Soc. Wash. 58: 265–270.

Evans, H.E. 1957a. Ethological studies on digger wasps of the genus *Astata* (Hymenoptera: Sphecidae). J. N.Y. Entomol. Soc. 65: 159–185.

Evans, H.E. 1957b. Studies on the Comparative Ethology of Digger Wasps of the Genus *Bembix*. Ithaca, N.Y.: Cornell University Press.

Evans, H.E. 1958a. The evolution of social life in wasps. Proc. 10th Int. Congr. Entomol. 2: 449–457.

Evans, H.E. 1958b. Observations on the nesting behavior of *Larropsis distincta* (Smith) (Hymenoptera, Sphecidae). Entomol. News 59: 197–200.

Evans, H.E. 1958c. Studies on the nesting behavior of digger wasps of the tribe Sphecini. Part I. Genus *Priononyx* Dahlbohm. Ann. Entomol. Soc. Amer. 51: 177–186.

Evans, H.E. 1959. Observations on the nesting behavior of digger wasps of the genus *Ammophila*. Amer. Midl. Nat. 62: 449–473.

Evans, H.E. 1960. A study of *Bembix u-scripta*, a crepuscular digger wasp. Psyche 67: 45–61.

Evans, H.E. 1962. The evolution of prey-carrying mechanisms in wasps. Evolution 16: 468–483.

Evans, H.E. 1963. Wasp Farm. Garden City, N.Y.: Anchor Press.

Evans, H.E. 1964. Observations on the nesting behavior of *Moniacera asperata* (Fox) with comments on communal nesting in solitary wasps. Insectes Sociaux 11: 71–78.

Evans, H.E. 1965. Simultaneous care of more than one nest by *Ammophila azteca* Cameron (Hymenoptera, Sphecidae). Psyche 72: 8–23.

Evans, H.E. 1966a. The accessory burrows of digger wasps. Science 152: 465–471.

Evans, H.E. 1966b. The behavior patterns of solitary wasps. Ann. Rev. Entomol. 11: 123–154.

Evans, H.E. 1966c. The Comparative Ethology and Evolution of the Sand Wasps. Cambridge, Mass.: Harvard University Press.

Evans, H.E. 1968. Studies on neotropical Pompilidae (Hymenoptera) IV. Examples of dual sex-linked mimicry in *Chirodamus*. Psyche 75: 1–22.

Evans, H.E. 1969a. Notes on the nesting behavior of *Pisonopsis clypeata* and *Belomicrus forbesii* (Hymenoptera, Sphecidae). J. Kans. Entomol. Soc. 42: 117–125.

Evans, H.E. 1969b. Phoretic copulation in Hymenoptera. Entomol. News 80: 113–124.

Evans, H.E. 1969c. Studies on neotropical Pompilidae (Hymenoptera) V. *Austrochares* Banks. Psyche 76: 18–28.

Evans, H.E. 1970a. Ecological-behavioral studies of the wasps of Jackson Hole, Wyoming. Bull. Mus. Comp. Zool. 140: 451–511.

Evans, H.E. 1970b. A new genus of ant-mimicking spider wasps from Australia (Hymenoptera, Pompilidae). Psyche 77: 303–307.

Evans, H.E. 1971. Observations on the nesting behavior of wasps of the tribe Cercerini. J. Kans. Entomol. Soc. 44: 500–523.

Evans, H.E. 1973a. Burrow sharing and nest transfer in the digger wasp *Philanthus gibbosus* (Fabricius). Anim. Behav. 21: 302–308.

Evans, H.E. 1973b. Notes on the nests of *Montezumia* (Hymenoptera, Eumenidae). Entomol. News 84: 285–290.

Evans, H.E. 1977. Extrinsic vs. intrinsic factors in the evolution on insect sociality. Bioscience 27: 613–617.

Evans, H.E. 1978. The Bethylidae of America North of Mexico. Mem. Amer. Entomol. Inst. 27: 1–332.

Evans, H.E. 1982. Two new species of Australian *Bembix* sand wasps, with notes on other species of the genus. Austr. Entomol. Mag. 9: 7–12.

Evans, H.E. 1987. Observations on the prey and nests of *Podalonia occidentalis* Murray (Hymenoptera: Sphecidae). Pan-Pac. Entomol. 63: 130–134.

Evans, H.E. 1989. The mating and predatory behavior of *Mellinus rufinodus* (Hymenoptera: Sphecidae). Pan-Pac. Entomol. 65: 414–417.

Evans, H.E., and J.E. Gillaspy. 1964. Observations on the ethology of digger wasps of the genus *Steniolia* (Hymenoptera: Sphecidae: Bembecini). Amer. Midl. Nat. 72: 257–280.

Evans, H.E., and A.W. Hook. 1982. Communal nesting in the digger wasp *Cerceris australis* (Hymenoptera: Sphecidae). Austr. J. Zool. 30: 557–568.

Evans, H.E., and A.W. Hook. 1986a. Nesting behavior of Australian *Cerceris* digger wasps, with special reference to nest reutililization and nest sharing (Hymenoptera, Sphecidae). Sociobiology 11: 275–302.

Evans, H.E., and A.W. Hook. 1986b. Prey selection by Australian wasps of the genus *Cerceris* (Hymenoptera: Sphecidae). J. Nat. Hist. 20: 1297–1307.

Evans, H.E., and E.G. Linsley. 1960. Notes on a sleeping aggregation of solitary bees and wasps. Bull. S. Calif. Acad. Sci. 59: 30–37.

Evans, H.E., and R.W. Matthews. 1971. Nesting behaviour and larval stages of some Australian nyssonine sand wasps (Hymenoptera: Sphecidae). Austr. J. Zool. 19: 293–310.

Evans, H.E., and R.W. Matthews. 1973a. Behavioural observations on some Australian spider wasps (Hymenoptera: Pompilidae). Trans. R. Entomol. Soc. Lond. 125: 45–55.

Evans, H.E., and R.W. Matthews. 1973b. Observations on the nesting behavior of *Trachypus petiolatus* (Spinola) in Columbia and Argentina. J. Kans. Entomol. Soc. 46: 165–175.

Evans, H.E., and R.W. Matthews. 1973c. Systematics and nesting behavior of Australian *Bembix* sand wasps. Mem. Amer. Entomol. Inst. 20: 1–387.

Evans, H.E., and R.W. Matthews. 1974. Notes on the nests and prey of two species of ground-nesting Eumenidae from So. America (Hymenoptera). Entomol. News 85: 149–153.

Evans, H.E., and K.M. O'Neill. 1978. Alternative mating strategies in the digger wasp *Philanthus zebratus* Cresson. Proc. Natl. Acad. Sci. U.S.A. 75: 1901–1903.

Evans, H.E., and K.M. O'Neill. 1985. Male territoriality in four species of the tribe Cercerini (Sphecidae: Philanthinae). J. N.Y. Entomol. Soc. 93: 1033–1040.

Evans, H.E., and K.M. O'Neill. 1986. The reproductive and nesting biology of *Bembecinus nanus strenuus* (Mickel) (Hymenoptera, Sphecidae). Proc. Entomol. Soc. Wash. 88: 628–633.

Evans, H.E., and K.M. O'Neill. 1988. The Natural History and Behavior of North American Beewolves. Ithaca, N.Y.: Cornell University Press.

Evans, H.E., and K.M. O'Neill. 1991. Beewolves. Sci. Amer. 265: 70–76.

Evans, H.E., and A. Shimizu. 1996. The evolution of nest building and communal nesting in Ageniellini (Insecta: Hymenoptera: Pompilidae). J. Nat. Hist. 30: 1633–1648.

Evans, H.E., and M.J. West-Eberhard. 1970. The Wasps. Ann. Arbor: University of Michigan Press.

Evans, H.E., and C.M. Yoshimoto. 1955. An annotated list of pompilid wasps taken at Blackjack Creek, Pottawotomie County, Kansas (Hymenoptera). J. Kans. Entomol. Soc. 28: 16–19.

Evans H.E., and C.M. Yoshimoto. 1962. Ecology and nesting behavior of the Pompilidae (Hymenoptera) of the northeastern United States. Misc. Publ. Entomol. Soc. Amer. 3: 65–119.

Evans, H.E., C.S. Lin, and C.M. Yoshimoto. 1953. A biological study of *Anoplius apiculatus autumnalis* (Banks) and its parasite, *Evagetes mohave* (Banks) (Hymenoptera: Pompilidae). J.N.Y. Entomol. Soc. 51: 61–78.

Evans, H.E., C. Kugler, and W.L. Brown Jr. 1980a. Rediscovery of *Scolebythus madacassus*, with a description of the male and of the female sting apparatus (Hymenoptera: Scolebythidae). Psyche 86: 45–51.

Evans, H.E., F.E. Kurczewski, and J. Alcock. 1980b. Observations on the nesting behaviour of seven species of *Crabro* (Hymenoptera: Sphecidae). J. Nat. Hist. 14: 865–882.

Evans, H.E., R.W. Matthews, and A. Hook. 1980c. Notes on the nests and prey of six species of *Pison* in Australia (Hymenoptera: Sphecidae). Psyche 87: 221–230.

Evans, H.E., K.M. O'Neill, and R.P. O'Neill. 1986. Nesting site changes and nocturnal clustering in the sand wasp *Bembecinus quinquespinosus* (Hymenoptera: Sphecidae). J. Kans. Entomol. Soc. 59: 280–286.

Fabre, J.H. 1915. The Hunting Wasps. New York: Dodd, Mead, and Co.

Fabre, J.H. 1918. The Wonders of Instinct. New York: The Century Co.

Fabre, J.H. 1919. The Mason Wasps. New York: Dodd, Mead, and Co.

Fabre, J.H. 1921. More Hunting Wasps. New York: Dodd, Mead, and Co.

Fales, H.M., T.M. Jaouni, J.O. Schmidt, and M.S. Blum. 1980. Mandibular gland allomones of *Dasymutilla occidentalis* and other mutillid wasps. J. Chem. Ecol. 6: 895–903.

Fattig, P.W. 1943. The Mutillidae or velvet ants of Georgia. Emory Univ. Mus. Bull. 1: 1–24.

Fenton, F.A. 1918. The parasites of leafhoppers. With special reference to the biology of the Anteoninae. Ohio J. Sci. 18: 177–212.

Ferguson, C.S., and J.H. Hunt. 1989. Near-nest behavior of a solitary mud-daubing wasp, *Sceliphron caementarium* (Hymenoptera: Sphecidae). J. Ins. Behav. 2: 315–323.

Ferguson, W.E. 1962. Biological charasteristics of the mutillid subgenus *Photopsis* Blake and their systematic values (Hymenoptera). Univ. Calif. Publ. Entomol. 27: 1–92.

Field, J. 1989a. Alternative nesting tactics in a solitary wasp. Behaviour 110: 219–243.

Field, J. 1989b. Intraspecific parasitism and nesting success in the solitary wasp *Ammophila sabulosa*. Behaviour 110: 23–46.

Field, J. 1992a. Guild structure in solitary spider-hunting wasps (Hymenoptera: Pompilidae) compared to null model predictions. Ecol. Entomol. 17: 198–208.

Field, J. 1992b. Intraspecific parasitism as an alternative reproductive tactic in nest-building wasps and bees. Biol. Rev. 67: 79–126.

Field, J. 1992c. Intraspecific parasitism and nest defence in the solitary pompilid wasp *Anoplius viaticus*. J. Zool. (Lond.) 228: 341–350.

Field, J. 1992d. Patterns of nest provisioning and parental investment in the solitary digger wasp *Ammophila sabulosa*. Ecol. Entomol. 17: 43–51.

Field, J. 1994. Selection of host nests by intraspecific nest-parasitic digger wasps. Anim. Behav. 48: 113–118.

Field, J., and W.A. Foster. 1995. Nest co-occupation in the digger wasp *Cerceris arenaria*: cooperation or usurpation? Anim. Behav. 50: 99–112.

Finlayson, L.H. 1950a. The biology of *Cephalonomia waterstoni* Gahan (Hym., Bethylidae), a parasite of *Laemophloeus* (Col., Cucujidae). Bull. Entomol. Res. 41: 79–97.

Finlayson, L.H. 1950b. Host preference of *Cephalonomia waterstoni* Gahan, a bethylid parasitoid of *Laemophloeus* species. Behaviour 2: 225–316.

Fisher, R.A. 1930. The Genetical Theory of Natural Selection. Oxford: Oxford University Press.

Flanders, S.E. 1956. The mechanisms of sex-ratio regulation in (Parasitic) Hymenoptera. Insectes Soc. 3: 325–334.

Flinn, P.W. 1991. Temperature-dependent functional response of the parasitoid *Cephalonomia waterstoni* (Gahan) (Hymentoptera: Bethylidae) attacking rusty grain beetle larvae (Coleoptera: Cucujidae). Environ. Entomol. 20: 872–876.

Flinn, P.W., and D.W. Hagstrum. 1995. Simulation model of *Cephalonomia waterstoni* (Hymenoptera: Bethylidae) parasitizing the rusty grain beetle (Coleoptera: Cucujidae). Environ. Entomol. 24: 1608–1615.

Flinn, P.W., D.W. Hagstrum, and W.H. McGaughey. 1996. Suppression of beetles in stored wheat by augmentative release of parasitic wasps. Environ. Entomol. 25: 505–511.

Fordham, F. 1946. Pollination of *Calochilus campestris*. Victorian Nat. 62: 199–201.

Frank, J.H., J.P. Parkman, and F.D. Bennett. 1995. *Larra bicolor* (Hymenoptera: Sphecidae), a biological control agent of *Scapteriscus* mole crickets (Orthoptera: Gryllotalpidae), established in northern Florida. Florida Entomol. 78: 619–623.

Frank, S.A. 1995. Sex allocation in solitary bees and wasps. Amer. Nat. 146: 316–323.

Fraizer, T. 1997. A dynamic model of mating behaviour in digger wasps: the energetics of male-male competition mimic size-dependent thermal constraints. Behav. Ecol. Sociobiol. 41: 423–434.

Freeman, B.E. 1966. Notes on conopid flies, including insect host, plant and phoretic relationships (Diptera: Conopidae). J. Kans. Entomol. Soc. 39: 123–131.

Freeman, B.E. 1973. Preliminary studies of the population dynamics of *Sceliphron assimile* Dahlbom (Hymenoptera: Sphecidae) in Jamaica. J. Anim. Ecol. 42: 173–182.

Freeman, B.E. 1981a. The dynamics in Trinidad of *Trypoxylon palliditarse* Saussure: a Thompsonian population? J. Anim. Ecol. 50: 563–572.

Freeman, B.E. 1981b. Parental investment, maternal size, and population dynamics of a solitary wasp. Amer. Nat. 117: 357–362.

Freeman, B.E. 1982. The comparative distribution and population dynamics in Trinidad of *Sceliphron fistularium* (Dahlbom) and *S. asiaticum* (L.) (Hymenoptera: Sphecidae). Biol. J. Linn. Soc. 17: 343–363.

Freeman, B.E., and K. Ittyeipe. 1976. Field studies on the cumulative response of *Melittobia* sp. (*Hawaiiensis* complex) (Eulophidae) to varying host density. J. Anim. Ecol. 45: 415–423.

Freeman, B.E., and K. Ittyeipe. 1993. The natural dynamics of the eulophid parasitoid *Melittobia australica*. Ecol. Entomol. 18: 129–140.

Freeman, B.E., and D.B. Jayasingh. 1975. Population dynamics of *Pachodynerus nasidens* (Hymenoptera) in Jamaica. Oikos 26: 86–91.

Freeman, B.E., and B. Johnston. 1978a. The biology in Jamaica of the adults of the sphecid wasp *Sceliphron assimile* Dahlbom. Ecol. Entomol. 2: 39–52.

Freeman, B.E., and B. Johnston. 1978b. Gregarious roosting in the sphecid wasp *Sceliphron assimile*. Ann. Entomol. Soc. Amer. 71: 435–441.

Freeman, B.E., and C.A. Taffe. 1974. Population dynamics and nesting behaviour of *Eumenes colona* (Hymenoptera) in Jamaica. Oikos 25: 388–394.

Frick, K.E. 1962. Ecological studies on the alkali bee, *Nomia melanderi*, and its bombyliid parasite, *Heterostylum robustum*, in Washington. Ann. Entomol. Soc. Amer. 55: 5–15.

Fricke, J.M. 1991. Trap-nest bore diameter preferences among sympatric *Passaloecus* spp. (Hymenoptera: Sphecidae). Great Lakes Entomol. 24: 123–125.

Fricke, J.M. 1992a. Factors influencing length and volume of cells provisioned by two *Passaloecus* species (Hymenoptera: Sphecidae). Great Lakes Entomol. 25: 107–114.

Fricke, J.M. 1992b. The influence of tree species on frequency of trap-nest use by *Passaloecus* species (Hymenoptera: Sphecidae). Great Lakes Entomol. 25: 51–53.

Fricke, J.M. 1993. Aphid prey of *Passaloecus cuspidatus* (Hymenoptera: Sphecidae). Great Lakes Entomol. 26: 31–34.

Fricke, J.M. 1995. Economics of cells partitions and closures by *Passaloecus cuspidatus* (Hymenoptera: Sphecidae). Great Lakes Entomol. 28: 221–223.

Frost, S.W. 1944. Notes on the habits of *Monobia quadridens* (Linn.). Entomol. News 60: 10–15.

Funasaki, G.Y., P. Lai, L.M. Nakahara, J.W. Beardsley, and A.K. Ota. 1988. A review of biological control introductions in Hawaii: 1890 to 1985. Proc. Haw. Entomol. Soc. 28: 105–160.

Fye, R.E. 1965. The biology of the Vespidae, Pompilidae, and Sphecidae (Hymenoptera) from trap nests in northwestern Ontario. Can. Entomol. 97: 744.

Gaimari, S.D., and R.D. Martins. 1996. Nesting behavior and nest distributions of *Ammophila gracilis* Lepeletier (Hymenoptera: Sphecidae) in Brazil. J. Hym. Res. 5: 240–248.

Garcia, M.V.B., and J. Adis. 1993. On the biology of *Penepodium goryanum* (Lepeletier) in wooden trap nests (Hymenoptera, Sphecidae). Proc. Entomol. Soc. Wash. 95: 547–553.

Garcia, M.V.B., and J. Adis. 1995. Nesting behaviour of *Trypoxylon (Trypargilum) rogenhoferi* Kohl (Hymenoptera, Sphecidae) in várzea inundation forest of central Amazonia. Amazoniana 13: 259–282 [in Portuguese].

Gauld, I., and B. Bolton (eds.). 1988. The Hymenoptera. British Museum (Natural History) and Oxford University Press, London and Oxford.

Gayubo, S.F., J.D. Asis, and J. Tormos. 1992. A new species of *Palarus* Latrielle from Spain with a comparative study on nesting behavior and larvae in the genus (Hymenoptera: Sphecidae). Ann. Entomol. Soc. Amer. 85: 26–33.

Gess, F.W. 1980a. Prey and nesting sites of some sympatric species of *Cerceris* (Hymenoptera: Sphecidae) with a review and discussion of the prey diversity in the genus. Ann. Cape Prov. Mus. (Nat. Hist.) 13: 85–93.

Gess, F.W. 1980b. Some aspects of the ethology of *Dasyproctus westermanni* (Dahlbom) (Hymenoptera: Sphecidae: Crabroninae) in the eastern Cape Province of South Africa. Ann. Cape Prov. Mus. (Nat. Hist.) 13: 95–106.

Gess, F.W. 1981. Some aspects of an ethological study of the aculeate wasps and bees of a karroid area in the vicinity of Grahamstown, South Africa. Ann. Cape Prov. Mus. (Nat. Hist.) 14: 1–80.

Gess, F.W. 1984. Some aspects of the ethology of *Ampulex bantuae* Gess (Hymenoptera: Sphecidae: Ampulicinae) in the eastern Cape Province of South Africa. Ann. Cape Prov. Mus. (Nat. Hist.) 16: 23–40.

Gess, F.W., and S.K. Gess. 1974. An ethological study of *Dichragenia pulchricoma* (Arnold) (Hymenoptera: Pompilidae), a southern African spider-hunting wasp which builds a turreted, subterranean nest. Ann. Cape Prov. Mus. (Nat. Hist.) 9: 187–214.

Gess, F.W., and S.K. Gess. 1975. Ethological studies of *Bembecinus cingulifer* (Smith) and *B. oxydorus* (Handl.) (Hymenoptera: Sphecidae), two southern African turret-building wasps. Ann. Cape Prov. Mus. (Nat. Hist.) 11: 21–46.

Gess, F.W., and S.K. Gess. 1980. Ethological studies of *Jugurtia confusa* Richards, *Ceramius capicola* Brauns, *C. linearis* Klug and *C. lichtensteinii* (Klug) (Hymenoptera: Masaridae) in the eastern Cape Province of South Africa. Ann. Cape Prov. Mus. (Nat. Hist.) 13: 63–83.

Gess, F.W., and S.K. Gess. 1982. Ethological studies of *Isodontia simoni* (du Buysson), *I. pelopoeiformis*, and *I. stanleyi* (Kohl) (Hymenoptera: Sphecidae: Sphecinae) in the eastern Cape Province of South Africa. Ann. Cape Prov. Mus. (Nat. Hist.) 14: 151–171.

Gess, F.W., and S.K. Gess. 1986. Ethological notes on *Ceramius bicolor* (Thunberg), *C. clypeatus* Richards, *C. nigripennis* Saussure and *C. socius* Turner (Hymenoptera: Masaridae) in the western Cape Province of South America. Ann. Cape Prov. Mus. (Nat. Hist.) 16: 161–178.

Gess, F.W., and S.K. Gess. 1988a. A contribution to the knowledge of the ethology of the genera *Parachilus* Giordani Soika and *Paravespa* Radoszkowski (Hymenoptera: Eumenidae) in southern Africa. Ann. Cape Prov. Mus. (Nat. Hist.) 18: 57–81.

Gess, F.W., and S.K. Gess. 1988b. A contribution to the knowledge of the taxonomy and the ethology of the genus *Masarina* Richards (Hymenoptera: Masaridae). Ann. Cape Prov. Mus. (Nat. Hist.) 16: 351–362.

Gess, F.W., and S.K. Gess. 1988c. A further contribution to the knowledge of the genus *Ceramius* Latreille (Hymenoptera: Masaridae) in the southern and western Cape Province of South Africa. Ann. Cape Prov. Mus (Nat. Hist) 18: 1–30.

Gess, F.W., and S.K. Gess. 1990. A fourth contribution to the knowledge of the ethology of the genus *Ceramius* Latreille (Hymenoptera: Vespoidea: Masaridae) in southern Africa. Ann. Cape Prov. Mus. (Nat. Hist) 18: 183–202.

Gess, F.W., and S.K. Gess. 1991. Some aspects of the ethology of five species of Eumenidae (Hymenoptera) in southern Africa. Ann. Cape Prov. Mus. (Nat. Hist.) 18: 245–270.

Gess, F.W., and S.K. Gess. 1992. Ethology of three southern African ground nesting Masarinae, two *Celonites* species and a silk-spinning *Quartinia* species, with a discussion of nesting by the subfamily as a whole (Hymenoptera: Vespidae). J. Hym. Res. 1: 145–155.

Gess, F.W., S.K. Gess, and A.J.S. Weaving. 1982. Some aspects of the ethology of *Chalybion (Hemichalybion) spinolae* (Lepeletier) (Hymenoptera: Sphecidae: Sphecinae) in the eastern Cape Province of South Africa. Ann. Cape Prov. Mus. (Nat. Hist.) 14: 139–149.

Gess, F.W., S.K. Gess, and R.W. Gess. 1995. An Australian masarine, *Rolandia angulata* (Richards) (Hymenoptera: Vespidae): nesting and evaluation of association with *Goodenia* (Goodeniaceae). J. Hym. Res. 4: 25–32.

Gess, S.K. 1992. Biogeography of the masarine wasps (Hymenoptera Vespidae: Masarinae), with particular emphasis on the southern African taxa and on correlations between masarine and forage plant distributions. J. Biogeog. 19: 491–503.

Gess, S.K. 1996. The Pollen Wasps. Cambridge, Mass.: Harvard University Press.

Gess, S.K., and F.W. Gess. 1989a. Flower visiting by masarid wasps in southern Africa (Hymenoptera: Vespoidea: Masaridae). Ann. Cape Prov. Mus. (Nat. Hist.) 13: 95–134.

Gess, S.K., and F.W. Gess. 1989b. Notes on the nesting behaviour in *Bembix bubalus* Handlirsch in southern Africa with emphasis on nest sharing and reaction to nest parasites (Hymenoptera: Sphecidae). Ann. Cape Prov. Mus. (Nat. Hist.) 18: 151–160.

Gess, S.K., and F.W. Gess. 1994. Potential pollinators of the Cape Group of Crotalarieae (*sensu* Polhill) (Fabales: Papilionaceae), with implications for seed production in cultivated rooibos tea. Afr. Entomol. 2: 97–106.

Ghazoul, J., and P.G. Willmer. 1994. Endothermic warm-up in two species of sphecid wasp and its relation to behaviour. Physiol. Entomol. 19: 103–108.

Gifford, J.R. 1965. *Goniozus indicus* as a parasite of the sugarcane borer. J. Econ. Entomol. 58: 799–800.

Gillaspy, J.E., H.E. Evans, and C.S. Lin. 1962. Observations on the behavior of digger wasps of the genus *Stictiella* (Hymenoptera: Sphecidae) with a partition of the genus. Ann. Entomol. Soc. Amer. 55: 559–566.

Giraldeau, L.-A., C. Soos, and G. Beauchamp. 1994. A test of the producer-scrounger foraging game in captive flocks of spice finches, *Lonchura punctulata.* Behav. Ecol. Sociobiol. 34: 251–256.

Given, B.B. 1954. Evolutionary trends in the Thynninae (Hymenoptera: Tiphiidae) with special reference to the feeding habits of Australian species. Trans. R. Entomol. Soc. Lond. 105: 1–10.

Godfray, H.C.J. 1994. Parasitoids: Behavioral and Evolutionary Ecology. Princeton, N.J.: Princeton University Press.

Goertzen, R., and R.L. Doutt. 1975. The ovicidal propensity of *Goniozus.* Ann. Entomol. Soc. Amer. 68: 869–870.

Gordh, G. 1976. *Goniozus gallicola* Fouts, a parasite of moth larvae, with notes on other bethylids (Hymenoptera: Bethylidae; Lepidoptera: Gelechiidae). USDA Tech. Bull. 1524: 1–27.

Gordh, G., and H.E. Evans. 1976. A new species of *Goniozus* imported into California from Ethiopia for the biological control of pink bollworm and some notes on the taxonomic status of *Parasierola* and *Goniozus* (Hymenoptera: Bethylidae). Proc. Entomol. Soc. Wash. 78: 479–489.

Gordh, G., and B. Hawkins. 1981. *Goniozus emigratus* (Rohwer), a primary external parasite of *Paramyelois transitella* (Walker), and comments on bethylids attacking Lepidoptera (Hymenoptera: Bethylidae; Lepidoptera: Pyralidae). J. Kans. Entomol. Soc. 54: 787–803.

Gordh, G., and R.E. Medved. 1986. Biological notes on *Goniozus pakmanus* Gordh (Hymenoptera: Bethylidae), a parasite of pink bollworm, *Pectinophora gossypiella* (Saunders) (Lepidoptera: Gelechiidae). J. Kans. Entomol. Soc. 59: 723–734.

Gordh, G., and L. Móczár. 1990. A catalog of the world Bethylidae (Hymenoptera: Aculeata). Mem. Amer. Entomol. Inst. 46: 1–364.

Gordh, G., J.B. Woolley, and R.A. Medved. 1983. Biological studies on *Goniozus legneri* Gordh (Hymenoptera: Bethylidae) a primary external parasite of the navel orangeworm *Amyelois transitella*. Mem. Amer. Entomol. Inst. 20: 433–468.

Goulet, H., and J.T. Huber. 1993. Hymenoptera of the World: An Identification Guide to Families. Research Branch, Agriculture Canada.

Publication 1894/E. Ottawa: Centre for Land and Biological Resources Research.

Graham, D.Y., and D.E. Conlong. 1988. Improved laboratory rearing of *Eldana saccharina* (Lepidoptera: Pyralidae) and its indigenous parasitoid Goniozus natalensis (Hymenoptera: Bethylidae). Proc. Sth. Afr. Sug. Tech. Assoc. 62: 116–119.

Green, R.F., G. Gordh, and B.A. Hawkins. 1982. Precise sex ratios in highly inbred parasitic wasps. Amer. Nat. 120: 653–665.

Griffiths, N.T., and H.C.J. Godfray. 1988. Local mate competition, sex ratio and clutch size in bethylid wasps. Behav. Ecol. Sociobiol. 22: 211–217.

Grissell, E.E. 1975. Ethology and larva of *Pterocheilus texanus* (Hymenoptera: Eumenidae). J. Kans. Entomol. Soc. 48: 244–253.

Grissell, E.E. 1979. Nesting biology of *Pluto littoralis* (Malloch) (Hymenoptera: Sphecidae). 52: 269–275.

Gross, M.R. 1996. Alternative reproductive strategies and tactics: diversity within sexes. Trends Ecol. Evol. 11: 92–98.

Guglielmino, A., and E.G. Virla. 1998. Post-embryonic development of *Gonatopus lunatus* Klug (Hymenoptera: Dryinidae: Gonatopodinae), with remarks on its biology. Ann. Soc. Entomol. Fr. (N.S.) 34: 321–333.

Gurney, A.B. 1953. Notes on the biology and immature stages of the cricket parasite of the genus *Rhopalosoma*. Proc. U.S. Nat. Mus. 103: 19–34.

Gwynne, D.T. 1978. Male territoriality in the bumblebee wolf, *Philanthus bicinctus* (Mickel) (Hymenoptera: Sphecidae): observations on the behaviour of individual males. Z. Tierpsychol. 47: 89–103.

Gwynne, D.T. 1979. Nesting biology of the spider wasps (Hymenoptera: Pompilidae) which prey on burrowing wolf spiders (Araneae: Lycosidae, *Geolycosa*). J. Nat. Hist. 13: 681–692.

Gwynne, D.T. 1980. Female defense polygyny in the bumblebeewolf, *Philanthus bicinctus* (Hymenoptera: Sphecidae). Behav. Ecol. Sociobiol. 7: 213–225.

Gwynne, D.T. 1981. Nesting biology of the bumblebee wolf *Philanthus bicinctus* Mickel (Hymenoptera: Sphecidae). Amer. Midl. Nat. 105: 130–138.

Gwynne, D.T. 1984. Male mating effort, confidence of paternity, and insect sperm competition. Pages 117–150 in R.L. Smith (ed.), Sperm Competition and the Evolution of Animal Mating Systems. New York: Academic Press.

Gwynne, D.T., and G.N. Dodson. 1983. Non-random provisioning by the digger wasp, *Palmodes laeviventris* (Hymenoptera: Sphecidae). Ann. Entomol. Soc. Amer. 76: 434–436.

Gwynne, D.T., and H.E. Evans. 1975. Nesting behavior of *Larropsis chilopsidis* and *L. vegeta* (Hymenoptera: Sphecidae: Larrinae). Psyche 82: 275–282.

Gwynne, D.T., and K.M. O'Neill. 1980. Territoriality in digger wasps results in sex-biased predation on males (Hymenoptera: Sphecidae, *Philanthus*). J. Kans. Entomol. Soc. 53: 220–224.

Habeck, D.H., R.T. Arbogast, and L.D. Cline. 1974. Biology and immature stages of *Schinia mitis* (Grote). J. Lep. Soc. 28: 152–157.

Hadlington, P., and F. Hoschke. 1959. Observations on the ecology of phasmatid *Ctenomorphodes tessulata* (Gray). Proc. Linn. Soc. N.S. Wales 84: 146–159.

Haeseler, V. 1980. Nectar robbing by solitary wasps (Hymenoptera: Vespoidea: Eumenidae). Entomol. Gen. 6: 49–55.

Hager, B.J., and F.E. Kurczewski. 1985. Cleptoparasitism of *Ammophila harti* (Fernald) (Hymenoptera: Sphecidae) by *Senotainia vigilans* Allen, with observations on *Phrosinella aurifacies* Downes (Diptera: Sarcophagidae). Psyche 92: 451–462.

Hager, B.J., and F.E. Kurczewski. 1986. Nesting behavior of *Ammophila harti* (Fernald) (Hymenoptera: Sphecidae). Amer. Midl. Nat. 116: 7–24.

Halffter, G. 1997. Subsocial behavior in Scarabaeinae beetles. Pages 237–259 in J.C. Choe and B.J Crespim (eds.), Social Behavior in Insects and Arachnids. Cambridge: Cambridge University Press.

Hamilton, W.D. 1967. Extraordinary sex ratios. Science 156: 477–478.

Hamm, A.H., and O.W. Richards. 1926. The biology of the British Crabronidae. Trans. Entomol. Soc. Lond. 74: 297–331.

Hamm, A.H., and O.W. Richards. 1930. The biology of the British fossorial wasps of the families Mellinidae, Gorytidae, Philanthidae, Oxybelidae, and Trypoxylidae. Trans. Entomol. Soc. Lond. 78: 95–131.

Handel, S.N., and R. Peakall. 1993. Thynnine wasps discriminate among heights when seeking mates: tests with a sexually deceptive orchid. Oecologia 95: 241–245.

Hardy, I.C.W. 1992. Non-binomial sex allocation and brood sex ratio variances in the parasitoid Hymenoptera. Oikos 65: 143–158.

Hardy, I.C.W., and T.M. Blackburn. 1991. Brood guarding in a bethylid wasp. Ecol. Entomol. 16: 55–62.

Hardy, I.C.W., and J.M. Cook. 1995. Brood sex ratio of variance, developmental mortality and virginity in a gregarious parasitoid wasp. Oecologia 103: 162–169.

Hardy, I.C.W., and P.J. Mayhew. 1998. Sex ratio, sexual dimorphism and mating structure in bethylid wasps. Behav. Ecol. Sociobiol. 42: 383–395.

Hardy, I.C.W., N.T. Griffiths, and H.C.J. Godfray. 1992. Clutch size in a parasitoid wasp: a manipulation experiment. J. Anim. Ecol. 61: 121–129.

Harris, A.C. 1998. Nesting behaviour, life history and description of the mature larva of the beetle predator, *Podagritus parrotti* Leclercq (Hymenoptera: Sphecidae: Crabroninae). J. Royal Soc. New Zealand 28: 591–604.

Hartmann, C.G. 1944a. How *Odynerus* suspends her egg. Psyche 51: 1–4.

Hartmann, C.G. 1944b. A note on the habits of *Trypoxylon politum* Say (Hymenoptera: Sphecidae). Entomol. News 55: 7–8.

Haskell, A.T. 1955. Further observations on the occurrence of *Sphex aegyptius* with swarms of the desert locust. Entomol. Mon. Mag. 91: 284–285.

Hastings, J. 1986. Provisioning by female western cicada-killer wasps, *Sphecius grandis* (Hymenoptera: Sphecidae): influence of body size and emergence time on individual provisioning success. J. Kans. Entomol. Soc. 59: 262–268.

Hastings, J. 1989a. The influence of size, age, and residency status on territory defense in male western cicada killer wasps (*Sphecius grandis*, Hymenoptera: Sphecidae). J. Kans. Entomol. Soc. 62: 363–373.

Hastings, J. 1989b. Protandry in western cicada killer wasps (*Sphecius grandis*, Hymenoptera: Sphecidae): an empirical study of emergence time and mating opportunity. Behav. Ecol. Sociobiol. 25: 255–260.

Heather, N.W. 1965. Occurrence of Cleptidae (Hymenoptera) parasites in eggs of *Ctenomorphodes tessulatus* (Gray) (Phasmida: Phasmidae) in Queensland. J. Entomol. Soc. Queensland 4: 86–87.

Heaversedge, R.C. 1968. Variation in the size of insect parasites of puparia of *Glossina* spp. Bull. Entomol. Res. 58: 153–159.

Heaversedge, R.C. 1969. Brief notes on the reproductive morphology of *Mutilla glossinae* Turner (Hymenoptera) and the development of its immature stages. J. Entomol. Soc. Sth. Afr. 32: 485–488.

Heaversedge, R.C. 1970. Developmental periods of insect parasites of *Glossina morsitans orientalis* Vanderplank (Diptera: Muscidae). J. Entomol. Soc. Sth. Afr. 33: 352–354.

Heim de Balsac, H. 1935. Ecologie de *Pedinomma rufescens* Westwood; sa présence dans les nids des micromammiféres (Hym. Embolemidae). Rev. Fran. d'Entomol. 2: 109–112.

Heimpel, G.E. 1994. Virginity and the cost of insurance in highly inbred Hymenoptera. Ecol. Entomol. 19: 299–302.

Heimpel, G.E., and T.R. Collier. 1996. The evolution of host-feeding behaviour in insect parasitoids. Biol. Rev. 71: 373–400.

Heinrich, B. 1993. The Hot-Blooded Insects. Cambridge, Mass.: Harvard University Press.

Helms, K.R. 1994. Sexual size dimorphism and sex ratios in bees and wasps. Amer. Nat. 143: 418–434.

Hermann, H.R., and J.M. Gonzalez. 1986. Venom apparatus of *Trypoxylon clavatum clavatum* (Hymenoptera: Sphecidae). J. Kans. Entomol. Soc. 59: 213–218.

Hertz, P.E., R.B. Huey, and R.D. Stevenson. 1993. Evaluating temperature regulation by field-active ectotherms: the fallacy of the inappropriate question. Amer. Nat. 142: 796–818.

Hicks, C.H. 1929. *Pseudomasaris edwardsii* Cresson, another pollen-provisioning wasp, with further notes on *P. vespoides* (Cresson). Can. Entomol. 61: 121–125.

Hicks, C.H. 1933a. Note on the relationship of an ichneumonid to certain digger wasps. Pan-Pac. Entomol. 9: 49–52.

Hicks, C.H. 1933b. Observations on a chrysidid parasite and its host (Hymenoptera: Chrysididae, Megachilidae). Entomol. News 44: 206–209.

Hicks, C.H. 1927. *Pseudomasaris vespoides* (Cresson), a pollen provisioning wasp. Can. Entomol. 59: 75–79.

Hilchie, G.J. 1982. Evolutionary aspects of geographical variation in color and prey in the beewolf species *Philanthus albopilosus* Cresson. Quaest. Entomol. 18: 91–126.

Hill, D.S. 1994. Agricultural Entomology. Portland, Oreg.: Timber Press.

Hine, J.S. 1906. A preliminary report on the horseflies of Louisiana with a discussion of remedies and natural enemies. Circ. State Crop Pest Commission Louisiana 6: 1–43.

Hirschfelder, H. 1952. Zur Biologie des Bienenwolfes (*Philanthus triangulum* F.). Z. Bienenf. 1: 1–3.

Hölldobler, B., and E.O. Wilson. 1990. The Ants. Boston: Harvard University Press.

Holloway, J.K. 1931. Temperature as a factor in the activity and development of the Chinese strain of *Tiphia popilliavora* (Rohw.) in New Jersey and Pennsylvania. J. N.Y. Entomol. Soc. 39: 555–564.

Hook, A.W. 1984. Notes on the nesting and mating behavior of *Trypoxylon (Trypargilum) spinosum* (Hymenoptera: Sphecidae). J. Kans. Entomol. Soc. 57: 534–535.

Hook, A.W. 1987. Nesting behavior of Texas *Cerceris* digger wasps with emphasis on nest reutilization and nest sharing (Hymenoptera: Sphecidae). Sociobiology 13: 93–118.

Hook, A.W., and H.E. Evans. 1991. Prey and parasites of *Cerceris fumipennis* (Hymenoptera: Sphecidae) from central Texas, with description of the larva of *Dasymutilla scaevola* (Hymenoptera: Mutillidae). J. Kans. Entomol. Soc. 64: 257–264.

Hook, A.W., and R.W. Matthews. 1980. Nesting biology of *Oxybelus sericeus* with a discussion of nest guarding by male sphecid wasps (Hymenoptera). Psyche 87: 21–37.

Houston, T.F. 1984. Bionomics of a pollen-collecting wasp, *Paragia tricolor* (Hymenoptera: Vespidae: Masarinae). Rec. W. Austr. Mus. 11: 141–151.

Houston, T.F. 1986. Biological notes on the pollen wasp *Paragia (Cygnaea) vespiformis* (Hymenoptera: Vespidae: Masarinae) with description of a nest. Austr. Entomol. Mag. 12: 115–124.

Houston, T.F. 1995. Notes on the ethology of *Rolandia maculata* (Hymenoptera: Vespidae; Masarinae), a pollen wasp with a psammophore. Rec. W. Austr. Mus. 17: 343–349.

Howard, L.O. 1901. The Insect Book. New York: Doubleday, Page and Co.

Howard, R.W., and P.W. Flinn. 1990. Larval trails of *Cryptolestes ferrugineus* (Coleoptera: Cucujidae) as kairomonal host-finding cues for the parasitoid *Cephalonomia waterstoni* (Hymenoptera: Bethylidae). Ann. Entomol. Soc. Amer. 83: 239–245.

Hudson, W.G., J.H. Frank, and J.L. Castner. 1988. Biological control of *Scapteriscus* spp. mole crickets (Orthoptera: Gryllotalpidae) in Florida. Bull. Entomol. Soc. Amer. 34: 192–198.

Hull, F.M. 1973. Bee Flies of the World: The Genera of the Family Bombyliidae. Washington, D.C.: Smithsonian Institution Press.

Hungerford, H.B. 1937. *Pseudomasaris occidentalis* (Cresson) in Kansas. J. Kans. Entomol. Soc. 10: 133–134.

Hunt, J. 1993. Survivorship, fecundity, and recruitment in a mud dauber wasp, *Sceliphron assimile* (Hymenoptera: Sphecidae). Ann. Entomol. Soc. Amer. 86: 51–59.

Hurd, P.D., Jr., and J.S. Moure. 1961. Some notes on sapygid parasitism in the nests of carpenter bees belonging to the genus *Xylocopa* Latreille (Hymenoptera: Aculeata). J. Kans. Entomol. Soc. 34: 20–22.

Hyslop, J.A. 1916. *Pristocera armifera* (Say) parasitic on *Limonius agonus* (Say). Proc. Entomol. Soc. Wash. 18: 169–170.

Itino, T. 1986. Comparison of life table between the solitary eumenid wasp *Anterhynchium flavomarginatum* and the subsocial eumenid wasp *Orancistrocerus drewseni* to evaluate the adaptive significance of maternal care. Res. Popul. Ecol. 28: 185–199.

Itino, T. 1988. The spatial patterns of parasitism of eumenid wasps, *Anterhynchium flavomarginatum* and *Orancistrocerus drewseni* by the miltogrammine fly *Amobia distorta*. Res. Popul. Ecol. 30: 1–12.

Itino, T. 1992. Differential diet breadths and species coexistence in leafroller-hunting eumenid wasps. Res. Popul. Ecol. 34: 203–211.

Itô, Y. 1978. Comparative Ecology. Cambridge: Cambridge University Press.

Iwata, K. 1932. Biology of *Homonotus iwatai* Yasumatsu. Annot. Zool. Jap. 13: 305–317.

Iwata, K. 1933. Studies on the nesting habits and parasites of *Megachile sculpturalis* Smith (Hymenoptera, Megachilidae). Mushi 6: 6–24.

Iwata, K. 1936. Biology of two Japanese species of *Methoca* with the description of a new species (Hymenoptera, Thynnidae). Kontyû 10: 57–89.

Iwata, K. 1942. Comparative studies on the habits of solitary wasps. Tenthredo 4: 1–146.

Iwata, K. 1949. Biology of *Goniozus japonicus* Ashmead, a parasite of the persimmon leaf-roller, *Dichocrocis chlorophanta* Butler. Tech. Bull. Fac. Agric. Kagawa Univ. 1: 58–60.

Iwata, K. 1955. The comparative anatomy of the ovary in Hymenoptera. Part I. Aculeata. Mushi 29: 17–34.

Iwata, K. 1960. The comparative anatomy of the ovary in Hymenoptera. Supplement on Aculeata with descriptions of ovarian eggs of certain species. Acta Hymenopterol. 1: 205–211.

Iwata, K. 1961. Farther biological observations on *Goniozus japonicus* Ashmead (Hymenoptera: Bethylidae). Mushi 35: 91–97.

Iwata, K. 1964. Egg gigantism in subsocial Hymenoptera, with ethological discussion on tropical bamboo carpenter bees. Pages 399–435 in T. Kira and T. Umesao (eds.), Nature and Life in Southeast Asia, vol. 3. Kyoto: Fauna and Flora Research Society.

Iwata, K. 1965. The comparative anatomy of the ovary in Hymenoptera (records on 64 species of Aculeata in Thailand with descriptions of ovarian eggs). Mushi 38: 101–109.

Iwata, K. 1976. Evolution of Instinct: Comparative Ethology of the Hymenoptera. New Delhi, India: Amerind Publishing [translation of 1971 Japanese edition].

Iwata, K., and S.F. Sakagami. 1966. Gigantism and dwarfism in bee eggs in relation to the modes of life, with notes on the number of ovarioles. Japan. J. Ecol. 16: 4–16.

Iwata, K., and M. Tanihata. 1963. Biological observations on *Larra amplipennis* (Smith) in Kagawa, Japan (Hymenoptera: Sphecidae). Trans. Shikoku Entomol. Soc. 7: 101–105.

Janvier, H. 1928. Recherches biologiques sur les predateurs du Chili. Ann. Sci. Nat. Zool. 11: 67–207.

Janvier, H. 1933. Etude biologique de quelques Hymenopteres du Chili. Ann. Sci. Nat. Zool. 16: 210–356.

Janvier, H. 1960. Recherches sur les Hymenopteres nidifiants aphidivores. Ann. Sci. Nat. Zool. 12: 281–321.

Jayakar, S.D. 1963. "Proterandry" in solitary wasps. Nature 198: 208–209.

Jayakar, S.D., and H. Spurway. 1966. Re-use of cells and brother sister mating in the Indian species *Stenodynerus miniatus* (Sauss.) (Vespidae: Eumeninae). J. Bombay Nat. Hist. Soc. 63: 378–398.

Jayasingh, D.B. 1980. A new hypothesis on cell provisioning in solitary wasps. Biol. J. Linn. Soc. 13: 167–170.

Jayasingh, D.B., and B.E. Freeman. 1980. The comparative population dynamics of eight solitary bees and wasps (Aculeata; Apocrita; Hymenoptera) trapnested in Jamaica. Biotropica 12: 214–219.

Jayasingh, D.B., and C.A. Taffe. 1982. The biology of the eumenid mud-wasp *Pachodynerus nasidens* in trapnests. Ecol. Entomol. 7: 283–289.

Jellison, W.L. 1982. Concentrations of mutillid wasps (Hymenoptera: Mutillidae). Entomol. News 93: 27–28.

Jennings, D.T., and M.W. Houseweart. 1984. Predation by eumenid wasps (Hymenoptera: Eumenidae) on spruce budworm (Lepidoptera: Tortricidae) and other lepidopterous larvae in spruce-fir forests of Maine. Ann. Entomol. Soc. Amer. 77: 39–45.

Jenks, G.E. 1938. Marvels of metamorphosis: a scientific "G-man" pursues rare trapdoor spider parasites for three years with a spade and a candid camera. Natl. Geogr. 74: 807–828.

Jervis, M.A. 1979. Courtship, mating and "swarming" in *Aphelopus melaleucus* (Dalman) (Hymenoptera: Dryinidae). Entomol. Gaz. 30: 191–193.

Jervis, M.A. 1980a. Ecological studies on the parasite complex associated with typhlocybine leafhoppers (Homoptera, Cicadellidae). Ecol. Entomol. 5: 123–136.

Jervis, M.A. 1980b. Life history studies on *Aphelopus* species (Hymenoptera, Dryinidae) and *Chalarus* species (Diptera, Pipunculidae), primary parasites of typhlocybine leafhoppers (Homoptera, Cicadellidae). J. Nat. Hist. 14: 769–780.

Jervis, M.A., and N.A.C. Kidd. 1991. The dynamic significance of host-feeding by insect parasitoids—what modellers ought to consider. Oikos 62: 97–99.

Jervis, M.A., N.A.C. Kidd, and A. Sahragard. 1987. Host-feeding in Dryinidae: its adaptive significance and its consequences for parasitoid-host population dynamics. Proc. 6th Auchenorrhyncha Meeting, Turin, Italy, pp. 591–596.

Jervis, M.A., B.A. Hawkins, and N.A.C. Kidd. 1996. The usefulness of destructive host feeding parasitoids in classical biological control: theory and observation conflict. Ecol. Entomol. 21: 41–46.

Jones, D.L., and B. Gray. 1974. The pollination of *Calochilus holtzei* F. Muell. Amer. Orchid. Soc. Bull. 43: 604–606.

Kapadia, M.N., and V.P. Mittal. 1986. Biology of *Parasierola nephantidis* Muesbeck and its importance in the control of *Opsinia arensolla* Walker under Mahuva (Gujarat State) conditions. GAU Res. J. 12: 29–34.

Karsai, I. 1989. Factors affecting diurnal activities of solitary wasps (Hymenoptera: Sphecidae and Pompilidae). Entomol. Gener. 14: 223–232.

Kaston, B.J. 1959. Notes on pompilid wasps that do not dig burrows to bury their spider prey. Bull. Brook. Entomol. Soc. 54: 103–113.

Kearns, C.W. 1934. A hymenopterous parasite (*Cephalonomia gallicola* Ashm.) new to the cigarette beetle (*Lasioderma serricorne* Fab.). J. Econ. Entomol. 27: 801–806.

Kerr, W.E. 1962. Genetics of sex determination. Ann. Rev. Entomol. 7: 157–176.

Kidd, N.A.C., and M.A. Jervis. 1989. The effects of host-feeding behaviour on the dynamics of parasitoid-host interactions, and the implications for biological control. Res. Popul. Ecol. 31: 235–274.

Kimsey, L.S. 1978. Nesting and mating behavior in *Dynatus nigripes spinolae* (Lepeletier) (Hymenoptera: Sphecidae). Pan-Pac. Entomol. 54: 65–68.

Kimsey, L.S. 1980. Notes on the biology of some Panamanian Pompilidae with a description of a communal nest (Hymenoptera). Pan-Pac. Entomol. 56: 98–100.

Kimsey, L.S. 1991. Relationships among the tiphiid wasp subfamilies (Hymenoptera). Syst. Entomol. 16: 427–438.

Kimsey, L.S., and R.M. Bohart. 1990. The Chrysidid Wasps of the World. New York: Oxford University Press.

Kimsey, L.S., R.B. Kimsey, and C.A. Toft. 1981. Life history of *Bembix inyoensis* in Death Valley (Hymenoptera: Sphecidae). J. Kans. Entomol. Soc. 54: 665–672.

King, J.L., and J.K. Holloway. 1930. *Tiphia popilliavora* Rohwer, a parasite of the Japanese beetle. USDA Circular 145: 1–11.

Kishitani, Y. 1961. Observations on the egg-laying habit of *Goniozus japonicus* Ashmead (Hymenoptera, Bethylidae). Kontyû 29: 175–179.

Kislow, C.J., and R.W. Matthews. 1977. Nesting behavior of *Rhopalum atlanticum* Bohart (Hymenoptera: Sphecidae: Crabroninae). J. Ga. Entomol. Soc. 12: 85–89.

Klein, J.A., and N.E. Beckage. 1990. Comparative suitability of *Trogoderma variabile* and *T. glabrum* (Coleoptera: Dermestidae) as hosts for the ectoparasite *Laelius pedatus* (Hymenoptera: Bethylidae). Ann. Entomol. Soc. Amer. 83: 809–816.

Klein, J.A., D.K. Ballard, K.S. Lieber, W.E. Burkholder, and N.E. Beckage. 1991. Host developmental stage and size as factors affecting parasitization of *Trogoderma variabile* (Coleoptera: Dermestidae) by *Laelius pedatus* (Hymenoptera: Bethylidae). Ann. Entomol. Soc. Amer. 84: 72–78.

Knisley, C.B. 1987. Habitats, food resources and natural enemies of a community of larval *Cicindela* in southeastern Arizona (Coleoptera: Cicindelidae). Can. J. Zool. 65: 1191–1200.

Knisley, C.B., D.L. Reeves, and T.S. Gregory. 1989. Behavior and development of the wasp *Pterombrus rufiventris hyalinatus* Krombein (Hymenoptera: Tiphiidae), a parasite of larval tiger beetles (Coleoptera: Cicindelidae). Proc. Entomol. Soc. Wash. 9: 179–184.

Kornhauser, S.I. 1919. The sexual characteristics of the membracid, *Thelia bimaculata* (Fabr.). J. Morphol. 32: 531–636.

Krombein, K.V. 1938. Notes on the biology of *Pseudomethoca frigida* (Smith) (Hymenoptera: Mutillidae). Bull. Brooklyn Entomol. Soc. 33: 14–15.

Krombein, K.V. 1948. Liberation of oriental scoliid wasps in the United States from 1920 to 1946 (Hymenoptera: Scoliidae, Tiphiidae). Ann. Entomol. Soc. Amer. 41: 58–62.

Krombein, K.V. 1955. Miscellaneous prey records of solitary wasps I. (Hymenoptera: Aculeata). Bull. Brooklyn Entomol. Soc. 50: 13–17.

Krombein, K.V. 1958. Miscellaneous prey records of solitary wasps. III. Proc. Biol. Soc. Wash. 71: 21–26.

Krombein, K.V. 1960. Additions to the Amesiginae and Adelphinae (Hymenoptera, Chrysididae). Trans. Amer. Entomol. Soc. 86: 27–39.

Krombein, K.V. 1961. Some symbiotic relations between saproglyphid mites and solitary vespid wasps. J. Wash. Acad. Sci. 51: 89–93.

Krombein, K.V. 1964. Natural history of Plummers Island, Maryland. XVIII. The hibiscus wasp, an abundant rarity, and its associates (Hymenoptera: Sphecidae). Proc. Biol. Soc. Wash. 77: 73–112.

Krombein, K.V. 1967. Trap-Nesting Wasps and Bees: Life Histories, Nests, and Associates. Washington, D.C.: Smithsonian Press.

Krombein, K.V. 1968. Studies in the Tiphiidae, X: *Hylomesa*, a new genus of myzinine wasp parasitic on larvae of longicorn beetles. Proc. U.S. Natl. Mus. Smithson. Inst. 124: 1–22.

Krombein, K.V. 1978. Biosystematic studies of Ceylonese wasps, III. Life history, nest and associates of *Paraleptomenes mephitus* (Cameron) (Hymenoptera: Eumenidae). J. Kans. Entomol. Soc. 51: 721–734.

Krombein, K.V. 1981. Biosystematic studies of Ceylonese wasps, VIII: a monograph of the Philanthidae (Hymenoptera: Sphecoidea). Smithson. Contrib. Zool. 343: 1–75.

Krombein, K.V. 1983. Biosystematic studies of Ceylonese wasps, XI: a monograph of the Amiseginae and Loboscelidiinae (Hymenoptera: Chrysididae). Smithson. Contrib. Zool. 376: 1–79.

Krombein, K.V. 1984. Biosystematic studies of Ceylonese wasps, XIV: a revision of *Carinostigmus* Tsuneki (Hymenoptera: Sphecoidea: Pemphredonidae). Smithson. Contrib. Zool. 396: 1–37.

Krombein, K.V. 1992. Host relationships, ethology and systematics of *Pseudomethoca* Ashmead (Hymenoptera: Mutillidae, Andrenidae, Halictidae and Anthophoridae). Proc. Entomol. Soc. Wash. 94: 91–102.

Krombein, K.V., and F.E. Kurczewski. 1963. Biological notes on three Floridian wasps. Proc. Entomol. Soc. Wash. 76: 139–152.

Krombein, K.V., and B. Norden. 1996. Behavior of nesting *Episyron conterminus posterus* (Fox) and its cleptoparasite *Ephuta slossonae* (Fox) (Hymenoptera: Pompilidae, Mutillidae). Proc. Entomol. Soc. Wash. 98: 188–194.

Krombein, K.V., and B. Norden. 1997. Nesting behavior of *Krombeinictus nordenae* Leclerq, a sphecid wasp with vegetarian larvae (Hymenoptera: Sphecidae: Crabroninae). Proc. Entomol. Soc. Wash. 99: 42–49.

Krombein, K.V., P.D. Hurd, D.R. Smith, and B.D. Burks. 1979a. Catalog of Hymenoptera in America North of Mexico, vol. 2, Apocrita (Aculeata). Washington, D.C.: Smithsonian Institution Press.

Krombein, K.V., P.D. Hurd, D.R. Smith, and B.D. Burks. 1979b. Catalog of Hymenoptera in America North of Mexico, Vol. 3, Indexes. Washington, D.C.: Smithsonian Institution Press.

Kullenberg, B. 1956. Field experiments with chemical sexual attractants on aculeate Hymenoptera males. Zool. Bidrag Uppsala 31: 253–354.

Kullenberg, B. 1973. New observations on the pollination of *Ophrys* L. (Orchidaceae). Zoon, Suppl. 1: 9–14.

Kurczewski, F.E. 1963a. Biological notes on *Campsomeris plumipes confluenta* (Say) (Hymenoptera: Scoliidae). Entomol. News 74: 21–24.

Kurczewski, F.E. 1963b. A first Florida record and note on the nesting of *Trypoxylon (Trypargilum) texense* Saussure (Hymenoptera: Sphecidae). Fla. Entomol. 46: 243–245.

Kurczewski, F.E. 1966a. Behavioral notes on two species of *Tachytes* that hunt pygmy mole-crickets (Hymenoptera: Sphecidae, Larrinae). J. Kans. Entomol. Soc. 39: 147–155.

Kurczewski, F.E. 1966b. A host record for the scoliid wasp *Campsomeris plumipes confluenta.* J. Kans. Entomol. Soc. 39: 156.

Kurczewski, F.E. 1966c. *Tachysphex terminatus* preying on Tettigoniidae—an unusual record (Hymenoptera: Sphecidae, Larrinae). J. Kans. Entomol. Soc. 39: 317–322.

Kurczewski, F.E. 1967a. *Campsomeris plumipes confluenta* (Scoliidae) preying on the gold beetle, *Cotalpa lanigera* (Scarabaeidae), in Kansas. J. Kans. Entomol. Soc. 40: 208–209.

Kurczewski, F.E. 1967b. *Hedychridium fletcheri* (Hymenoptera: Chrysididae, Elampinae), a probable parasite of *Tachysphex similis* (Hymenoptera: Sphecidae, Larrinae). J. Kans. Entomol. Soc. 40: 278–284.

Kurczewski, F.E. 1969. Comparative ethology of female digger wasps in the genera *Miscophus* and *Nitelopterus* (Hymenoptera: Sphecidae, Larrinae). J. Kans. Entomol. Soc. 42: 470–509.

Kurczewski, F.E. 1972. Observations on the nesting behavior of *Diploplectron peglowi* Krombein. Proc. Entomol. Soc. Wash. 74: 385–397.

Kurczewski, F.E. 1987a. Nesting behavior of *Tachysphex laevifrons* and *T. crassiformis*, with a note on *T. krombeini* (Hymenoptera: Sphecidae). Proc. Entomol. Soc. Wash. 89: 715–730.

Kurczewski, F.E. 1987b. A review of nesting behavior in *Tachysphex pompiliformis* group, with observations on five species (Hymenoptera: Sphecidae). J. Kans. Entomol. Soc. 60: 118–126.

Kurczewski, F.E. 1998. Territoriality and mating behavior of *Sphex pennsylvanicus* L. (Hymenoptera: Sphecidae). J. Hym. Res. 7: 74–83.

Kurczewski, F.E., and R.E. Acciavatti. 1968. A review of the nesting behaviors of the nearctic species of *Crabro*, including observations on *C. advenus* and *C. latipes* (Hymenoptera: Sphecidae). J. N.Y. Entomol. Soc. 76: 196–212.

Kurczewski, F.E., and N.B. Elliot. 1978. Nesting behavior and ecology of *Tachysphex pechumani* Krombein (Hymenoptera: Sphecidae). J. Kans. Entomol. Soc. 54: 765–780.

Kurczewski, F.E., and H.E. Evans. 1972. Nesting behavior and description of the larva of *Bothynostethus distinctus* Fox. Psyche 79: 88–103.

Kurczewski, F.E., and S.E. Ginsberg. 1971. Nesting behavior and description of *Tachysphex (Tachyplena) validus*. J. Kans. Entomol. Soc. 44: 113–131.

Kurczewski, F.E., and E.J. Kurczewski. 1968a. Host records for some North American Pompilidae (Hymenoptera) with a discussion of factors in prey selection. J. Kans. Entomol. Soc. 41: 1–33.

Kurczewski, F.E., and E.J. Kurczewski. 1968b. Host records for some North American Pompilidae (Hymenoptera): first supplement. J. Kans. Entomol. Soc. 41: 367–382.

Kurczewski, F.E., and E.J. Kurczewski. 1972. Host records for some North American Pompilidae, second supplement. Tribe Pepsini. J. Kans. Entomol. Soc. 45: 181–193.

Kurczewski, F.E., and E.J. Kurczewski. 1973. Host records for some North American Pompilidae, third supplement. Tribe Pompilini. J. Kans. Entomol. Soc. 46: 65–81.

Kurczewski, F.E., and E.J. Kurczewski. 1984. Mating and nesting behavior of *Tachytes intermedius* (Viereck) (Hymenoptera: Sphecidae). Proc. Entomol. Soc. Wash. 86: 176–184.

Kurczewski, F.E., and C.J. Lane. 1974. Observations on the nesting behavior of *Mimesa (Mimesa) basirufa* Packard and *M. (M.) cressoni* Packard (Hymenoptera: Sphecidae). Proc. Entomol. Soc. Amer. 76: 375–384.

Kurczewski, F.E., and R.C. Miller. 1984. Observations on the nesting of three species of *Cerceris* (Hymenoptera: Sphecidae). Fla. Entomol. 67: 146–155.

Kurczewski, F.E., and R.C. Miller. 1986. Observations on some nests of *Crossocerus (Blepharipus) A. annulipes* (Lepeletier and Brullé) (Hymenoptera: Sphecidae). Proc. Entomol. Soc. Wash. 88: 157–162.

Kurczewski, F.E., and D.J. Peckham. 1970. Nesting behavior of *Anacrabro ocellatus ocellatus* (Hymenoptera: Sphecidae). Ann. Entomol. Soc. Amer. 63: 1419–1424.

Kurczewski, F.E., and D.J. Peckham. 1982. Nesting behavior of *Lyroda subita* (Say) (Hymenoptera: Sphecidae). Proc. Entomol. Soc. Wash. 84: 149–156.

Kurczewski, F.E., and M.G. Spofford. 1986. Observations on the nesting behaviors of *Tachytes parvus* Fox and *T. obductus* Fox (Hymenoptera: Sphecidae). Proc. Entomol. Soc. Wash. 88: 13–24.

Kurczewski, F.E., and D.L. Wochaldo. 1998. Relationship of cell depth and soil moisture in *Oxybelus bipunctatus* (Hymenoptera: Sphecidae). Entomol. News 109: 7–14.

Kurczewski, F.E., N.A. Burdick, and G.C. Gaumer. 1969. Observations on the nesting behavior of *Crossocerus (C.) maculiclypeus* (Fox) (Hymenoptera: Sphecidae). J. N.Y. Entomol. Soc. 77: 92–104.

Kurczewski, F.E., E.J. Kurczewski, and M.G. Spofford. 1988. Nesting behavior of *Aporinellus wheeleri* Bequaert and *A. taeniolatus* (Dalla Torre) (Hymenoptera: Pompilidae). Proc. Entomol. Soc. Wash. 90: 294–306.

Lai, P.-Y. 1988. Biological control: a positive point of view. Proc. Haw. Entomol. Socl 28: 179–190.

Laing, D.J. 1979. Studies on populations of the tunnel web spider *Porrhotthele antipodiana* (Mygalomorphae: Dipluridae). Tuatara 24: 1–21.

Lamborn, W.A. 1915. Second report on *Glossinia* investigations in Nyasaland. Bull. Entomol. Res. 6: 249–265.

Lamborn, W.A. 1916. Third report on *Glossinia* investigations in Nyasaland. Bull. Entomol. Res. 7: 29–50.

Landes, D.A., M.S. Obin, A.B. Cady, and J.H. Hunt. 1987. Seasonal and latitudinal variation in spider prey of the mud dauber *Chalybion californicum* (Hymenoptera: Sphecidae). J. Arachnol. 15: 249–256.

Lane, R.S., J.R. Anderson, and E. Rogers. 1986. Nest provisioning and related activities of the sand wasp, *Bembix americana comata* (Hymenoptera: Sphecidae). Pan-Pac. Entomol. 62: 258–268.

LaRivers, I. 1945. The wasp *Chorion laeviventris* as a natural population control of the mormon cricket. Amer. Midl. Nat. 33: 743–763.

Larsen, O.N., G. Gleffe, and J. Tengö. 1986. Vibration and sound communication in solitary bees and wasps. Physiol. Entomol. 11: 287–296.

Larsson, F.K. 1986. Increased nest density of the digger wasp *Bembix rostrata* as a response to parasites and predators (Hymenoptera: Sphecidae). Entomol. Gener. 12: 71–75.

Larsson, F.K. 1989a. Female body size relationships with fecundity and egg size in two solitary species of fossorial Hymenoptera (Colletidae and Sphecidae). Entomol. Gener. 15: 167–171.

Larsson, F.K. 1989b. Temperature-induced alternative male mating tactics in a tropical digger wasp. J. Ins. Behav. 2: 849–852.

Larsson, F.K. 1990. Thermoregulation and activity patterns of the sand wasp *Steniolia longirostra* (Say) (Hymenoptera: Sphecidae) in Costa Rica. Biotropica 22: 65–68.

Larsson, F.K. 1991. Some take it cool, some like it hot—a comparative study of male mate searching tactics in two species of Hymenoptera (Colletidae and Sphecidae). J. Therm. Biol. 16: 45–51.

Larsson, F.K., and J. Tengö. 1989. It is not always good to be large: some female fitness components in a temperate digger wasp, *Bembix rostrata* (Hymenoptera: Sphecidae). J. Kans. Entomol. Soc. 62: 490–495.

Lashomb, J.H., and A.L. Steinhauer. 1975. Observations of *Zethus spinipes* Say (Hymenoptera: Eumenidae). Proc. Entomol. Soc. Wash. 77: 164.

Lavigne, R.J., and F.R. Holland. 1969. Comparative behavior of eleven species of Wyoming robber flies (Diptera: Asilidae). Sci. Mono. 18: 1–61.

Legaspi, B.A.C., B.M. Shepard, and L.P. Almazan. 1987. Oviposition behavior and development of *Goniozus triangulifer* Kieffer (Hymenoptera: Bethylidae). Environ. Entomol. 16: 1283–1286.

Legner, E.F. 1983. Patterns of field diapause in the navel orangeworm (Lepidoptera: Phycitidae) and three imported parasites. Ann. Entomol. Soc. Amer. 76: 503–506.

Legner, E.F., and G. Gordh. 1992. Lower navel orangeworm (Lepidoptera: Phycitidae) population densities following establishment of *Goniozus legneri* (Hymenoptera: Bethylidae) in California. J. Econ. Entomol. 85: 2153–2160.

Legner, E.F., and A. Silveira-Guido. 1983. Establishment of *Goniozus emigratus* and *Goniozus legneri* (Hym: Bethylidae) on navel orangeworm, *Amyelois transitella* (Lep: Phycitidae) in California and biological control potential. Entomophaga 28: 97–106.

Legner, E.F., and R.W. Warkentin. 1988. Parasitization of *Goniozus legneri* (Hymenoptera: Bethylidae) at increasing parasite and host, *Amyelois transitella* (Lepidoptera: Phycitidae), densities. Ann. Entomol. Soc. Amer. 81: 774–776.

Lin, N. 1963. Territorial behavior in the cicada killer wasp *Sphecius speciosus* (Hymenoptera: Sphecidae). Behaviour 20: 115–133.

Lin, N. 1966. Copulatory behavior of the cicada killer wasp, *Sphecius speciosus*. Anim. Behav. 14: 130–131.

Lin, N. 1967. Role differentiation in copulating cicada killer wasps. Science 157: 1334–1335.

Lin, N. 1978a. Defended hunting territories and hunting behavior of females of *Philanthus gibbosus* (Hymenoptera: Sphecidae). Proc. Entomol. Soc. Wash. 80: 234–239.

Lin, N. 1978b. Sequential hypermaxalation in the digger wasp *Diodontus franclemonti* Krombein (Hymenoptera: Sphecidae). J. Kans. Entomol. Soc. 51: 235–238.

Lin, N. 1979a. Differential prey selection for the sex of offspring in the cicada killer *Sphecius speciosus* (Hymenoptera: Sphecidae). Proc. Entomol. Soc. Wash. 81: 269–275.

Lin, N. 1979b. The weight of cicada killer wasps, *Sphecius speciosus*, and the weight of their prey. J. Wash. Acad. Sci. 69: 159–163.

Lin, N., and C.D. Michener. 1972. Evolution of sociality in insects. Q. Rev. Biol. 47: 131–159.

Linsley, E.G. 1962. Sleeping aggregations of solitary Hymenoptera—II. Ann. Entomol. Soc. Amer. 55: 148–164.

Linsley, E.G., and J.W. MacSwain. 1942. The parasites, predators, and inquiline associates of *Anthophora linsleyi*. Amer. Midl. Nat. 27: 402–417.

Linsley, E.G., and J.W. MacSwain. 1954. Observations on the habits and prey of *Eucerceris ruficeps* Scullen (Hymenoptera: Sphecidae). Pan-Pac. Entomol. 30: 11–14.

Linsley, E.G., and J.W. MacSwain. 1956. Some observations on the nesting habits and prey of *Cerceris californica* Cresson (Hymenoptera: Sphecidae). Ann. Entomol. Soc. Amer. 49: 71–84.

Linsley, E.G., J.W. MacSwain, and R.F. Smith. 1955. Observations on the mating habits of *Dasymutilla formicalis* (Hymenoptera: Mutillidae). Can. Entomol. 87: 411–413.

Lomholdt, O. 1982. On the origin of the bees (Hymenoptera: Apidae, Sphecidae). Entomol. Scand. 13: 185–190.

Longair, R.W. 1981. Sex-ratio variations in xylophilous Hymenoptera. Evolution 35: 597–600.

Longair, R.W. 1984. Male mating behavior in solitary wasps (Hymenoptera: Vespidae). Ph.D. dissertation, Colorado State University.

Longair, R.W. 1985a. Male behavior of *Euparagia richardsii* Bohart (Hymenoptera: Vespidae). Pan-Pac. Entomol. 61: 318–320.

Longair, R.W. 1985b. Mating behavior at floral resources in two species of *Pseudomasaris* (Hymenoptera: Vespidae). Proc. Entomol. Soc. Wash. 89: 759–769.

Longair, R.W. 1987. Mating behavior at floral resources in two species of *Pseudomasaris* (Hymenoptera: Vespidae: Masarinae). Proc. Entomol. Soc. Wash. 89: 759–769.

Longair, R.W., J.H. Cane, and L. Elliott. 1987. Male competition and mating behavior within mating aggregations of *Glenostictia satan* Gillaspy (Hymenoptera: Sphecidae). J. Kans. Entomol. Soc. 60: 264–272.

Low, B.S., and W.T. Wcislo. 1992. Male foretibial plates and mating in *Crabro cribrellifer* (Packard) (Hymenoptera: Sphecidae), with a survey of expanded male forelegs in Apoidea. Ann. Entomol. Soc. Amer. 85: 219–223.

Ma, M., W.E. Burkholder, and S.D. Carlson. 1978. Supra-anal organ: A defensive mechanism of the furniture carpet beetle, *Anthrenus flavipes* (Coleoptera: Dermestidae). Ann. Entomol. Soc. Amer. 71: 718–723.

Malyshev, S.I. 1968. Genesis of the Hymenoptera and the Phases of Their Evolution. London: Methuen.

Manley, D.G. 1975. Notes on the courtship and mating of *Dasymutilla* Ashmead (Hymenoptera: Mutillidae) in California. Southwest. Nat. 21: 552–554.

Manley, D.G., and S. Taber III. 1978. A mating aggregations of *Dasymutilla foxi* in southern Arizona. Pan-Pac. Entomol. 231–235.

Markin, G.P., and A.R. Gittins. 1967. Biology of *Stenodynerus claremontensis* (Cameron) (Hymenoptera: Vespidae). Univ. Idaho Coll. Agric. Res. Bull. 74: 1–25.

Marston, N. 1964. The biology of *Anthrax limatulus fur* (Osten Sacken), with a key to and description of pupae of some species in the *Anthrax albofasciatus* and *trimaculatus* groups (Diptera: Bombyliidae). J. Kans. Entomol. Soc. 37: 89–105.

Martin, R.F. 1971. The cañon wren (*Catherpes mexicanus*) raiding food storage of a trypoxylid wasp. Auk 88: 677.

Martins, E.P. 1996. Phylogenies and the Comparative Methods in Animal Behavior. New York: Oxford University Press.

Martins, R.P. 1991. Nesting behavior and prey of *Poecilopompilus algidus fervidus* and *Tachypompilus xanthopterus* (Hymenoptera: Pompilidae). J. Kans. Entomol. Soc. 64: 231–236.

Masters, W.M. 1979. Insect defensive stridulation: its defensive role. Behav. Ecol. Sociobiol. 5: 187–200.

Matthes-Sears, W., and J. Alcock. 1985. An experimental study of the attractiveness of artificial perch territories to male tarantula-hawk wasps, *Hemipepsis ustulata* (Hymenoptera: Pompilidae). Psyche 92: 255–263.

Matthews, R.W. 1968. Nesting biology of the social wasp *Microstigmus comes* (Hymenoptera: Sphecidae, Pemphredonidae). Psyche 75: 23–45.

Matthews, R.W. 1970. A new thrips-hunting *Microstigmus* from Costa Rica (Hymenoptera: Sphecidae, Pemphredonidae). Psyche 77: 120–126.

Matthews, R.W. 1997. Unusual sex allocation in a solitary parasitoid wasp, *Sphaeropthalma pensylvanica* (Hymenoptera: Mutillidae). Great Lakes Entomol. 30: 51–54.

Matthews, R.W. 1991. Evolution of social behavior in the sphecid wasps. Pages 570–602 in K.G. Ross and R.W. Matthews (eds.), The Social Biology of Wasps. Ithaca, N.Y.: Cornell University Press.

Matthews, R.W., and H.E. Evans. 1970. Biological notes on two species of *Sericophorus* from Australia (Hymenoptera: Sphecidae). Psyche 77: 413–429.

Matthews, R.W., A. Hook, and J.W. Krispyn. 1979. Nesting behavior of *Crabro argusinus* and *C. hilaris* (Hymenoptera: Sphecidae). Psyche 86: 149–166.

Mauss, V. 1996. Contribution to the bionomics of *Ceramius tuberculifer* Saussure (Hymenoptera, Vespidae, Masarinae). J. Hym. Res. 5: 22–37.

Mayhew, P.J. 1997. Fitness consequences of ovicide in a parasitoid wasp. Entomol. Exp. Appl. 84: 115–126.

Mayhew, P.J. 1998. Offspring size-number strategy in the bethylid parasitoid *Laelius pedatus*. Behav. Ecol. 9: 54–59.

Mayhew, P.J., and H.C.J. Godfray. 1997. Mixed sex allocation strategies in a parasitoid wasp. Oecologia 110: 218–221.

Mayhew, P.J., and I.C.W. Hardy. 1998. Non-siblicidal behavior and the evolution of clutch size in bethylid wasps. Amer. Nat. 151: 409–424.

McColloch, J.W., W.P. Hayes, and H.R. Bryson. 1928. Hibernation of certain scarabaeids and their *Tiphia* parasites. Ecology 9: 34–42.

McCorquodale, D.B. 1986. Digger wasp (Hymenoptera: Sphecidae) provisioning flights as a defence against a nest parasite, *Senotainia trilineata* (Diptera: Sarcophagidae). Can. J. Zool. 64: 1620–1627.

McCorquodale, D.B. 1988a. Prey carriage on the sting by *Sericophorus relucens* (Hymenoptera: Sphecidae: Larrinae). J. N.Y. Entomol. Soc. 96: 121–122.

McCorquodale, D.B. 1988b. Relatedness among nestmates in a primitively social wasp, *Cerceris antipodes* (Hymenoptera: Sphecidae). Behav. Ecol. Sociobiol. 23: 401–406.

McCorquodale, D.B. 1989a. Nest defense in single- and multi-female nests of *Cerceris antipodes* (Hymenoptera: Sphecidae). J. Ins. Behav. 2: 267–275.

McCorquodale, D.B. 1989b. Nest sharing, nest switching, longevity and overlap of generations in *Cerceris antipodes* (Hymenoptera: Sphecidae). Insectes Soc. 36: 42–50.

McCorquodale, D.B. 1989c. Soil softness, nest initiation, and nest sharing, in the wasp *Cerceris antipodes* (Hymenoptera: Sphecidae). Ecol. Entomol. 14: 196.

McCorquodale, D.B., and I.D. Naumann. 1988. A new Australian species of communal ground nesting wasp, in the genus *Spilomena* Shuckard (Hymenoptera: Sphecidae: Pemphredoninae). J. Aust. Entomol. Soc. 27: 221–231.

McDaniel, C.A., R.W. Howard, K.M. O'Neill, and J.O. Schmidt. 1987. The chemistry of the male mandibular gland secretions of *Philanthus basilaris* Cresson and *Philanthus bicinctus* (Mickel) (Hymenoptera: Sphecidae). J. Chem. Ecol. 113: 227–235.

McDaniel, C.A., J.O. Schmidt, and R.W. Howard. 1992. Mandibular gland secretions of the male beewolves *Philanthus crabroniformis*, *P. barbatus*, and *P. pulcher* (Hymenoptera: Sphecidae). J. Chem. Ecol. 18: 27–37.

McIntosh, M. 1996. Nest-substrate preferences of the twig-nesters *Ceratina acantha*, *Ceratina nanula* (Apidae) and *Pemphredon lethifer* (Sphecidae). J. Kans. Entomol. Soc. 69 (suppl.): 216–231.

McQueen, D.J. 1979. Interactions between the pompilid wasp *Anoplius relativus* (Fox) and the burrowing wolf spider *Geolycosa domifex* (Hancock). Canad. J. Zool. 57: 542–550.

Meddis, R. 1987. Sleep. Pages 512–517 in D. McFarland (ed.), The Oxford Companion to Animal Behavior. Oxford: Oxford University Press.

Medler, J.T. 1964. Biology of *Rygchium foraminatum* in trap-nests in Wisconsin (Hymenoptera: Vespidae). Ann. Entomol. Soc. Amer. 57: 56–60.

Medler, J.T. 1965. Biology of *Isodontia (Murrayella) mexicana* in trap-nests in Wisconsin (Hymenoptera: Sphecidae). Ann. Entomol. Soc. Amer. 58: 137–142.

Medler, J.T. 1967. Biology of *Trypoxylon* in trap nests in Wisconsin. Amer. Midl. Nat. 78: 344–358.

Melander, A.L., and C.T. Brues. 1903. Guests and parasites of the burrowing bee *Halictus*. Biol. Bull. 5: 4–7, 24–25.

Mellor, J.E.M. 1927. A note on the mutillid *Ephutoma continua* Fabr. and on *Bembix mediterranea* Hdl. in Egypt. Bull. Soc. Roy. d'Egypte 20: 69–79.

Mertins, J.W. 1980. Life history and behavior of *Laelius pedatus*, a gregarious bethylid ectoparasitoid of *Anthrenus verbasci*. Ann. Entomol. Soc. Amer. 73: 686–693.

Mertins, J.W. 1982. Occurrence of *Anthrenus fuscus* Olivier (Coleoptera: Dermestidae) in Iowa. Entomol. News 93: 139–142.

Mertins, J.W. 1985. *Laelius utilis* (Hym.: Bethylidae), a parasitoid of *Anthrenus fuscus* (Col.: Dermestidae) in Iowa. Entomophaga 30: 65–68.

Michener, C.D. 1971. Notes on crabronine wasp nests. J. Kans. Entomol. Soc. 44: 405–407.

Michener, C.D. 1974. The Social Behavior of Bees. Cambridge, Mass.: Harvard University Press.

Michener, C.D. 1979. Biogeography of the bees. Ann. Mo. Bot. Gardens 66: 277–347.

Mickel, C.E. 1928. Biological and taxonomic investigations on the mutillid wasps. Bull. U.S. Natl. Mus. 143: 1–351.

Mickel, C.E. 1938. Photopsid mutillids collected by Dr. K.A. Salman at Eagle Lake, California. Pan-Pac. Entomol. 14: 178–185.

Miller, R.C., and F.E. Kurczewski. 1972. A review of nesting behavior in the genus *Entomognathus* with notes on *E. memorialis* Banks. Psyche 79: 61–78.

Miller, R.C., and F.E. Kurczewski. 1975. Comparative behavior of wasps in the genus *Lindenius* (Hymenoptera: Sphecidae, Crabroninae). J. N.Y. Entomol. Soc. 83: 82–120.

Miller, R.C., and F.E. Kurczewski. 1976. Comparative nesting behaviors of *Crabro rufibasis* and *Crabro arcadiensis* (Hymenoptera: Sphecidae, Crabroninae). Fla. Entomol. 59: 267–286.

Milliron, H.E. 1950. The identity of a cleptid egg parasite of the common walking stick, *Diapheromera femorata* Say (Hymenoptera: Cleptidae). Proc. Entomol. Soc. Wash. 52: 47.

Miotk, V.P. 1979. Biology and ecology of *Odynerus spinipes* (L.) and *O. reniformis* (Gmel.) inhabiting loess-waals in the Kaiserstuhl-area (Hymenoptera: Eumenidae). Zool. Jb. Syst. 106: 374–405.

Móczár, L. 1961. On the habits of *Stilbum cyanurum cyanurum* Forst. (Hymenoptera, Chrysididae). Ann. Hist. Mus. Natl. Hung. 53: 463–466.

Móczár, L., M. Andó, and L. Gallé. 1973. Microclimate and the activity of *Paragymnomerus spiricornis* (Spinola) (Hymenoptera: Eumenidae). Acta Biol. Szeged. 19: 147–160.

Molumby, A. 1995. Dynamics of parasitism in the organ-pipe wasp, *Trypoxylon politum*: effects of spatial scale on parasitoid functional response. Ecol. Entomol. 20: 159–168.

Molumby, A. 1997. Why make daughters larger? Maternal sex allocation and sex-dependent selection for body size in a mass-provisioning wasp, *Trypoxylon politum*. Behav. Ecol. 8: 279–287.

Morgan, D.J.W., and J.M. Cook. 1994. Extremely precise sex ratios in small clutches of a bethylid wasp. Oikos 71: 423–430.

Moya-Raygoza, G., and J. Trujillo-Arriaga. 1993. Evolutionary relationships between *Dalbulus* leafhopper (Homoptera: Cicadellidae) and its dryinid (Hymenoptera: Dryinidae) parasitoids. J. Kans. Entomol. Soc. 66: 41–50.

Mueller, U.G., A.F. Warneke, T.U. Grafe, and P.R. Ode. 1992. Female size and nest defense in the digger wasp *Cerceris fumipennis* (Hymenoptera: Sphecidae: Philanthinae). J. Kans. Entomol. Soc. 65: 44–92.

Müller, A. 1996. Convergent evolution of morphological specializations in central European bee and honey wasp species as an adaptation to the uptake of pollen from nototribic flowers (Hymenoptera, Apoidea and Masaridae). Biol. J. Linn. Soc. 57: 235–252.

Muma, M.H., and W.F. Jeffers. 1945. Studies of the spider prey of several mud-dauber wasps. Ann. Entomol. Soc. Amer. 38: 245–255.

Myers, J.G. 1927. A sarcophagid "parasite" of solitary wasps: *Pachyopthalmus* parasitizing *Ancistrocerus*. Entomol. Mon. Mag. 63: 190–196.

Naumann, I.D. 1987. A new megalyrid (Hymenoptera: Megalyridae) parasitic on a sphecid wasp in Australia. J. Austr. Entomol. Soc. 26: 215–222.

Naumann, I.D., and J.C. Cardale. 1987. Notes on the behavior and nests of an Australian masarid wasp *Paragia (Paragia) decipiens decipiens* Shuckhard (Hymenoptera: Vespoidea, Masaridae). Austr. Entomol. Mag. 13: 59–65.

Nazarova, S., and S.B. Baratov. 1982. The predatory wasps (Hymenoptera: Sphecidae, Vespidae) of Tadzhikistan and their role in regulating the numbers of Tabanidae. Entomol. Rev. 60: 95–101.

Neff, J.L., and B.B. Simpson. 1985. Hooked setae and narrow tubes: foretarsal pollen collection by *Trimeria buyssoni* (Hymenoptera: Masaridae). J. Kans. Entomol. Soc. 58: 730–732.

Nelson, J.W. 1986. Ecological notes on male *Mydas xanthopterus* (Loew) (Diptera: Mydidae) and their interactions with *Hemipepsis ustulata* Dahlbohm (Hymenoptera: Pompilidae). Pan-Pac. Entomol. 62: 316–322.

Nentwig, W. 1985. A mimicry complex between mutillid wasps (Hymenoptera: Mutillidae) and spiders (Araneae). Studies Neotrop. Fauna Environ. 20: 113–116.

Newton, R.C. 1956. Digger wasps, *Tachsphex* spp., as predators of a range grasshopper in Idaho. J. Econ. Entomol. 49: 615–619.

Newcomer, E.J. 1930. Notes on the habits of a digger wasp and its inquiline flies. Ann. Entomol. Soc. Amer. 23: 552–563.

Nielsen, E. 1936. The biology of *Homonotus sanguinolentus* Fabr. Entomol. Meddel. 19: 385–404.

Nilsson, L.A. 1978. Pollination ecology of *Epipactus palustris* (Orchidaceae). Bot. Notiser 131: 355–368.

Nilsson, L.A., L. Jonsson, L. Rason, and E. Randrianjohany. 1986. The pollination of *Cymbidiella flabellata* (Orchidaceae) in Madagaskar: a system operated by sphecid wasps. Nord. J. Bot. 6: 411–422.

Obin, M.S. 1982. Spiders living at wasp nesting sites: what constrains predation by mud-daubers? Psyche 89: 321–335.

O'Brien, M.F. 1982. *Trypargilum tridentatum* (Packard) in trap nests in Oregon (Hymenoptera: Sphecidae: Trypoxyloninae). Pan-Pac. Entomol. 58: 288–290.

O'Brien, M.F., and F.E. Kurczewski. 1982a. Ethology and overwintering of *Podalonia luctuosa* (Hymenoptera: Sphecidae). Great Lakes Entomol. 15: 261–275.

O'Brien, M.F., and F.E. Kurczewski. 1982b. Further observations on the ethology of *Alysson conicus* Provancher (Hymenoptera: Sphecidae). Proc. Entomol. Soc. Wash. 84: 225–231.

O'Brien, M.F., and F.E. Kurczewski. 1982c. Nesting and overwintering behavior of *Liris argentata* (Hymenoptera: Larridae). J. Georgia Entomol. Soc. 17: 60–68.

Ohgushi, R. 1954. On the plasticity of the nesting habit of a hunting wasp, *Pemphredon lethifer fabricii* (Müller). Mem. Coll. Sci. Univ. Kyoto, Series B 21: 45–48.

Olberg, G. 1959. Das Verhalten der Solitären Wespen Mitteleuropas (Vespidae, Pompilidae, Sphecidae). Berlin: Deutscher Verlag Wissenschaften.

Olmi, M. 1984a. Revision of the Dryinidae, p. 1. Mem. Amer. Entomol. Inst. 37: 1–946.

Olmi, M. 1984b. Revision of the Dryinidae, p. 2. Mem. Amer. Entomol. Inst. 37: 947–1913.

O'Neill, K.M. 1979. Territorial behavior in males of *Philanthus psyche* (Hymenoptera: Sphecidae). Psyche 86: 19–43.

O'Neill K.M. 1983a. The significance of body size in territorial interactions of male beewolves (Hymenoptera: Sphecidae, *Philanthus*). Anim. Behav. 31: 404–411.

O'Neill, K.M. 1983b. Territoriality, body size, and spacing in males of the beewolf *Philanthus basilaris* (Hymenoptera: Sphecidae). Behaviour 86: 395–421.

O'Neill, K.M. 1985. Egg size, prey size, and sexual size dimorphism in digger wasps (Hymenoptera: Sphecidae). Can. J. Zool. 63: 2187–2193.

O'Neill, K.M. 1990. Female nesting behavior and male territoriality in *Aphilanthops subfrigidus* Dunning (Hymenoptera: Sphecidae). Pan-Pac. Entomol. 66: 19–23.

O'Neill, K.M. 1992a. Patch-specific foraging by the robber fly *Megaphorus willistoni*. Environ. Entomol. 21: 1333–1340.

O'Neill, K.M. 1992b. Temporal and spatial dynamics of predation in a robber fly population (Diptera: Asilidae, *Efferia staminea*). Can. J. Zool. 70: 1546–1552.

O'Neill, K.M. 1994. The male mating strategy of the ant *Formica subpolita* Mayr (Hymenoptera: Formicidae): swarming, mating, and predation risk. Psyche 101: 93–108.

O'Neill, K.M. 1995. Digger wasps (Hymenoptera: Sphecidae) and robber flies (Diptera: Asilidae) as predators of grasshoppers (Orthoptera: Acrididae) on Montana rangeland. Pan-Pac. Entomol. 71: 248–250.

O'Neill, K.M. 1997. Multispecies mating swarms of *Formica* (Hymenoptera: Formicidae) in southwestern Montana. J. Hym. Res. 6: 336–343.

O'Neill, K.M., and L. Bjostad. 1987. The male mating strategy in the bee *Nomia nevadensis* (Hymenoptera: Halictidae): leg structure and mate-guarding. Pan-Pac. Entomol. 63: 207–217.

O'Neill, K.M., and H.E. Evans. 1981. Predation on conspecific males by females of the beewolf *Philanthus basilaris* Cresson (Hymenoptera: Sphecidae). J. Kans. Entomol. Soc. 54: 553–556.

O'Neill, K.M., and H.E. Evans. 1982. Patterns of prey use in four sympatric species of *Philanthus* (Hymenoptera: Sphecidae) with a review of prey selection in the genus. J. Nat. Hist. 16: 791–801.

O'Neill, K.M., and H.E. Evans. 1983a. Alternative male mating tactics in *Bembecinus quinquespinosus* (Hymenoptera: Sphecidae): correlations with size and color variation. Behav. Ecol. Sociobiol. 14: 39–46.

O'Neill, K.M., and H.E. Evans. 1983b. Body size and alternative mating tactics in the beewolf *Philanthus zebratus* (Hymenoptera: Sphecidae). Biol. J. Linn. Soc. 20: 175–184.

O'Neill, K.M., and H.E. Evans. 1999. Observations on nests and prey of *Podalonia valida* (Hymenoptera: Sphecidae). Proc. Entomol. Soc. Wash. 101: 312–315.

O'Neill, K.M., and R.P. O'Neill. 1988. Thermal stress and microhabitat selection in territorial males of *Philanthus psyche* (Hymenoptera: Sphecidae). J. Therm. Biol. 13: 15–20.

O'Neill, K.M., and C. Seibert. 1996. Foraging behavior of the robber fly *Megaphorus willistoni* (Cole) (Diptera: Asilidae). J. Kans. Entomol. Soc. 69: 317–325.

O'Neill, K.M., and S.W. Skinner. 1990. Ovarian egg size and number in relation to female size in five species of parasitoid wasp. J. Zool. 220: 115–122.

O'Neill, K.M., H.E. Evans, and R.P. O'Neill. 1989. Phenotypic correlates of mating success in the sand wasp *Bembecinus quinquespinosus* (Hymenoptera: Sphecidae). Can. J. Zool. 67: 2557–2568.

O'Neill, K.M., H.E. Evans, and L.J. Bjostad. 1991. Territorial behavior in males of three North American species of bumble bees (Hymenoptera: Apidae, *Bombus*). Can. J. Zool. 69: 604–613.

Paetzel, M.M. 1973. Behavior of the male *Trypoxylon rubrocinctum* (Hymenoptera: Sphecidae). Psyche 86: 19–43.

Palmer, M. 1976. Notes on the biology of *Pterombrus piceus* Krombein (Hymenoptera: Tiphiidae). Proc. Entomol. Soc. Wash. 78: 369–375.

Parker, D.E. 1936. *Chrysis shanghaiensis* Smith, a parasite of the oriental moth. J. Agric. Res. 52: 449–458.

Parker, F.D. 1967. Notes on the nests of three species of *Pseudomasaris* Ashmead (Hymenoptera: Masaridae). Pan-Pac. Entomol. 43: 213–216.

Parker, F.D., and R.M. Bohart. 1966. Host-parasite associations in some twig-nesting Hymenoptera from western North America. Pan-Pac. Entomol. 42: 91–98.

Parker, F.D., and R.M. Bohart. 1968. Host-parasite associations in some twig-nesting Hymenoptera from western North America. Part II. Pan-Pac. Entomol. 44: 1–6.

Parker, F.D., V.J. Tepedino, and D.L. Vincent. 1980. Observations on the provisioning behavior of *Ammophila aberti* Haldeman. Psyche 87: 249–258.

Parker, J.B. 1917. A revision of the bembecine wasps of America north of Mexico. Proc. U.S. Nat. Mus. 52: 1–155.

Parkman, J.P., J.H. Frank, T.J. Walker, and D.J. Schuster. 1996. Classical biological control of *Scapteriscus* spp. (Orthoptera: Gryllotalpidae) in Florida. Environ. Entomol. 25: 1415–1420.

Pasteur, G. 1994. Jean Henri Fabre. Sci. Amer. 268: 74–80.

Paul, A.V.N., B.V. David, and S. Jayraj. 1979. Effect of two hosts on the development and reproduction of the parasite, *Parasierola nephantidis* Mues. (Bethylidae: Hymenoptera). Madras Agric. J. 66: 140.

Peakall, R. 1990. Responses of male *Zaspilothynnus trilobatus* Turner wasps to females and the sexually deceptive orchid it pollinates. Func. Ecol. 4: 159–167.

Peckham, D.J. 1977. Reduction of miltogrammine cleptoparasitism by male *Oxybelus subulatus* (Hymenoptera: Sphecidae). Ann. Entomol. Soc. Amer. 70: 823–828.

Peckham, D.J. 1985. Ethological observations on *Oxybelus* (Hymenoptera: Sphecidae) in southwestern New Mexico. Ann. Entomol. Soc. Amer. 78: 865–872.

Peckham, D.J. 1991. Consequences relating to the inclusion of female sarcophagids (Diptera) as prey by *Oxybelus sparideus* (Hymenoptera: Sphecidae). Ann. Entomol. Soc. Amer. 84: 170–173.

Peckham, D.J., and A.W. Hook. 1980. Behavioral observations on *Oxybelus* in southwestern North America. Ann. Entomol. Soc. Amer. 73: 557–567.

Peckham, D.J., and A.W. Hook. 1994. Nesting behavior of *Enchemicrum australe* (Hymenoptera: Sphecidae). Ann. Entomol. Soc. Amer. 87: 972–977.

Peckham, D.J., and F.E. Kurczewski. 1978. Nesting behavior of *Chlorion aerarium*. Ann. Entomol. Soc. Amer. 71: 758–761.

Peckham, D.J., F.E. Kurczewski, and D.B. Peckham. 1973. Nesting behavior of nearctic species of *Oxybelus* (Hymenoptera: Sphecidae). Ann. Entomol. Soc. Amer. 66: 647–661.

Peckham, G.W., and E.G. Peckham. 1898. On the instincts and habits of the solitary wasps. Bull. Wis. Geol. Nat. Hist. Surv. 2: 1–245.

Pedigo, L. 1996. Entomology and Pest Management, 2d ed. Upper Saddle River, N.J.: Prentice Hall.

Peña, N., and M. Shepard. 1986. Seasonal incidence of parasitism of brown planthoppers, *Nilaparvata lugens* (Homoptera: Delphacidae), green planthoppers, *Nephotettix* spp., *Sogatella furcifera* (Homoptera: Cicadellidae) in Laguna Province, Philippines. Environ. Entomol. 15: 263–267.

Perkins, R.C.L. 1905. Leaf-hoppers and their natural enemies (Pt. I. Dryinidae). Bull. Haw. Sugar Planters' Assoc. 1: 1–69.

Peter, C., and B.V. David. 1991. Biology of *Goniozus sensorius* Gordh (Hymenoptera: Bethylidae) a parasitoid of the pumpkin caterpillar, *Diaphania indica* (Saunders) (Lepidoptera: Pyralidae). Ins. Sci. Applic. 12: 339–345.

Petters, R.M., and R.V. Mettus. 1980. Decreased diploid male viability in the parasitic wasp, *Bracon hebetor*. J. Hered. 71: 353–356.

Pfennig, D.W., and H.K. Reeve. 1989. Neighbor recognition and context-dependent aggression in a solitary wasp, *Sphecius speciosus* (Hymenoptera: Sphecidae). Ethology 80: 1–18.

Pfennig, D.W., and H.K. Reeve. 1993. Nepotism in a solitary wasp as revealed by DNA fingerprinting. Evolution 47: 700–704.

Philippi, T., and W.G. Eberhard. 1986. Foraging behavior of *Stictia signata* (Hymenoptera: Sphecidae). J. Kans. Entomol. Soc. 59: 604–608.

Piek, T. 1985. Insect venoms and toxins. Pages 595–636 in G.A. Kerkut and L.I. Gilbert (eds.), Comprehensive Insect Physiology, Biochemistry, and Pharmacology. New York: Pergamon Press.

Piek, T., and W. Spanjer. 1986. Chemistry and pharmacology of solitary wasp venoms. Pages 161–307 in T. Piek (ed.), Venoms of the Hymenoptera: Biochemical, Pharmacological and Behavioural Aspects. New York: Academic Press.

Piek, T., J.H. Visser, and R.L. Veenendaal. 1984. Change in behaviour of the cockroach, *Periplaneta americana*, after being stung by the sphecid wasp *Ampulex compressa*. Entomol. Exp. Appl. 35: 195–203.

Piel, O. 1933. *Monema flavescens* Wkr. and its parasites (Lepidoptera, Heterogeneidae). Lingnan Sci. J. (suppl.) 12: 173–201.

Ponomarenko, N.G. 1971. Some peculiarities of development of Dryinidae. Proc. 13th Int. Congr. Entomol. Moscow 1968. 1: 281–282.

Powell, D. 1938. The biology of *Cephalonomia tarsalis* (Ash.), a vespoid wasp (Bethylidae: Hymenoptera) parasitic on the sawtoothed grain beetle. Ann. Entomol. Soc. Amer. 31: 44–49.

Powell, J.A. 1963. Biology and behavior of nearctic wasps of the genus *Xylocelia*, with special reference to *X. occidentalis* (Fox) (Hymenoptera: Sphecidae). Wasmann J. Biol. 21: 155–176.

Powell, J.A. 1967. Behavior of ground nesting wasps of the genus *Nitelopterus*, particularly *N. californicus*. J. Kans. Entomol. Soc. 40: 331–346.

Proctor, M., P. Yeo, and A. Lack. 1996. The Natural History of Pollination. Portland, Oreg.: Timber Press.

Punzo, F. 1994a. The biology of the spider wasp, *Pepsis thisbe* (Hymenoptera: Pompilidae) from Trans Pecos Texas. I. Adult morphometrics, larval development and the ontogeny of larval feeding patterns. Psyche 101: 229–242.

Punzo, F. 1994b. The biology of the spider wasp, *Pepsis thisbe* (Hymenoptera: Pompilidae) from Trans Pecos Texas. II. Temporal patterns of activity and hunting behavior with special reference to the effects of experience. Psyche 101: 243–256.

Qi, Y., and W.E. Burkholder. 1990. Attraction of larval kairomone of *Trogoderma* spp. to the parasitoid *Laelius pedatus* (Hymenoptera: Bethylidae). Contr. Shanghai Inst. Entomol. 9: 52–55.

Qi, Y., J.F. Andersen, J. Philips, and W.E. Burkholder. 1990. Isolation and identification of *Trogoderma variabile* (Coleoptera: Dermestidae) larval kairomone for the female parasitoid *Laelius pedatus* (Hymenoptera: Bethylidae). Contr. Shanghai Inst. Entomol. 9: 59–66.

Quicke, D.L.J. 1997. Parasitic Wasps. New York: Chapman and Hall.

Raatinkainen, M. 1961. *Dicondylus helleni* n. sp. (Hym., Dryinidae) a parasite of *Calligypona sordidula* (Stål) and *C. excisa* (Mel.). Ann. Entomol. Fenn. 27: 126–137.

Radović, I.T. 1985. Morphology and adaptive value of the sting apparatus of digger wasps (Hym., Sphecidae). Acta Entomol. Jugoslavica 21: 61–74.

Radović, I.T., and S. Sušić. 1997. Morphological characteristics of the sting and prey carriage mechanism in *Sericophorus relucens* F. Smith (Hymenoptera: Sphecidae: Larrinae). Proc. Entomol. Soc. Wash. 99: 537–540.

Rathmayer, W. 1962. Paralysis caused by the digger wasp *Philanthus*. Nature 196: 1148–1151.

Rau, P. 1928. Field studies in the behavior of the non-social wasps. Trans. Acad. Sci. St. Louis 25: 325–489.

Rau, P. 1948. A note on the nesting habits of the wasp, *Pemphredon inornatus* Say. Ann. Entomol. Soc. Amer. 41: 326.

Rau, P., and N. Rau. 1916. The sleep of insects: an ecological study. Ann. Entomol. Soc. Amer. 9: 227–274.

Rau, P., and N. Rau. 1918. Wasp Studies Afield. New York: Dover Publications.

Raw, A. 1975. Territoriality and scent marking by *Centris* males (Hymenoptera, Anthophoridae) in Jamaica. Behaviour 54: 311–321.

Rayor, L.S. 1996. Attack strategies of predatory wasps (Hymenoptera: Pompilidae; Sphecidae) on colonial orb web–building spiders (Araneidae: *Metepeira incrassata*). J. Kans. Entomol. Soc. 69 (suppl.): 67–75.

Readshaw, J.L. 1965. A theory of phasmatid outbreak release. Aust. J. Zool. 13: 475–490.

Rehnberg, B.G. 1987. Selection of prey by *Trypoxylon politum* (Say) (Hymenoptera: Sphecidae). Can. Entomol. 119: 189–194.

Reinhard, E.G. 1929. The Witchery of Wasps. New York: Century Co.

Ribi, W.A., and L. Ribi. 1979. Natural history of the Australian digger wasp *Sphex cognatus* Smith (Hymenoptera, Sphecidae). J. Nat. Hist. 13: 693–701.

Richards, O.W. 1939. The British Bethylidae (s.l.) (Hymenoptera). Trans. R. Soc. Lond. 89: 185–344.

Richards, O.W. 1953. The classification of the Dryinidae (Hym.), with descriptions of new species. Trans. R. Soc. Lond. 104: 51–70.

Richards, O.W. 1962. A revisional study of the masarid wasps (Hymenoptera, Vespoidea). London: British Museum (Natural History).

Richards, O.W. 1963. The species of *Pseudomasaris* Ashmead. Univ. Calif. Publ. Entomol. 27: 283–310.

Richards, O.W. 1978. The Social Wasps of the Americas Excluding the Vespinae. London: British Museum.

Richards, O.W., and A.H. Hamm. 1939. The biology of the British Pompilidae (Hymenoptera). Trans. Soc. Brit. Entomol. 6: 51–114.

Riek, E.F. 1955. Australian cleptid (Hymenoptera: Chrysidoidea) egg parasites of Cresmododea (Phasmodea). Austr. J. Sci. 3: 118–130.

Riek, E.F. 1970. Hymenoptera. Pages 867–869 in D.F. Waterhouse (ed.), The Insects of Australia. Melbourne: Melbourne University Press.

Rilett, R.O. 1949. The biology of *Cephalonomia waterstoni* Gahan. Can. J. Res. 27: 93–111.

Ristich, S.S. 1956. The host relationship of a miltogrammid fly *Senotainia trilineata* (VDW). Ohio J. Sci. 56: 271–274.

Rivers, R.L., Z.B. Mayo, and T.J. Helms. 1979. Biology, behavior and description of *Tiphia berbereti* (Hymenoptera: Tiphiidae). J. Kans. Entomol. Soc. 52: 362–372.

Roble, S.M. 1985. Submergent capture of *Dolomedes triton* (Araneae, Pisauridae) by *Anoplius depressipes* (Hymenoptera, Pompilidae). J. Arachnol. 13: 392.

Rocha, I.R.D., and A. Raw. 1982. Dinâmica das populações da vespa solitária *Zeta argillacea* (Linnaeus, 1758) (Hymenoptera: Eumenidae). An. Soc. Entomol. Brasil 11: 57–78.

Roig-Alsina, A., and C.D. Michener. 1993. Phylogenetic studies of long-tongued bees. Univ. Kans. Sci. Bull. 55: 123–162.

Rosenheim, J.A. 1987a. Host location and exploitation by the cleptoparasitic wasp *Argochrysis armilla*: the role of learning (Hymenoptera: Chrysididae). Behav. Ecol. Sociobiol. 21: 401–406.

Rosenheim, J.A. 1987b. Nesting behavior and bionomics of a solitary ground-nesting wasp, *Ammophila dysmica* (Hymenoptera: Sphecidae): influence of parasite pressure. Ann. Entomol. Soc. Amer. 80: 739–749.

Rosenheim, J.A. 1988. Parasite presence acts as a proximate cue in the nest-site selection process of a solitary digger wasp *Ammophila dysmica* (Hymenoptera: Sphecidae). J. Ins. Behav. 1: 333–342.

Rosenheim, J.A. 1989. Behaviorally mediated spatial and temporal refuges from a cleptoparasite, *Argochrysis armilla* (Hymenoptera: Chrysididae), attacking a ground-nesting wasp, *Ammophila dysmica* (Hymenoptera: Sphecidae). Behav. Ecol. Sociobiol. 25: 335–348.

Rosenheim, J.A. 1990a. Aerial prey caching by solitary ground-nesting wasps: a test of the predator defense hypothesis. J. Insect. Behav. 3: 241–250.

Rosenheim, J.A. 1990b. Density-dependent parasitism and the evolution of aggregated nesting in the solitary Hymenoptera. Ann. Entomol. Soc. Amer. 83: 277–286.

Rosenheim, J.A. 1993. Single-sex broods and the evolution of non-siblicidal parasitoid wasps. Amer. Nat. 141: 90–104.

Rosenheim, J.A., and J.K. Grace. 1987. Biology of a wood-nesting wasp, *Mimumesa mixta* (W. Fox) (Hymenoptera: Sphecidae), and its parasite, *Elampus viridicyaneus* Norton (Hymenoptera: Chrysididae). Proc. Entomol. Soc. Wash. 89: 351–355.

Roubaud, E. 1911. The natural history of the solitary wasps of the genus *Synagris*. Ann. Rep. Smithson. Inst. 1910: 507–525.

Roubaud, E. 1916. Recherches biologiques sur les guêpes solitaires et sociales d'Afrique. La genèse de la vie sociale et l'évolution de l'instinct maternel chez les vespides. Ann. Sci. Nat. Zool. (Sér. 10) 1: 1–160.

Rubink, W.L. 1978. The use of edaphic factors as cues for nest-site selection by sand wasps (Hymenoptera: Sphecidae). Ph.D. dissertation, Colorado State University.

Rubink, W.L. 1982. Spatial patterns in a nesting aggregation of solitary wasps: evidence for the role of conspecifics in nest-site selection. J. Kans. Entomol. Soc. 55: 52–56.

Rubink, W.L., and H.E. Evans. 1979. Notes on the nesting behavior of the bethylid wasp, *Epyris eriogoni* Kieffer, in southern Texas. Psyche 86: 313–319.

Rubink, W.L., and K.M. O'Neill. 1980. Observations on the nesting behavior of three species of *Plenoculus* Fox (Hymenoptera:Sphecidae). Pan-Pac. Entomol. 56: 187–196.

Rude, C.S. 1937. Parasites of pink bollworm in northern Mexico. J. Econ. Entomol. 30: 838–842.

Sahragard, A., M.A. Jervis, and N.A.C. Kidd. 1991. Influence of host availability on rates of oviposition and host-feeding, and on longevity in *Dicondylus indianus* Olmi (Hym., Dryinidae), a parasitoid of the rice brown planthopper, *Nilaparvata lugens* Stål (Hem., Delphacidae). J. Appl. Entomol. 112: 153–162.

Salt, G. 1931. A further study of the effects of stylopization on wasps. J. Exp. Zool 59: 133–166.

Scarborough, A.G. 1981. Ethology of *Eudioctria tibialis* Banks (Diptera: Asilidae) in Maryland: prey, predator behavior, and enemies. Proc. Entomol. Soc. Wash. 83: 258–268.

Schaber, B.D. 1985. Observations on nesting behavior and turret construction by *Odynerus dilectus* (Hymenoptera: Eumenidae). Can. Entomol. 117: 1159–1161.

Schaefer, C.H. 1962. Life history of *Conophthorus radiatae* (Coleoptera: Scolytidae) and its principal parasite, *Cephalonomia utahensis* (Hymenoptera: Bethylidae). Ann. Entomol. Soc. Amer. 55: 569–577.

Schmidt, J.O. 1978. *Dasymutilla occidentalis*: a long-lived aposematic wasp (Hymenoptera: Mutillidae). Entomol. News 89: 135–136.

Schmidt, J.O. 1990. Hymenopteran venoms: striving toward the ultimate defense against vertebrates. Pages 387–420 in D.L. Evans and J.O. Schmidt (eds.), Insect Defenses: Adaptive Mechanisms and Strategies of Prey and Predators. Albany, N.Y.: State University of New York Press.

Schmidt, J.O., and M.S. Blum. 1977. Adaptations and responses of *Dasymutilla occidentalis* (Hymenoptera: Mutillidae) to predators. Entomol. Exp. Appl. 21: 99–111.

Schmidt, J.O., K.M. O'Neill, H.M. Fales, C.A. McDaniel, and R.W. Howard. 1985. Volatiles from the mandibular glands of male beewolves (Hymenoptera: Sphecidae, *Philanthus*) and their possible roles. J. Chem. Ecol. 11: 895–901.

Schmidt, J.O., C.A. McDaniel, and R.T. Simonthomas. 1990. Chemistry of male mandibular gland secretions of *Philanthus triangulum*. J. Chem. Ecol. 16: 2135–2143.

Schmieder, R.G. 1933. The polymorphic forms of *Melittobia chalybii* Ashmead and the determining factors involved in their production. Biol. Bull. 65: 338–354.

Schöne, H., and J. Tengö. 1981. Competition of males, courtship behaviour and chemical communication in the digger wasp *Bembix rostrata* (Hymenoptera, Sphecidae). Behaviour 77: 44–66.

Schöne, H., and J. Tengö. 1991. Homing in the digger wasp *Bembix rostrata* (Hymenoptera, Sphecidae)—release direction and weather conditions. Ethology 87: 160–164.

Schöne, H., and J. Tengö. 1992. Insolation, air temperature, and behavioural activity in the digger wasp *Bembix rostrata* (Hymenoptera, Sphecidae). Entomol. Gener. 17: 259–264.

Schöne, H., A.C. Harris, H. Schöne, and P.A. Mahalski. 1993. Homing after displacement in open or closed containers by the digger wasp *Argogorytes carbonarius* (Hymenoptera, Sphecidae). Ethology 95: 152–156.

Schöne, H., H. Schöne, W.-D. Kühme, and L. Kühme. 1994. Arena headings, vanishing bearings, and homing times by *Argogorytes carbonarius*

(Hymenoptera, Sphecidae) after displacement in open or closed containers. Ethology 98: 291–297.

Schwarz, E.A. 1901. Sleeping trees of Hymenoptera. Proc. Entomol. Soc. Wash. 4: 24–26.

Schwarz, H.F. 1929. Honey wasps. Nat. Hist. (N.Y.) 29: 421–426.

Seger, J. 1983. Partial bivoltinism may cause alternating sex ratio biases that favor eusociality. Nature 301: 59–62.

Sexton, O. 1964. Differential predation by the lizard, *Anolis carolinensis* upon unicoloured and polycoloured insects after an interval of no contact. Anim. Behav. 12: 101–110.

Shafer, G.D. 1949. The Ways of a Mud Dauber. Stanford, Calif.: Stanford University Press.

Sheehan, W. 1984. Nesting biology of the sand wasp *Stictia heros* (Hymenoptera: Sphecidae: Nyssoninae) in Brazil. Rev. Biol. Trop. 29: 105–113.

Sheldon, J.K. 1968. The nesting behavior and larval morphology of *Pison koreense* (Radoszkowski) (Hymenoptera: Sphecidae). Psyche 75: 107–117.

Sheldon, J.K. 1970. Sexual dimorphism in the head structure of Mutillidae Hymenoptera: a possible behavioral explanation. Entomol. News 81: 57–61.

Shields, O. 1967. Hilltopping. J. Res. Lep. 6: 69–178.

Shimizu, A. 1989. An ethological study of *Agenioideus ishikawai* (Hymenoptera, Pompilidae). Jap. J. Entomol. 57: 654–662.

Shimizu, A. 1992. Nesting behavior of the semi-aquatic spider wasp, *Anoplius eous*, which transports its prey on the surface film of water (Hymenoptera, Pompilidae). J. Ethol. 10: 85–102.

Shimizu, A. 1994. Phylogeny and classification of the family Pompilidae (Hymenoptera). TMU Bull. Nat. Hist. 2: 1–142.

Simonthomas, R.T., and E.P.R. Poorter. 1972. Notes on the behaviour of males of *Philanthus triangulum* (F.) (Hymenoptera, Sphecidae). Tijdschr. Entomol. 115: 141–152.

Simonthomas, R.T., and A.M.J. Simonthomas. 1972. Some observations on the behaviour of females of *Philanthus triangulum* (F.) (Hymenoptera, Sphecidae). Tijdschr. Entomol. 115: 123–139.

Simonthomas, R.T., and A.M.J. Simonthomas. 1977. A Pest of the Honeybee in the Apiculture of the Dakhla Oasis, Egypt. Amsterdam: Pharmacological Laboratory, University of Amsterdam.

Simonthomas, R.T., and A.M.J. Simonthomas. 1980. *Philanthus triangulum* and its recent eruption as a predator of honeybees in an Egyptian oasis. Bee World 61: 97–107.

Simonthomas, R.T., and R.L. Veenendaal. 1974. Observations on the reproduction behaviour of *Crabro peltarius* (Schreber) (Hymenoptera, Sphecidae). Neth. J. Zool. 24: 58–66.

Smith, A.P. 1975. Life strategy and mortality factors of *Sceliphron laetum* (Smith) (Hymenoptera: Sphecidae) in Australia. Austral. J. Ecol. 4: 181–186.

Smith, A.P. 1978. An investigation of the mechanisms underlying nest construction in the mud wasp *Paralastor* sp. (Hymenoptera: Eumenidae). Anim. Behav. 26: 232–240.

Smith, A.P., and J. Alcock. 1980. A comparative study of the mating systems of Australian eumenid wasps (Hymenoptera). Z. Tierpsychol. 53: 41–60.

Smith, C.E. 1935. *Larra analis* Fabricius, a parasite of the mole cricket *Gryllotalpa hexadactyla* Perty. Proc. Entomol. Soc. Wash. 37: 65–82.

Smith, G.S., J.C.S. Allison, and N.W. Pammenter. 1994. Bio-assay study of response by a parasitoid to frass and feeding substrates of its host, the stalk borer *Eldana saccharina*. Ann. Appl. Biol. 125: 439–446.

Smith, K.G. 1986. Downy woodpecker foraging on mud-dauber wasp nests. Southwest. Nat. 31: 134.

Snelling, R.R. 1963. A host of *Macrosiagon cruentum* (Genmar) in Georgia (Coleoptera: Rhipiphoridae). Pan-Pac. Entomol. 39: 87–88.

Snoddy, E.L. 1968. Simuliidae, Ceratopogonidae, and Chloropidae as prey of *Oxybelus emarginatum*. Ann. Entomol. Soc. Amer. 61: 1029–1030.

Spangler, H. 1973. Vibration aids in soil manipulation in Hymenoptera. J. Kans. Entomol. Soc. 46: 157–160.

Spangler, H., and D.G. Manley. 1978. Sounds associated with the mating behavior of a mutillid wasp. Ann. Entomol. Soc. Amer. 71: 389–392.

Spofford, M.G., F.E. Kurczewski, and D.J. Peckham. 1986. Cleptoparasitism of *Tachysphex terminatus* (Hymenoptera: Sphecidae) by three species of Miltogrammini (Diptera: Sarcophagidae). Ann. Entomol. Soc. Amer. 79: 350–358.

Stage, G.I. 1960. First North American host record of the adventive wasp, *Chrysis fuscipennis* Brulle (Hymenoptera: Chrysididae). Pan-Pac. Entomol. 36: 191–195.

Starr, C.K. 1985. A simple pain scale for field comparison of hymenopteran stings. J. Entomol. Sci. 20: 225–232.

Steiner, A.L. 1971. Behavior of the hunting wasp *Liris nigra* V.d.L. (Hymenoptera, Larrinae) in particular or in unusual situations. Can. J. Zool. 49: 1401–1415.

Steiner, A.L. 1975. Description of territorial behavior of *Podalonia valida* (Hymenoptera: Sphecidae) females in southeast Arizona, with remarks on digger wasp territorial behavior. Quaest. Entomol. 11: 113–137.

Steiner, A.L. 1976. Digger wasp predatory behavior (Hymenoptera, Sphecidae) II. Comparative study of closely related wasps (Larrinae: *Liris nigra*, Palearctic; *L. argentata* and *L. aequalis*, Nearctic) that all paralyze crickets (Orthoptera, Gryllidae). Z. Tierpsychol. 42: 343–380.

Steiner, A.L. 1978. Evolution of prey-carrying mechanisms in digger wasps: possible role of a functional link between prey-paralyzing and carrying studied in *Oxybelus uniglumis* (Hymenoptera, Sphecidae, Crabroninae). Quaest. Entomol. 14: 393–409.

Steiner, A.L. 1979. Digger wasp predatory behavior (Hymenoptera, Sphecidae): fly hunting and capture by *Oxybelus uniglumis* (Crabroninae: Oxybelini); a case of extremely concentrated stinging pattern and prey nervous system. Can. J. Zool. 57: 953–962.

Steiner, A.L. 1981. Digger wasp predatory behavior (Hymenoptera, Sphecidae) IV. Comparative study of some distantly related Orthoptera-hunting wasps (Sphecinae vs. Larrinae), with emphasis on *Prionyx parkeri* (Sphecini). Z. Tierpsychol. 57: 305–339.

Steiner, A.L. 1982. Use of the proboscis for prey-piercing and sucking by sphecid wasps of the genus *Prionyx* (Hymenoptera: Sphecidae), a case of convergent evolution. Pan-Pac. Entomol. 58: 129–134.

Steiner, A.L. 1983a. Predatory behavior of digger wasps (Hymenoptera, Sphecidae) VI. Cutworm hunting and stinging by the ammophiline wasp *Podalonia luctuosa* (Smith). Wash. Entomol. Soc. 41: 1–16.

Steiner, A.L. 1983b. Predatory behavior of solitary wasps V. Stinging of caterpillars by *Euodynerus foraminatus* (Hymenoptera: Eumenidae): Weakening of the complete four-sting pattern. Biol. Behav. 8: 11–26.

Steiner, A.L. 1984. Observations on the possible use of habitat cues and token stimuli by caterpillar-hunting wasps: *Euodynerus foraminatus* (Hymenoptera, Eumenidae). Quaest. Entomol. 20: 25–33.

Steiner, A.L. 1986. Stinging behaviour of solitary wasps. Pages 63–160 in T. Piek (ed.), Venoms of the Hymenoptera: Biochemical, Pharmacological and Behavioural Aspects. New York: Academic Press.

Steiner, H.M. 1936. New nymphal-adult parasite of white apple leafhopper. J. Econ. Entomol. 29: 632–633.

Steiner, H.M. 1938. Effects of orchard practices on natural enemies of the white apple leafhopper. J. Econ. Entomol. 31: 232–240.

Steiner, K.E., V.B. Whitehead, and S.D. Johnson. 1994. Floral and pollinator divergence in two sexually deceptive South African orchids. Amer. J. Bot. 81: 185–194.

Stephens, D.W., and J.R. Krebs. 1986. Foraging Theory. Princeton, N.J.: Princeton University Press.

Stiling, P.D. 1994. Interspecific interactions and community structure in planthoppers and leafhoppers. Pages 449–516 in R.E. Denno and T.J. Perfect (eds.), Planthoppers: Their Ecology and Management. New York: Chapman and Hall.

Stone, G.N. 1989. Endothermy and thermoregulation in solitary bees. Ph.D. dissertation, University of Oxford.

Stone, G.N., and P.G. Willmer. 1989. Warm-up rates and body temperatures in bees: the importance of body size, thermal regime and phylogeny. J. Expt. Biol. 147: 303–328.

Stoutamire, W.P. 1983. Wasp-pollinated species of *Caladenia* (Orchidaceae) in southwestern Australia. Aust. J. Bot. 31: 383–394.

Strohm, E., and K.E. Linsenmair. 1997a. Female size affects provisioning and sex allocation in a digger wasp. Anim. Behav. 54: 23–34.

Strohm, E., and K.E. Linsenmair. 1997b. Low resource availability causes extremely male-biased investment ratios in the European beewolf, *Philan-*

thus triangulum F. (Hymenoptera: Sphecidae). Proc. R. Soc. Lond. B 264: 423–429.

Stubblefield, J.W., J. Seger, J.W. Wenzel, and M.M. Heisler. 1993. Temporal, spatial, sex-ratio and body size heterogeneity of prey species taken by the beewolf *Philanthus sanbornii* (Hymenoptera: Sphecidae). Phil. Trans. R. Soc. Lond. 339: 397–423.

Swaminathan, S., and T.N. Ananthakrishan. 1984. Population trends of some monophagous and polyphagous fulgorids in relation to biotic and abiotic factors (Insecta: Homoptera). Proc. Indian Acad. Sci. (Anim. Sci.) 93: 1–8.

Tachikawa, T., and M. Yukinari. 1974. Parasites of *Goniozus japonica* Ashmead (Hymenoptera: Bethylidae) in Shikoku. Trans. Shikoku Entomol. Soc. 12: 45–46.

Taffe, C.A. 1978. Temporal distribution of mortality in a field population of *Zeta abdominale* (Hymenoptera) in Jamaica. Oikos 31: 106–111.

Taffe, C.A. 1979. The ecology of two West Indian species of mud-wasps (Eumenidae: Hymenoptera). Biol. J. Linn. Soc. 11: 1–17.

Tallamy, D.W., and C. Schaeffer. 1997. Maternal care in the Hemiptera: ancestry, alternatives, and current adaptive value. Pages 94–115 in J.C. Choe and B.J Crespim (eds.), Social Behavior in Insects and Arachnids. Cambridge: Cambridge University Press.

Tengö, J., H. Schöne, and J. Chmurzyński. 1990. Homing in the digger wasp *Bembix rostrata* (Hymenoptera, Sphecidae) in relation to sex and stage. Ethology 86: 47–56.

Tepedino, V.J. 1979. Notes on the flower-visiting habits of *Pseudomasaris vespoides* (Hymenoptera: Masaridae). Southwest. Nat. 24: 380–381.

Thornhill, R., and J. Alcock. 1983. The Evolution of Insect Mating Systems. Cambridge, Mass.: Harvard University Press.

Thorpe, W.H. 1950. A note on detour experiments with *Ammophila pubescens* Curt (Hymenoptera; Sphecidae). Behaviour 2: 257–263.

Tinbergen, N. 1932. Über die Orientierung des Bienenwolfes (*Philanthus triangulum* Fabr.). Z. Vgl. Physiol. 16: 305–335.

Tinbergen, N. 1935. Über die Orientierung des Bienenwolfes. II. Die Bienenjagd. Z. Vgl. Physiol. 21: 699–716.

Tinbergen, N. 1951. The Study of Instinct. London: Oxford University Press.

Tinbergen, N. 1958. Curious Naturalists. New York: Doubleday.

Tinbergen, N. 1972. The Animal in Its World: Field Studies. Cambridge, Mass.: Harvard University Press.

Toft, C.A. 1987. Activity budgets, habitat use and body size in two coexisting species of sand wasps (*Microbembex*: Sphecidae, Hymenoptera). Ecol. Entomol. 12: 427–438.

Torchio, P.F. 1970. The ethology of the wasp, *Pseudomasaris edwardsii* (Cresson), and a description of its immature forms (Hymenoptera: Vespoidea, Masaridae). L.A. Co. Mus. Contrib. Sci. 202: 1–32.

Torchio, P.F. 1972a. *Sapyga pumila* Cresson, a parasite of *Megachile rotundata* (F.) (Hymenoptera: Sapygidae; Megachilidae). I: Biology and description of immature stages. Melanderia 10: 1–22.

Torchio, P.F. 1972b. *Sapyga pumila* Cresson, a parasite of *Megachile rotundata* (F.) (Hymenoptera: Sapygidae; Megachilidae). II: Methods for control. Melanderia 10: 23–30.

Torchio, P.F. 1974. Mechanisms involved in the pollination of *Penstemon* visited by the masarid wasp *Pseudomasaris vespoides* (Cresson) (Hymenoptera: Vespoidea). Pan-Pac. Entomol. 50: 226–234.

Torchio, P.F. 1979. An eight-year field study involving control of *Sapyga pumila* Cresson (Hymenoptera: Sapygidae), a wasp parasite of the alfalfa leafcutter bee, *Megachile pacifica* Panzer. J. Kans. Entomol. Soc. 52: 412–419.

Trexler, J.C. 1985. Density-dependent parasitism by a eulophid parasitoid: tests of an intragenerational hypothesis. Oikos 44: 415–422.

Triapitzin, V.A. 1964. Encyrtidae (Hymenoptera)—parasites of Dryinidae (Hymenoptera) in USSR. Zool. Zhurn. 43: 142–145 [in Russian, with English summary].

Trivers, R.L. 1972. Parental investment and sexual selection. Pages 136–179 in B. Campbell (ed.), Sexual Selection and the Descent of Man. Chicago: Aldine.

Trivers, R.L., and D.E. Willard. 1973. Natural selection of parental ability to vary the sex ratio of offspring. Science 179: 90–92.

Truc, C., and J. Gervet. 1983. Influence of the reactions of the prey on the stinging patterns of digger wasps. Experientia 39: 1320–1322.

Tsuneki, K. 1950. Experimental analysis of the sensory cues working in the return to the nest of the Pompilidae (solitary Hymenoptera). Ann. Zool. Jap. 23: 75–84.

Tsuneki, K. 1952. Ethological studies on the Japanese species of *Pemphredon* (Hymenoptera: Sphecidae), with notes on their parasites, *Elampus* spp. (Hym., Chrysididae). J. Fac. Sci. Hokkaido Univ., Ser. VI. Zool. 11: 57–75.

Tsuneki, K. 1956. Ethological studies of *Bembix niponica* Smith, with emphasis on the psychobiological analysis of behaviour inside of the nest (Hymenoptera: Sphecidae). I. Biological part. Mem. Fac. Lib. Arts. Fukui Univ. (Ser. II, Nat. Sci.) 6: 77–179.

Tsuneki, K. 1959. Contributions to the knowledge of the Cleptinae and Pseninae faunae of Japan and Korea. Mem. Fac. Lib. Arts, Fukui Univ. (Ser. II, Nat. Sci.) 9: 1–78.

Tsuneki, K. 1961. Colour vision and figure discriminating capacity of the solitary diplopterous wasp *Odynerus frauenfeldi* Saussure. Mem. Fac. Lib. Arts, Fukui Univ. (Ser. II, Nat. Sci.) 11: 103–160.

Tsuneki, K. 1963. Comparative studies on the nesting biology of the genus *Sphex* (s.l.) in East Asia. Mem. Fac. Lib. Arts, Fukui Univ. (Ser. II, Nat. Sci.) 13: 13–78.

Tsuneki, K. 1965. The nesting biology of *Stizus pulcherrimus* F. Smith (Hym., Sphecidae) with special reference to the geographic variation. Etizenia, Occ. Publ. Biol. Lab. Fukui Univ. 10: 1–21.

Tsuneki, K. 1968. The biology of some Japanese spider wasps. Etizenia 34: 1–37.

Tsuneki, K. 1970. Gleanings on the bionomics of the East-Asiatic non-social wasps (Hymenoptera) VI. Some species of Trypoxyloninae. Etizenia 45: 1–20.

Turner, R.E. 1907. A revision of the Thynnidae of Australia. Proc. Linn. Soc. N.S.W. 32: 206–290.

Valdeyron-Fabre, L. 1955. Observations sur la biologie de *Brachytrupes megacephalus* Lef. En Tunisie. Rev. Path. Vég. Entomol. Agr. France 34: 136–158.

van Iersal, J.J.A. 1952. On the orientation of *Bembex rostrata* L. Trans 9th Int. Congr. Entomol. 1: 384–393.

van Iersel, J.J.A., and J. van den Assem. 1964. Aspects of orientation in the diggerwasp *Bembix rostrata*. Anim. Behav. Suppl. 1: 145–162.

Venkatraman, T.V., and M.J. Chacko. 1961. Some factors influencing the efficiency of *Goniozus marasmi* Kurian, a parasite of the maize and jowar leaf roller. Proc. Indian Acad. Sci. Sec. B 53: 275–283.

Vesey-Fitzgerald, D. 1940. Notes on Bembicidae and allied wasps from Trinidad. Proc. R. Entomol. Soc. London (A) 15: 37–39.

Villalobos, E.M., and T.E. Shelly. 1994. Observations on the mating behavior of male *Stictia heros* (Hymenoptera: Sphecidae). Fla. Entomol. 77: 99–104.

Vincent, D.L. 1978. A revision of the genus *Passaloecus* (Hymenoptera: Sphecidae) in America North of Mexico. Wasmann J. Biol. 36: 127–198.

Vinson, S.B., and G.W. Frankie. 1990. Territorial and mating behavior of *Xylocopa fimbriata* F. and *Xylocopa gualanensis* Cockerell from Costa Rica. J. Ins. Behav. 3: 13–32.

Vinson, S.B., H.J. Williams, G.W. Frankie, J.W. Wheeler, M.S. Blum, and R.E. Coville. 1982. Mandibular glands of male *Centris adani* (Hymenoptera: Anthophoridae): their morphology, chemical constituents, and function in scent marking and terrtorial behavior. J. Chem. Ecol. 8: 319–327.

Vouskassovitch, M.P. 1924. Sur la biologie de *Goniozus claripennis* Forst., parasite d'*Oenophthira pilleriana* Schiff. Bull. Soc. Hist. Toulousse 52: 225–246.

Waldbauer, G.P., and D.P. Cowan. 1985. Defensive stinging and Müllerian mimicry among eumenid wasps (Hymenoptera: Vespoidea: Eumenidae). Amer. Midl. Nat. 113: 198–199.

Waldbauer, G.P., and W.E. LaBerge. 1985. Phenological relationships of wasps, bumblebees, their mimics and insectivorous birds in northern Michigan. Ecol. Entomol. 10: 99–110.

Waldbauer, G.P., J.G. Sternburg, and C.T. Maier. 1977. Phenological relationships of wasps, bumblebees, their mimics, and insectivorous birds in an Illinois sand area. Ecology 58: 583–591.

Waloff, N. 1974. Biology and behaviour of some species of Dryinidae (Hymenoptera). J. Entomol. (A) 49: 97–109.

Waloff, N. 1975. The parasitoids of the nymphal and adult stages of leafhoppers (Auchenorrhyncha: Homoptera) of acidic grassland. Trans. R. Entomol. Soc. Lond. 126: 637–686.

Waloff, N. 1980. Studies on grassland leafhoppers (Auchenorrhyncha, Homoptera) and their natural enemies. Adv. Ecol. Res. 11: 81–215.

Waloff, N., and M.A. Jervis. 1987. Communities of parasitoids associated with leafhoppers and planthoppers in Europe. Adv. Ecol. Res. 17: 281–410.

Ward, D., and J.R. Henschel. 1992. Experimental evidence that a desert parasitoid keeps its host cool. Ethology 92: 135–142.

Wcislo, W.T. 1981. The roles of seasonality, host synchrony, and behaviour in the evolutions and distributions of nest parasites in Hymenoptera (Insecta), with special reference to bees (Apoidea). Biol. Rev. 62: 515–543.

Wcislo, W.T. 1984. Gregarious nesting of a digger wasp as a "selfish herd" response to a parasitic fly (Hymenoptera: Sphecidae; Diptera: Sarcophagidae). Behav. Ecol. Sociobiol. 15: 157–160.

Wcislo, W.T. 1986. Host nest discrimination by a cleptoparasitic fly, *Metopia campestris* (Fallén) (Diptera: Sarcophagidae: Miltogramminae). J. Kans. Entomol. Soc. 59: 82–86.

Wcislo, W.T., B.S. Low, and C.J. Karr. 1985. Parasite pressure and repeated burrow use by different individuals of *Crabro* (Hymenoptera: Sphecidae; Diptera: Sarcophagidae). Sociobiology 11: 115–125.

Wcislo, W.T., M.J. West-Eberhard, and W.G. Eberhard. 1988. Natural history and behavior of a primitively social wasp, *Auplopus semialatus*, and its parasite, *Irenangelus eberhardi* (Hymenoptera: Pompilidae). J. Ins. Behav. 1: 247–260.

Weaving, A.J.S. 1989. Habitat selection and nest construction behaviour in some Afrotropical species of *Ammophila* (Hymenoptera: Sphecidae). J. Nat. Hist. 23: 847–871.

Weaving, A.J.S. 1994. Nesting behaviour in three Afrotropical trap-nesting wasps, *Chalybion laevigatum* (Kohl), *Proepipona meadewaldoi* Bequaert and *Tricarinodynerus guernii* (Saussure), (Hymenoptera: Sphecidae, Eumenidae). Entomologist 113: 183–197.

Werren, J.H., and E.L. Charnov. 1978. Facultative sex ratios and population dynamics. Nature 272: 349–350.

West-Eberhard, M.J. 1984. Sexual selection, social communication, and species specific signals in insects. Pages 283–324 in T. Lewis (ed.), Insect Communication. London: Academic Press.

West-Eberhard, M.J. 1987. Observations of *Xenorhynchium nitidulum* (Fabricius) (Hymenoptera, Eumeninae), a primitively social wasp. Psyche 94: 317–323.

Wharton, R.A. 1989. Final instar larva of the embolemid wasp, *Ampulicomorpha confusa* (Hymenoptera). Proc. Entomol. Soc. Wash. 91: 509–512.

Wheeler, G.C. 1983. A mutillid mimic of an ant (Hymenoptera: Mutillidae and Formicidae). Entomol. News 94: 143–144.

Wheeler, W.M. 1919. The parasitic Aculeata: a study in evolution. Proc. Amer. Philos. Soc. 58: 1–40.

White, E. 1962. Nest-building and provisioning in relation to sex in *Sceliphron spirifex* L. (Sphecidae). J. Anim. Ecol. 31: 317–329.

White, R.T. 1943. Effect of milky disease on *Tiphia* parasites of Japanese beetle larvae. J. N.Y. Entomol. Soc. 51: 213–218.

Whitfield, J.B. 1992. Phylogeny of the non-aculeate Apocrita and the evolution of parasitism in the Hymenoptera. J. Hym. Res. 1: 3–14.

Whiting, P.W. 1935. Sex determination in bees and wasps. J. Hered. 26: 263–278.

Wickler, W. 1968. Mimicry in Plants and Animals. New York: McGraw-Hill.

Williams, F.X. 1914. Notes on the habits of some solitary wasps that occur in Kansas, with description of a new species. Univ. Kans. Sci. Bull. 18: 223–230.

Williams, F.X. 1916. Notes on the life-history of *Methoca stygia* Say. Psyche 23: 121–125.

Williams, F.X. 1919a. *Epyris extraneus* Bridwell (Bethylidae), a fossorial wasp that preys on the larva of the tenebrionid beetle, *Gonocephalum seriatum* (Boisduval). Proc. Haw. Entomol. Soc. 4: 55–63.

Williams, F.X. 1919b. A note on *Epactothynnus opaciventris* Turner, an Australian thynnid wasp. Psyche 26: 160–162.

Williams, F.X. 1919c. Philippine wasp studies. Part 2. Descriptions of new species and life history studies. Bull. Haw. Sug. Planters' Assoc., Entomol. Ser. 14: 19–186.

Williams, F.X. 1919d. Some observations on the leaf-hopper wasp, *Nesomimesa hawaiiensis* Perkins, at Pahala, Hawaii Feb. 11–April 25, 1918. Proc. Haw. Entomol. Soc. 4: 63–68.

Williams, F.X. 1928. *Pterombrus*, a wasp-enemy of the larva of tiger-beetles. Bull. Haw. Sug. Planters' Assoc., Entomol. Ser. 19: 144–151.

Williams, F.X. 1929. Notes on the habits of the cockroach-hunting wasps of the genus *Ampulex* sens. lat., with particular reference to *Ampulex (Rhinopsis) canaliculatus* Say. Proc. Haw. Entomol. Soc. 7: 315–329.

Williams, F.X. 1942. *Ampulex compressa* (Fabr.), a cockroach hunting wasp introduced from New Caledonia into Hawaii. Proc. Haw. Entomol. Soc. 11: 221–233.

Williams, F.X. 1956. Life history studies of *Pepsis* and *Hemipepsis* wasps in California (Hymenoptera, Pompilidae). Ann. Entomol. Soc. Amer. 49: 447–466.

Williams, G.C. 1966. Adaptation and Natural Selection. Princeton, N.J.: Princeton University Press.

Willmer, P.G. 1982. Microclimate and the environmental physiology of insects. Pages 1–60 in M.J. Berridge, J.E. Treherne, and V.B. Wigglesworth (eds.), Advances in Insect Physiology, New York: Academic Press.

Willmer, P.G. 1985a. Size effects on the hygrothermal balance and foraging patterns of a sphecid wasp, *Cerceris arenaria*. Ecol. Entomol. 10: 469–479.

Willmer, P.G. 1985b. Thermal ecology, size effects, and the origins of communal behavior in *Cerceris* wasps. Behav. Ecol. Sociobiol. 17: 151–160.

Willmer, P.G., and D.M. Unwin. 1981. Field analyses of insect heat budgets: reflectance, size, and heating rates. Oecologia 50: 250–255.

Wilson, E.O. 1971. The Insect Societies. Cambridge, Mass.: Harvard University Press.

Wilson, E.O., and D.J. Farish. 1973. Predatory behaviour in the ant-like wasp *Methoca stygia* (Say) (Hymenoptera: Tiphiidae). Anim. Behav. 21: 292–295.

Wilson, L.T., I. Carmeau, and D.L. Flaherty. 1991. *Aphelopus albopictus* Ashmead (Hymenoptera: Dryinidae): abundance, parasitism, and distribution in relation to leafhopper hosts in grapes. Hilgardia 59: 1–16.

Witethom, B., and G. Gordh. 1994. Development and life table of *Goniozus thailandensis* Gordh & Witethom (Hymenoptera: Bethylidae), a gregarious ectoparasitoid of a phycitine fruit borer (Lepidoptera: Pyralidae). J. Sci. Soc. Thailand 20: 101–114.

Wittenberger, J.F. 1981. Animal Social Behavior. Boston: Duxbury Press.

Wolcott, G.N. 1914. Notes on the life history and ecology of *Tiphia inornata* Say. J. Econ. Entomol. 7: 382–389.

Woyke, J. 1973. Reproductive organs of haploid and diploid drone honeybees. J. Apicult. Res. 12: 35–51.

Yamada, Y. 1955. Studies on the natural enemy of the woollen pest, *Anthrenus verbaci* Linné (*Allepyris microneurus* Kieffer) (Hymenoptera, Bethylidae). Mushi 28: 13–41.

Yamada, Y. 1987a. Characteristics of the oviposition of a parasitoid, *Chrysis shanghaiensis* (Hymenoptera: Chrysididae). Appl. Entomol. Zool. 22: 456–464.

Yamada, Y. 1987b. Factors determining the rate of parasitism by a parasitoid with a low fecundity, *Chrysis shanghaiensis* (Hymenoptera: Chrysididae). J. Anim. Ecol. 56: 1029–1042.

Yamada, Y. 1988. Optimal use of patches by parasitoids with a limited fecundity. Res. Popul. Ecol. 30: 235–249.

Yamada, Y. 1990. Role of a parasitoid with a low fecundity, *Praestochrysis shanghaiensis* (Hymenoptera: Chrysididae), in the population dynamics of its host. Res. Popul. Ecol. 32: 365–379.

Yanega, D. 1994. Aboreal, ant-mimicking mutillid wasps, *Pappognatha*; parasites of neotropical *Euglossa* (Hymenoptera: Mutillidae and Apidae). Biotropica 26: 465–468.

Yoshimoto, C.M. 1954. A study of the biology of *Priocnemis minorata* Banks (Hymenoptera, Pompilidae). Bull. Brooklyn Entomol. Soc. 49: 130–138.

Zalom, F.G., R.P. Meyer, and P.H. Mason. 1979. Sympatric associations of *Systrophus* spp. (Diptera: Bombyliidae) and *Ammophila* spp. (Hymenoptera: Sphecidae). Pan-Pac. Entomol. 55: 239–240.

Zeil, J. 1993a. Orientation flights of solitary wasps (*Cerceris*; Sphecidae; Hymenoptera) I. Description of flight. J. Comp. Physiol. 172: 189–205.

Zeil, J. 1993b. Orientation flights of solitary wasps (*Cerceris*; Sphecidae; Hymenoptera) II. Similarities between orientation and return flights and the use of motion parallax. J. Comp. Physiol. 172: 207–222.

Zeil, J., and A. Kelber. 1991. Orientation flights in ground-nesting wasps and bees share a common organization. Verh. Dtsch. Zool. Ges. 84: 371.

Zeil, J., A. Kelber, and R. Voss. 1996. Structure and function of learning flights in bees and wasps. J. Expt. Biol. 199: 245–252.

Zucchi, R., S. Yamane, and S.F. Sakagami. 1976. Preliminary notes on the habits of *Trimeria howardi*, a neotropical communal masarid wasp, with description of the mature larva (Hymenoptera: Vespoidea). Insect Matsumurana 8: 47–57.

Index

abdomen dragging, 252–253
Abispa, 159; *ephippium*, 234; *splendida*, 279
accessory burrows, 155, 208–209
Acroricnus, 191; *ambulator*, 192
Aculeata, 2, 4–5
Adelphe anisomorphe, 36
adult feeding on carbohydrates, 74, 78, 115, 142–143
Ageniella, 128, 158
Agenioideus, 92, 159, 328; *cinctellus*, 55–56, 327
Allepyris microneurus, 30
Allochares azureus, 56
Allocoelia bidens, 37
Alysson, 107, 158, 164; *conicus*, 293; *melleus*, 90, 177
Amisega kahlii, 36
Ammophila, 113, 124, 130, 157, 160, 166, 181, 206, 328; *aberti*, 130; *azteca*, 80, 162–163, 177–179, 325; *dysmica*, 80, 122, 197, 201–202, 214, 233, 273, 290; *ferrugineipes*, 163, 168; *gracilis*, 201; *harti*, 80, 163, 325; *infesta*, 116; *insignis*, 167–168; *marshi*, 212, *procera*, 324; *pubescens*, 325; *sabulosa*, 108, 121, 129, 201, 300, 310, 312
Amobia, 185, 187, 196–199; *floridensis*, 184, 200
Ampulex amoena, 61; *assimilis*, 60; *bantuae*, 58–59, 61; *canaliculata*, 60–61; *compressa*, 60, 62
Anacrabro ocellatus, 88
Ancistrocerus, 115, 117, 124, 159; *adiabatus*, 100, 220, 226, 236, 270, 301, 312–313; *antilope*, 98, 104, 116, 200, 227, 233, 277–278, 312–313, 317–318; *catskill*, 100, 199; *fukaianus*, 304
Ancistromma distincta, 85
Anoplius, 92, 104, 127, 156, 159, 289, 328; *apiculatus*, 95; *cleora*, 95; *depressipes*, 104–105; *eous*, 95, 104; *marginatus*, 96; *semirufus*, 96; *splendens*, 96; *tenebrosus*, 96; *viaticus*, 182
Anospilus orbitalis, 56
ant clamp, 112
ant lions, 195
Anteon brachycerum, 275; *pubicorne*, 21; *flavicorne*, 20
Antepipona scutellaris, 98
Anterhynchium, 159; *flavomarginatum*, 98, 188, 199, 202
Anthobosca chilensis, 50
Anthrax, 190, 193, 200; *distigmus*, 193
ants: as predators, 194, 197–198; relationship to solitary wasps, 8
Apenesia, 248
Aphelopus albopictus, 20; *atratus*, 20; *melaleucus*, 20, 275; *serratus*, 20
Aphilanthops, 92, 108, 158, 211, 253, 258; *frigidus*, 93; *subfrigidus*, 98, 103, 113, 246, 257, 262–263
Apocrita, 2
Apoidea, classification, 5, 7–8
Aporinellus wheeleri, 96
Aporus hirsutus, 57
Arachnospila, 328
Argochrysis, 197; *armilla*, 122–126, 201, 214
Argogorytes, 113, 129; *fargei*, 148; *mystaceus*, 146–148

Arpactophilus, 330
Asilidae, 194–195
Astata, 84, 157; *occidentalis*, 83, 156, 162–164; *unicolor*, 83, 114
Auplopus, 92, 110, 128, 158, 176, 324, 327; *argentifrons*, 332; *caerulescens*, 173; *esmerelda*, 332; *nyemitawa*, 128, 174–176; *semialatus*, 331–332
Austrochares gastricus, 206
Austrogorytes bellicosus, 90

Bathyplectes, 186
Batozonellus, 159; *annulatus*, 128, 293; *maculifrons*, 304
bats, 195
bees: foraging compared to Masarinae, 139–141, 149–151; relationship to solitary wasps, 7
Belomicrus, 157, 161; *forbesii*, 88
Bembecinus, 87, 158, 161, 325; *hirtulus*, 90; *japonicus*, 304–305; *nanus*, 90, 161, 163, 232, 236, 289; *neglectus*, 116, 329; *quinquespinosus*, 90, 161, 168, 177–178, 220, 227–232, 236, 250, 270, 290, 295, 319
Bembix, 87, 103, 107, 113, 117, 158, 165, 182, 194, 233, 277, 290–291, 329; *americana*, 91, 156, 177–179, 208, 293, 325; *amoena*, 91, 156, 325; *belfragei*, 156; *bubalus*, 91; *cinerea*, 156, 177, 235; *coonundura*, 168; *furcata*, 268; *hinei*, 156; *moma*, 91; *occidentalis*, 156, 235, 325; *olivacea*, 47; *pallidipicta*, 153–156, 325; *pruinosa*, 235, 329; *rostrata*, 114, 208, 230–232, 236, 250, 273–274, 279, 285–289, 293, 300; *sayi*, 91, 156, 325; *texana*, 91; *tuberculiventris*, 91; *u-scripta*, 91, 144, 293, 329
Bethylus fuscicornis, 29
Bicyrtes, 87, 157, 324; *quadrifasciata*, 91
biological control, 66–69
birds, 189, 195
Blaesoxipha, 185
Bocchus europaeus, 21, 23–24
body size: and alternative mating tactics, 269–270; and egg number, 310; and egg size, 310; and mating success, 220, 224, 229–230, 255–258, 266, 271; and thermoregulation, 291–292
Bombyliidae, 184–185, 190, 193, 196–199, 201
Bothynostethus distinctus, 86
Brachymeria, 190, 196, 198, 201
Braconidae, 186, 190
Bradynobaenidae, 9
brood parasites: defined, 120; incidental, 188–189; of solitary wasps, 184–187

Caenochrysis mucronata, 202
Calligaster, 176; *cyanoptera*, 107
Calliphoridae, 185
Campoplex, 186, 188
Campsomeris, 67; *annulata*, 52–53, 326; *dorsata*, 52; *marginella*, 52, 67; *plumipes*, 52
Campsoscolia, 148; *ciliata*, 148–149; *octomaculata*, 144; *tasmaniensis*, 148
Carinafoenus, 186
Celonites, 159; *abbreviatus*, 138, 140, 142; *latitarsis*, 138, 164; *peliostomi*, 138; *wahlenbergiae*, 138, 142, 173
Cephalonomia gallicola, 30, 226; *stephanoderis*, 31; *tarsalis*, 225–226; *utahensis*, 31; *waterstoni*, 26, 31–32, 68, 300
Ceramius, 142, 159, 164, 165; *bicolor*, 138; *binodis*, 258; *braunsi*, 140; *capicola*, 138, 142, 234; *clypeatus*, 138, 140, 144; *lichtensteinii*, 139, 162–164; *linearis*, 139; *nigripennis*, 139; *rex*, 139; *socius*, 139; *tuberculifer*, 138, 140, 233
Cerceris, 92, 110–111, 113, 158, 160, 181, 253, 258, 324; *antipodes*, 93, 156, 331–332; *arenaria*, 93, 167, 181, 287, 311–312; *australis*, 93, 332; *californica*, 93, 332; *echo*, 211; *flavofasciata*, 93; *fumipennis*, 93, 162–163, 182; *goddardi*, 332; *graphica*, 233; *halone*, 92, 116; *holconota*, 93; *rybyensis*, 94, 114; *sabulosa*, 164; *watlingensis*, 233
Ceroctis, 185
Ceropales, 127, 327; *maculata*, 127
Chalcididae, 190, 201

Chalybion, 79, 157; *caerulum*, 81, 295; *californicum*, 100, 103; *spinolae*, 81
chelae, 16, 23
Chirodamus longulus, 206
Chlorion, 79, 103, 157, 289, 302, 328; *aerarium*, 60, 81; *lobatum*, 62, 303, 305; *maxillosum*, 63
Chrestomutilla glossinae, 43, 45, 310
Chrysanthrax, 190, 193
Chrysidoidea, 5–7
Chrysis, 126, 199; *angolensis*, 37; *carinata*, 124; *coerulans*, 124; *fuscipennis*, 39–40, 123; *ignata*, 304; *inaequidens*, 200; *purpurata*, 124
Chrysura davidi, 37; *kyrae*, 37; *pacifica*, 37
Chrysurissa densa, 38
Cicindela, 185
Cicindelidae, 185
Cleptes purpuratus, 36; *semiauratus*, 37
cleptoparasitism, 189
Cleridae, 194
Clitemnestra gayi, 296
clutch size, 306–308
Clypeadon, 92, 111, 158, 253, 322; *laticinctus*, 93, 103, 121
color, and thermoregulation, 290–292
communal nesting, 213, 321, 330–333
conditional mating strategies, 268–272
Conopidae, 194
courtship, 222, 272–280
courtship feeding, 246–252
Crabro, 87, 157, 181, 276, 324; *advenus*, 163; *arcadiensis*, 88; *argusinus*, 87–88, 211; *cribrarius*, 293; *cribrellifer*, 201, 207, 232, 276–277; *deserticola*, 277; *digitatus*, 277; *maculiclypeus*, 88; *pallidus*, 277; *rufibasis*, 89; *spinuliferus*, 277; *tenuis*, 277
Crossocerus, 87, 157, 170–171; *annulipes*, 89; *capitosus*, 197
Crovettia theliae, 19, 25
Cryptocheilus, 158
cuckoo wasps, 122

Dasymutilla, 205, 289; *bioculata*, 45, 233, 274; *foxi*, 233, 276; *klugii*, 204; *nigripes*, 45; *occidentalis*, 45, 203; *scaevola*, 45; *vesta*, 43, 45, 235
Dasyproctus, 157, 170; *westermanni*, 89
delayed provisioning, 329
Dermestidae, 185, 194, 201
Diamma bicolor, 49, 51
Diapriidae, 190
Dichragenia, 128; *pulchricoma*, 95
Dicondylus bicolor, 275; *indianus*, 300
Dienoplus, 129
diet: spatial variation, 97–100; temporal variation, 100–101
Dinetus, 84, 157
Diodontus, 157, 187; *argentina*, 177–179; *occidentalis*, 82
Diogmites angustipennis, 194
Diomorus, 191, 196–198
Diploplectron, 84, 157; *peglowi*, 83
Dipogon, 158, 173, 324, 327; *sayi*, 95, 197
Discoelius japonicus, 98
Dissomphalus, 28, 248
Dolichurus stantoni, 62
Dryinus collaris, 21; *tarraconensis*, 21
Dryudella, 84
Duckeia cyanea, 36
Dynatus nigripes, 242

Ectemnius, 87, 157, 170–171; *paucimaculatus*, 89, 201; *sexcinctus*, 89; *spiniferus*, 89
Elampus viridicyaneus, 124, 319
Enchemicrum australe, 88
Encyrtidae, 190
endothermy, 283–285
Entomognathus memorialis, 89
Epactothynnus opaciventris, 51
Ephuta slossonae, 44
Ephutomorpha ignita, 44
Epinysson, 128
Epistenia, 191
Episyron, 110, 159; *arrogans*, 100, 104, 197, 202, 301; *quinquenotatus*, 96, 116, 162–163, 178
Epsilon, 224, 256
Epyris eriogoni, 31; *extraneus*, 31; *subangulatus*, 233
Eremochares dives, 79
Eucerceris 158, 253, 258; *arenaria* 254, 257; *flavocincta*, 255, 257, 263; *ruficeps*, 94

Eulophidae, 190
Eumenes, 159, 176; *alluaudi*, 301; *campaniformis*, 305; *colona*, 174, 181, 199, 202; *micado*, 304; *pedunculatus*, 143
Euodynerus, 117, 159; *annulatus*, 162–163; *dantica*, 98; *foraminatus*, 76–78, 104, 200, 202, 224, 227, 236, 257, 270, 273, 277, 301, 310–313, 317; *leucomelas*, 97, 199; *schwarzi*, 313
Euparagia richardsii, 234
Eurytoma, 191
Eusapyga rubripes, 134
Evagetes, 127, 328; *mohave*, 126–127; *subangulatus*, 233
evolutionary game theory, 180–181
Exeirus, 158; *lateritus*, 87
Exoprosopa, 190, 193; *fascipennis*, 192, 201
Exorista, 185

Fabriogenia, 158
fecundity: and egg size, 303–306; lifetime, 299–301
Fedtschenkia anthracina, 133
foraging behavior: Bethylidae, 25–32; Chrysididae (brood parasites), 122–126; Chrysididae (parasitoids), 33–40; Dryinidae, 16–25; Masarinae, 135–143; Mutillidae, 40–47; Pompilidae (brood parasites), 126–128; Pompilidae (parasitoids), 54–57; Pompilidae (predators), 103–105, Sphecidae (brood parasites), 128–131; Sphecidae (parasitoids), 58–64; Sphecidae (predators), 102–107; Sapygidae, 130–134; Scoliidae, 47, 49, 52–54; Tiphiidae, 47–51, 53–54
Formicidae, 186
fossil wasps, 4, 5, 319

Gasteruptiidae, 186, 194
Gayella, 159
Glenostictia, 107, 325; *satan*, 230–232, 236, 279; *scitula*, 87, 91, 114, 329
Glypta, 186
Gonatopus bicolor, 21, 23–24; *clavipes*, 18–24, 65, 319; *distinctus*, 22; *erythroides*, 19; *flavifemur*, 25; *helleni*, 22; *lunatus*, 22; *natalensis*, 66; *unicus*, 22
Goniozus, 225, 302, 315; *aethiops*, 226; *claripennis*, 29; *cellularis*, 29; *emigratus*, 29, 67, 226, 300; *gallicola*, 300, 323; *gordhi*, 226, 316; *indicus*, 225, 323; *jacintae*, 29, 66; *japonicus*, 28, 29, 323; *legneri*, 67, 226, 236, 274–275, 300, 307, 323; *longinervus*, 30; *marasmi*, 30; *natalensis*, 30, 32, 226; *nephantidis*, 30, 202, 316; *sensorius*, 30; *thailandensis*, 30; *triangulifer*, 300
Gorytes, 128, 158, 324; *canaliculatus*, 90, 114

Habritys, 191
Habrocytus, 191
haplodiploidy, 3
Haplogonatopus vitiensis, 68
Hedychridium, 210; *fletcheri*, 124; *solierellae*, 125
Hedychrum, 126; *intermedium*, 125, 211, 213, 224
Hemipepsis capensis, 149; *hilaris*, 148–149; *ustulata*, 95, 205, 220, 256, 264–268, 270, 279
Hemithynnus, 279
Heterogynaidae, 8
heterothermy, 284
Hilarella, 185, 187, 197; *hilarella*, 201
hilltopping, 264–268
homing and orientation, 72, 113–114, 321, 327
Homonotus iwatai, 54, 57; *sanguinolentus*, 57
Hoplisoides, 124, 130, 158; *nebulosus*, 90, 165; *placidus*, 131; *spilographus*, 178–179
host records: Bethylidae 28–31; Bradynobaenidae, 55; Chrysididae (brood parasites), 123–126; Chrysididae (parasitoids), 35–38; Dryinidae, 19–22; Embolemidae, 40; Mutillidae, 43–46; Plumariidae 40; Pompilidae (brood parasites), 127–128; Pompilidae (parasitoids), 55–57; Rhopalosomatidae, 55; Sclerogibbidae, 40; Scolebythidae, 40;

Scoliidae, 49, 52; Sphecidae (brood parasites), 129–130; Sphecidae (parasitoids), 60–64; Tiphiidae, 49–51
host plants, Masarinae, 137–139
host feeding, 16, 18, 23, 32, 47, 53, 59, 74, 75, 78, 115
host finding, 28–32, 123
Hylomesa, 49, 51

Ichneumonidae, 186, 191
inbreeding, 227
inquilines, 184
intercalary cells, 169, 173
Irenangelus, 127–128, 302, 327; *eberhardi*, 128, 331; *luzonensis*, 128; *pernix*, 301
Iridomimus, 206
Ischnurgops, 191
Ismarus, 190
Isodontia, 105, 157, 173; *auripes*, 80; *mexicana*, 80; *pelopoeiformis*, 80

Jugurtia, 159, 164, 166; *confusa*, 139, 233

Kennethiella, 195; *trisetosa*, 195
kin selection, 182, 225, 333
Krombeinictus nordenae, 149

Lackerbaueria, 185; *krombeini*, 187
Laelius, 315; *pedatus*, 25–28, 31, 226, 300, 308, 316; *trogodermatis*, 31; *utilis*, 26, 31
Larra, 60, 161, 302, 328; *amplipennis*, 63, 304; *analis*, 63; *bicolor*, 64, 233; *erebus*, 305; *luzonensis*, 67
Larropsis chilopsidis, 60, 64
larviposition, 187
leks, 261–262
Lepidophora, 185, 198; *lepidocera*, 184; *vetusta*, 201
Leskellia, 185
Lestica, 171
Lestiphorus, 130
Leucospididae, 191
Leucospis, 191
leveling, of nest mounds, 156, 208
Ligyra, 190, 193; *oenomaus*, 193, 201
Lindenius, 87, 181; *armaticeps*, 89; *columbianus*, 100
Liris, 103, 105; *argentata*, 296; *nigra*, 106
lizards, 195, 205, 227
Loboscelidia, 36
local mate competition, 314–316
Lonchodryinus ruficornis, 21
Lycogaster, 191
Lyroda, 84; *subita*, 86

Macrocentrus, 186
Macromerella, 158
Macromeris, 158, 324
Macrosiagon, 190, 194, 199; *pusillum*, 201
Masarina, 159; *familiaris*, 139–140, 142, 144
mass provisioning, 322, 329–330
mate rejection, 272–274
mating: among siblings, 224–227; frequency of, 218–219
Megachile rotundata, 132
Megadryinus magnificus, 21, 23–24
Megalothynnus klugii, 247–250
Megalyra, 191
Megalyridae, 191
Megaphorus willistoni, 194
Megascolia flaviphrons, 52
Megaselia, 185; *aletiae*, 184, 196, 199, 201
Melittobia, 189–192, 196–200, 203; *acasta*, 192; *australica*, 192; *chalybii*, 189–192
Mellinus, 103, 110, 158; *arvensis*, 87, 293; *rufinodus*, 246
Meloidae, 185
Messatoporus, 186
Meteorus, 186, 190
Methoca californica, 49, 50, 53, 326; *ichneumonoides*, 50, 326; *striatella*, 51; *stygia*, 51; *yasumatsui*, 51
Metopia, 185, 187, 197, 201, 208, 212; *argyrocephala*, 184; *campestris*, 201, 207, 289
Microbembex, 87, 158; *cubana*, 250; *monodonta*, 87, 114, 116, 163, 182, 233, 329
Microdontomerus, 191
Microstigmus, 84, 331; *comes*, 82, 304, 306; *thripoctenus*, 82, 84

Miltogrammini, 184–187, 196–201, 245
Mimesa, 157; *basirufa*, 82; *cressoni*, 82
mimicry, 206–207
Mimumesa, 124, 157, 170; *littoralis*, 83; *mixta*, 83
Minagenia osoria, 53, 57
Miscophus, 84, 157; *kansensis*, 86
Moniaecera asperata, 89, 332
Monobia, 159; *quadridens*, 98
Monobiacarus, 195
Monodontomerus, 191
Montezumia, 159; *cortesia*, 332
Mutilla attenuata, 44; *clythrae*, 43–44; *europaea*, 44, 305
Myrmecomimesis rubrifemur, 36; *semiglabra*, 36, 39
Myrmosa bradleyi, 44; *unicolor*, 44
Myrmosula parvula, 45
Myzinum, 51; *andrei*, 49; *navajo*, 144; *quinquecinctum*, 51, 294

Nemorilla, 185
Neochrysis, 198; *alabamensis*, 124
Neorhacodes, 191
Neozeloboria, 148; *proximus*, 249
Nesomimesa, 157; *hawaiiensis*, 164
nest: cell depth, 154; cell linings, 164; cell number per nest, 161–163; cleaning of cells, 329; closures, 156, 165–166, 209–210, 321; density, 153–154, 177, 206–208; depth, 154; digging, 153–155; sharing, 331; site selection, 153–154, 156–157, 168–170, 172–174, 176, 178; supersedure 179–182; turrets, 165; use of vibrations in digging, 161; use of water in digging, 161; usurpation, 179–182
nesting behavior, evolution of, 320–330
nests: mud, 173–177; in pre-existing cavities 171–173; in rotten wood, 170; in soil, 153–168; and soil moisture, 153, 168; in stems, 168–171; and temperature, 166–168; trap, 171–172; use of plant materials, 170–172, 176
Netelia, 186
Nitela, 84, 157
Nitelopterus californicus, 86; *evansi*, 178–179
Notocyphus tyrannicus, 56
Nysson, 128–131

Ocleroptera bipunctata, 90
Odynerus, 114, 159, 204; *dilectus*, 98, 104, 117
Omalus, 196–197; *aeneus*, 124
operational sex ratio, 220, 241, 244
Orancistrocerus, 159, 325; *drewseni*, 98, 199, 202
orientation flights, 113–114
Oryttus, 130
Osprynchotus, 191
overwintering, 296
oviposition, 15, 18–19, 35, 39, 42–43, 49, 54, 59, 114–116, 123, 126, 129, 131, 321
ovipositor, 4–5
Oxybelus, 111, 161, 165, 241, 280; *bipunctatus*, 88, 233, 273, 290; *emarginatus*, 88, 102–103; *exclamans*, 88; *fossor*, 88; *sericeus*, 88, 243–244, 256, 279; *similis*, 103; *sparideus*, 188; *subulatus*, 88, 243–244; *uniglumis*, 88, 105, 177–179
Oxyrrhexis, 186

Pachodynerus nasidens, 100, 104, 181, 199, 301
Palarus, 84, 110, 157, 194; *almariensis*, 86; *latifrons*, 103, 117; *variegatus*, 86
Palmodes, 105, 117, 157; *carbo*, 104; *laeviventris*, 79, 80, 103, 109, 117, 129, 195
Pappognatha myrmiciformis, 206
Parachilus major, 98
Paragia, 159, 164; *decipiens*, 138, 202; *tricolor*, 138, 140, 164; *vespiformis*, 138
Paragiaxenos, 195; *decipiens*, 202
Paragymnomerus spiricornis, 97
Paraleptomenes, 159, 325
Paralastor, 159, 162–163; *tricarinulatus*, 234
Paraleptomenes mephitus, 99; *miniatus*, 225–226
Parancistrocerus fulvipes, 99; *pensylvanicus*, 279
Parascleroderma, 28
parasites, of solitary wasps, 195–196

Parasitica, 2, 4, 35
parasitoids: defined, 13; of solitary wasps, 189–194
Paravespa mimus, 181
parental behavior: evolution in the Pompilidae, 326–328; evolution in the Sphecidae, 328–330; evolution in the Vespidae, 330; role of males, 240–245
parental investment: differences between the sexes, 309–314; and egg production, 299–303; and lifetime fecundity, 299–301
Pareumenes quadrispinosus, 325
Parnopes, 142; *edwardsii*, 38; *fulvicornis*, 38
Passaloecus, 108, 117, 125, 157; *cuspidatus*, 82, 171–173; *gracilis*, 197; *insignis*, 82; *ithacae*, 197
Pediobius, 190
Pemphredon, 125, 157, 170; *lethifer*, 82, 84, 103, 168–170, 181, 192, 197; *schuckardi*, 197
Penepodium, 79; *goryanum*, 81, 197
Pepsis, 110, 158, 203; *formosa*, 204; *thisbe*, 95, 104, 233
Perilampidae, 191
Perilampus, 191
Perithous, 191, 196–197; *divinator*, 192
Phalacrotophora, 185; *punctiapex*, 201
Phanagenia, 128, 158
pheromones, 254
Philanthus, 92, 103, 108, 125, 158, 165, 194, 201, 211, 253–264, 275, 290, 324; *albopilosus*, 156, 254, 291; *barbiger*, 263; *basilaris*, 254–255, 257–258, 260–261, 263; *bicinctus*, 100, 177–178, 219, 254, 256, 258–263; *crabroniformis*, 94, 156, 162–164, 254, 256, 260, 263; *gibbosus*, 201, 210, 331; *inversus*, 94; *lepidus*, 209; *multimaculatus*, 94, 255; *psyche*, 156, 162–163, 254–255, 257, 260, 262–263, 290; *pulchellus*, 94; *pulcher*, 156, 254–255, 257–258, 260, 263; *sanbornii*, 78, 94; *serrulatae*, 263; *triangulum*, 70–74, 108, 116, 118, 126, 214, 254, 263, 311–313; *zebratus*, 209, 233–236, 260, 269–271
phoretic copulation, 248
Phoridae, 185, 201
Photocryptus, 191–192, 196, 198
Phrosinella, 185, 187, 210
Physocephala, 194
Pimpla, 191; *spatulata*, 192
Pison, 157; *koreense*, 85; *rufipes*, 85; *strandi*, 242
Pisonopsis, 157; *clypeata*, 85
Platygastridae, 191
Plenoculus, 84, 157; *boregensis*, 233; *cockerelli*, 86
Pluto, 157; *littoralis*, 83
Podagritus parrotti, 89
Podalonia, 79, 102, 106, 117, 130, 157, 212, 324, 328; *canescens*, 148; *communis*, 178; *hirsuta*, 296; *luctuosa*, 81, 105, 109, 296; *occidentalis*, 201; *valida*, 81, 182, 303
Podium, 79, 157; *rufipes*, 81
Poecilopompilus, 159; *algidus*, 96, 100
Poemenia, 191, 196–197
pollen: carriage, 141; collection, 137, 140–141
pollination: by Masarinae, 144–145; by non-Masarinae, 143–150; deceptive, 145–150
Polochrum, 134
polyembryony, 19
Pompilus, 110, 127–128, 159; *cinereus*, 324; *scelestis*, 96, 160
Praestochrysis, 126; *lusca*, 124; *shanghaiensis*, 33–35, 38, 66, 68, 300
predators: defined, 13; of solitary wasps, 194–195
prey: caching, 114, 211–212; carriage, 72, 77, 108–112, 211–212, 321–322; of Eumeninae, 97–99; finding, 71, 75–77, 102–105; number per cell, 102; paralysis, 105–107; of Pompilidae, 92, 95–97; of Sphecidae, 79–94; storage, 112–113; theft, intraspecific, 121
Priocnemis, 92, 110, 158, 163, 324; *cornica*, 95; *germana*, 95; *minorata*, 95, 113
Prionyx, 102, 105, 116–117, 157, 212, 294, 324, 328; *atratus*, 79–80, 104, 128; *crudelis*, 104; *parkeri*, 106; *thomae*, 128
Pristocera, 248; *armifera*, 28, 31

progressive provisioning, 322, 329–330
Prorops nasuta, 31
Prosopigastra, 84
protandry, 219–220, 223–224, 227, 236, 259
Protomiltogramma, 185
psammophore, 161, 321
Psen, 157; *barthi*, 83
Psenulus, 157; *lukricus*, 304; *schencki*, 83, 197
Pseudisobrachium, 28
Pseudogonatopus distinctus, 275; *flavifemur*, 300; *hospes*, 68
Pseudolopyga carrilloi, 125–126
Pseudomalus auratus, 124
Pseudomasaris, 135, 159; *edwardsii*, 136–138, 140, 142, 174–175; *maculiphrons*, 138, 268; *marginalis*, 138; *phaceliae*, 138, 140; *vespoides*, 138, 140, 144–145, 234, 252, 276; *zonalis*, 138, 234, 252
Pseudomethoca frigida, 40–43, 274; *propinqua*, 233
Pseudophotopsis continua, 47
Pseudopompilus humboltdi, 55, 57
pseudostings, 204
Pseudoxenos, 195
Psorthaspis planata, 57
Pterocheilus, 159, 161; *texanus*, 99, 104, 162–163; *trichogaster*, 133
Pteromalidae, 191
Pterombrus, 49, 326; *cicindelus*, 51; *iheringi*, 51; *piceus*, 51; *rufiventris*, 51
Ptychoneura, 185, 187, 196–198
Pulverro, 157
pygidial plate, 161, 171, 321
Pymotes, 185; *ventricosus*, 187

Quartinia, 159; *vagepunctata*, 139, 142, 164, 234
Quartinoides, 140, 142; *capensis*, 142

rake spines, 160, 321
Raphiglossa, 159; *zethoides*, 97
removal experiments, 255
reproductive trade-offs, 299–306
Rhagigaster, 250; *aculeatus*, 249
Rhipiphoridae, 190, 194, 201
Rhopalosomatidae, 9
Rhopalum, 87, 125, 157; *atlanticum*, 89; *clavipes*, 197; *coarctatum*, 198
Rhyancanthrax, 193
Rhynchium marginellum, 99
robber flies, 194–195
Rolandia, 159, 161; *angulata*, 138, 142; *maculata*, 138, 140
Rubrica, 107, 117, 158, 329; *surinamensis*, 163

Sapyga centrata, 134; *confluenta*, 134; *pumila*, 130–133, 275; *quinquepunctata*, 134
Sarcophagidae, 185, 201
satellite flies, 187
sawflies, 3–4
Sceliphron, 79, 114, 124, 157, 175–177, 319; *asiaticum*, 198, 202; *assimile*, 181, 198, 301, 310–312; *caementarium*, 81, 173–175; *fistularium*, 198; *fuscum*, 145; *laetum*, 176, 198; *spirifex*, 310, 312
scent-marking, 252–254, 259, 262
Schedioides, 190
Sclerodermus, 306; *immigrans*, 28, 31–32, 226, 323
Scolia, 49, 287; *dubia*, 203, 295; *erythrosoma*, 305; *japonica*, 52–53, 326; *manilae*, 52, 233; *ruficornis*, 52
Senotainia, 185, 187
Senotainia sauteri, 187; *trilineata*, 184–188, 210–212; *vigilans*, 187
Sericophorus, 84, 111, 322; *victoriensis*, 164; *viridis*, 86
Sericopompilus apicalis, 96
sex allocation, 308–319; in bivoltine species, 317–318; and mating systems, 314–316
sex determination, 3
sex ratio, 308–319; and inbreeding, 318; and nest-hole diameter, 313, 319; precise, 315–316; problems with sampling, 318–319
sexual size dimorphism, 237, 248–250
Sierolomorphidae, 9
sleeping, 293–296
Solierella, 84, 125, 157, 172; *affinis*, 173; *blaisdelli*, 126; *peckhami*, 126
solitary wasp, defined, 1
sperm precedence, 219
spermatheca, 3
Sphaeropthalma orestes, 43, 46, 233;

pensylvanica, 46, 319; *unicolor*, 46
Sphecidae, as a paraphyletic family, 7–8
Sphecius, 111, 158, 181; *grandis*, 220–224, 257, 270; *speciosus*, 87, 90, 104, 108, 165, 182, 221–224, 236, 257, 312
Sphenometopa, 185
Sphex, 117, 157, 165, 181, 289, 324; *argentatus*, 80, 114, 164, 209; *cognatus*, 80; *crudelis*, 305; *flammatrichus*, 209; *ichneumoneus*, 100, 153, 180–181, 189, 289, 301; *nigellus*, 202, 304; *pennsylvanicus*, 256
Spilomena, 84, 157, 202; *subterranea*, 82, 84, 156, 332
Stangeella, 79, 157
Steniolia, 107, 142, 158, 325; *duplicata*, 329; *obliqua*, 91, 163, 177–179, 234, 273, 294, 329
Stenodynerus, 159; *claremontensis*, 331; *microstictus*, 162–163; *papagorum*, 177–179; *taos*, 246; *toltecus*, 313
Stictia, 117, 158, 165; *carolina*, 91, 103, 177, 233, 235; *heros*, 233, 271, 287; *signata*, 103, 168
Stictiella, 158, 324; *formosa*, 91, 164
Stigmus, 125, 157, 171
Stilbum cyanurum, 38–39, 126, 287, 305
stings and stinging, 14, 18, 23, 25–26, 32, 34, 42, 48–49, 53–55, 59, 71, 75, 77, 105–107; as a defense, 203–206
Stizoides, 128, 294; *renicinctus*, 128–129
Stizus, 158; *pulcherrimus*, 90; *ruficornis*, 106
Stomatomyia, 185
Strepsiptera, 195–196
sun dances, 235
Symmorphus, 97, 124, 159; *cristatus*, 99, 313
Synagris, 142, 159, 323; *cornuta*, 224; *spiniventris*, 330

Tachinidae, 185
Tachynomyia, 148, 250; *pilosula*, 249
Tachypompilus, 159; *analis*, 128; *xanthopterus*, 96
Tachysphex, 84, 105, 110, 117–118, 124, 157, 164, 324, 328; *albocinctus*, 290; *costai*, 106; *pechumani*, 85; *terminatus*, 85, 201, 214
Tachytes, 84, 108, 110, 157; *distinctus*, 273; *intermedius*, 85, 233; *mergus*, 104, 160; *minutus*, 104; *tricinctus*, 256; *validus*, 85
tarsal combs, 160
Taxigramma, 185
territory, 221–224, 237–246, 252–268
Tetrabaeus, 191
Tetradontochelys unicus, 300
Tetrastichus, 189–190
thermoregulation, 166–168, 282–293
Thynnoides, 148, 249; *bidens*, 148
Timulla vagans, 248, 250
Tiphia, 201; *agilis*, 53; *asericae*, 53; *berbereti*, 275; *koreana*, 53; *lucida*, 50; *nevadana*, 194; *ovinigris*, 53; *parallela*, 300; *popilliavora*, 47–50, 53, 68, 233, 300, 304, 312; *pullivora*, 201; *segregata*, 50, 67; *tegitiphaga*, 53; *notipolita*, 53; *vernalis*, 50, 53, 67
tool use, 166
Torymidae, 191
Toxophora, 190
Tracheliodes, 87; *amu*, 173; *quinquenotatus*, 116
Trachypus, 158; *denticolli*, 243; *petiolatus*, 94
trap nests, 171–172
Tricarinodynerus, 159; *guernii*, 173
Trigonalidae, 191
Trigonopsis, 79, 157; *cameroni*, 176, 231, 236, 256
Trimeria, 159; *buyssoni*, 140; *howardi*, 332
Trogoderma, 185, 198; *ornatum*, 201
Trypoxylon, 103, 124, 130, 157, 172, 175–177, 181, 280, 319; *attenuatum*, 85, 198; *figulus*, 85; *monteverdae*, 242, 245, 256; *nitidum*, 242; *palliditarse*, 198, 201, 203; *politum*, 49, 85, 100, 102, 174, 176, 198, 201, 237–245, 256, 278, 301, 310–312, 317; *rogenhoferi*, 198, 202; *rubricinctum*, 245; *spinosum*, 242; *superbum*, 85, 201, 242, 244–245; *tenoctitlan*, 86, 201–202, 243–245, 256; *texense*, 301; *tridentatum*, 86, 202, 243; *xanthandrum*, 244

venom, 14, 60, 105, 204
Vespacarus, 195
Vespoidea, 5, 8–9
vestibular cells, 173

Xenorhynchium, 159; *nitidulum*, 176, 213, 332
Xysma, 84

Zaspilothynnus, 148; *trilobata*, 146, 148, 150
Zeta abdominale, 199, 301; *argillacea*, 199; *canaliculatum*, 199
Zethus, 159, 176; *miniatus*, 181; *otomitus*, 99; *spinipes*, 104, 117
Zodion, 194
Zyzzyx, 142, 158